Keywords for Environmental Studies

Keywords for Environmental Studies

Edited by Joni Adamson, William A. Gleason, and David N. Pellow

NEW YORK UNIVERSITY PRESS *New York and* London

NEW YORK UNIVERSITY PRESS
New York and London
www.nyupress.org
© 2016 by New York University

References to Internet websites (URLs) were accurate at the time of writing.

Neither the author nor New York University Press is responsible for URLs that may have expired or changed since the manuscript was prepared.

Library of Congress Cataloging-in-Publication Data
Keywords for environmental studies / edited by Joni Adamson,
William A. Gleason, and David N. Pellow.
pages cm
Includes bibliographical references and index.
ISBN 978-0-8147-6296-7 (cl: alk. paper) —
ISBN 978-0-8147-6074-1 (pb: alk. paper)
1. Environmental protection—Terminology. 2. Ecology—Terminology.
I. Adamson, Joni, 1958– editor. II. Gleason, William A., 1961– editor.
III. Pellow, David N., 1969– editor.
TD169.3.K49 2016
363.7003--dc23   2015032230

New York University Press books are printed on acid-free paper, and their binding materials are chosen for strength and durability. We strive to use environmentally responsible suppliers and materials to the greatest extent possible in publishing our books.

Manufactured in the United States of America

10 9 8 7 6 5 4 3 2 1

Also available as an ebook

Contents

Foreword Lawrence Buell   vii

Acknowledgments   ix

Introduction Joni Adamson, William A. Gleason, and David N. Pellow   1

1. Agrarian Ecology Gary Paul Nabhan   7
2. Animal Stacy Alaimo   9
3. Anthropocene Jan Zalasiewicz, Mark Williams, and Colin N. Waters   14
4. Biodiversity Andy Dobson   17
5. Biomimicry Bryony Schwan   20
6. Biopolitics James J. Hughes   22
7. Bioregionalism Keith Pezzoli   25
8. Biosemiotics Timo Maran   29
9. Biosphere Tyler Volk   32
10. Built Environment William A. Gleason   35
11. Climate Change Andrew Ross   37
12. Conservation-Preservation William G. Moseley   41
13. Consumption Andrew Szasz   44
14. Cosmos Laura Dassow Walls   47
15. Culture Dianne Rocheleau and Padini Nirmal   50
16. Degradation Stephanie Foote   55
17. Democracy Sheila Jasanoff   57
18. Eco-Art Basia Irland   60
19. Ecocriticism Greg Garrard   61
20. Ecofascism Michael E. Zimmerman and Teresa A. Toulouse   64
21. Ecofeminism Greta Gaard   68
22. Ecology Reinmar Seidler and Kamaljit S. Bawa   71
23. Ecomedia Michael Ziser   75
24. Economy Robert Costanza   77
25. Ecopoetics Kate Rigby   79
26. Eco-terrorism David N. Pellow   82
27. Ecotourism Robert Melchior Figueroa   86
28. Education Mitchell Thomashow   89
29. Environment Vermonja R. Alston   93
30. Environmentalism(s) Joan Martinez-Alier   97
31. Environmental Justice Giovanna Di Chiro   100
32. Ethics Hava Tirosh-Samuelson   106
33. Ethnography Deborah Bird Rose   110
34. Evolution Dorion Sagan   113
35. Extinction Ursula K. Heise   118
36. Genome David E. Salt   121
37. Globalization Arthur P. J. Mol   125
38. Green Stephanie LeMenager and Teresa Shewry   128
39. Health Alissa Cordner, Phil Brown, and Rachel Morello-Frosch   130
40. History Stefania Barca   132
41. Humanities Joni Adamson   135
42. Imperialism Ashley Dawson   139
43. Indigeneity Kyle Powys Whyte   143
44. Landscape Dorceta E. Taylor   145
45. Natural Disaster Priscilla Wald   148
46. Nature Noel Castree   151
47. Nature Writing Karla Armbruster   156
48. Pastoral Sarah Phillips Casteel   158
49. Place Wendy Harcourt   161

50. Political Ecology Mario Blaser and
    Arturo Escobar   164
51. Pollution Serenella Iovino   167
52. Queer Ecology Catriona Sandilands   169
53. Religion Mary Evelyn Tucker and John Grim   172
54. Risk Society Robert J. Brulle   175
55. Scale Julie Sze   178
56. Species Quentin Wheeler   181
57. Sublime Patrick D. Murphy   183
58. Sustainability Julian Agyeman   186
59. Translation Carmen Flys-Junquera and
    Carmen Valero-Garcés   189
60. Urban Ecology Nik Heynen   192

Bibliography   195
About the Contributors   231

Foreword

Lawrence Buell

For as multicentric, interdisciplinary, and rapidly expanding an inquiry as environmental studies is, a work like this one couldn't be more timely. What better way to expand the horizons of those already on the way to immersing themselves in environmental studies than a collection of deeply informed, succinct, but nonreductive essays that range well beyond the horizons of any one scholar?

The collection's potential value seems all the greater given that the overwhelming majority of these sixty keywords are omnibus terms subject to dispute, ambiguation, reinterpretation, and revaluation over time. The rubric that stands almost in the exact middle of the series, "Environmentalism(s)," underscores this point: many if not most of the terms here discussed turn out to be plurals disguised as singulars, whether (as with "Environmentalism(s)," Martinez-Alier, this volume) because of sharp differences in demography and ideology among advocacy groups or because of an inherent semantic ductility that has been exploited in different ways by different stakeholders ("Sustainability," Agyeman, this volume). For the most part, the authors—who embody an auspicious blend of established expertise and fresh voices—adeptly negotiate the indeterminacies that go with their territories in such a way as to offer takeaway insights that resist the opposite pitfalls of timid fudging or overinsistence on one's pet ideas.

Keywords for Environmental Studies promises, then, both to satisfy reasonable desires for grounding and clarification and at the same time to put readers on notice that most of the concepts reviewed are, for better or worse, labile signifiers—arenas of discourse rather than fixed lexical building blocks—without unitary canonical definitions. With scattered exceptions like "Genome" and "Biomimicry," what's presented here, as the entry on "Democracy" puts it, is a reconnaissance of sets "of culturally inflected practices" rather than "a globally homogeneous norm that applies universally across all times and places."

None of what I've said so far about the kind of service a good guide of this kind can and should perform should be surprising. After all, the seminal late-twentieth-century precedent that stands behind this project went out of its way to make the same point: cultural theorist Raymond Williams's *Keywords: A Vocabulary of Culture and Society* (1976)—a pioneer study in media theory, in addition to much else. The polymathic Williams would have been impressed even if not unequivocally gratified by what to a time traveler from 1976 would have seemed the near-miraculous power of Internet publication's capacity for instant revision. The exploratory and reflexive character of many of the entries in *Keywords for Environmental Studies* would surely have pleased him too. But neither would he have failed to be struck by the sharp differences between his project and this one. The fact that it spotlights an area of inquiry he himself helped make central but engaged only at intervals until near the end of his long career might have pleased him more than otherwise, and so too the propensity of many if not most of these essays to assess environmental issues with regard to the inextricable linkages between human and other-than-human and with the claims of social justice as well as ecological health and survival conscience fully in mind. The humanist in him might also have taken issue, however, with the dispersed and specialized character of the present collection's keywords compared with his own. Virtually all Williams's 112 keywords of 1976 were cross-cutting, generalized terms of common usage—"Art," "Family," "Management," "Society," "Theory," "Tradition," and so forth. No one discipline or specialization had or has a monopoly on any, even if in a few cases insiders to a particular discourse might have lodged dubious claims of preemption ("Dialectic," "Evolution," "Hegemony," etc.). The same obviously can't be said for *Keywords for Environmental Studies*. Some of its rubrics are much more broadly cross-cutting than others; few readers are likely to interest themselves equally in the whole range; and it's impossible to imagine any one person writing credibly on all sixty of them.

But it is self-evidently bogus to hold up a genre's origination point as the measure of success a half-century later, as if the plays of Henrik Ibsen were to be taken as one's yardstick for evaluating those of Berthold Brecht (to mention another arena where Williams's work proved groundbreaking, and to compare oeuvres about as far apart in time as Williams's *Keywords* and this new compendium of the 2010s). The two books share only seven terms in common: "Consumption" (aka "Consumer" for Williams), "Culture," "Democracy," "Education" ("Educated"), "History," "Humanities" ("Humanity"), and "Nature." Not only does the more concentrated topical focus on "environment" justify this; what counts as "culture and society" has itself changed very markedly. In 1976, nearly a quarter of this collection's keywords hadn't yet been coined: "Anthropocene," "Biomimicry," "Biosemiotics," "Eco-Art," "Ecomedia," "Ecopoetics," "Eco-terrorism," "Ecotourism," "Environmental Justice," "Genome," "Indigeneity," "Queer Ecology," "Risk Society," and maybe more. Another sixth or more were neologisms at the edges of virtually everyone's radar screen—"Bioregion," "Ecocriticism," "Ecofeminism"—or somewhat recognizable, but with nothing like their present degree of portentousness and/or ramification: "Biosphere," "Climate Change," "(Environmental) Ethics," "Globalization," "Political Ecology," "Urban Ecology." In 1976, an increasing number of leading thinkers were contending that the state and fate of the planet would become the most crucial social issue for future generations—as science fiction had already been insinuating for generations before. Only during the 1980s did "the environment" start to become front-page news. Only in the twenty-first century has it begun to be widely accepted that the planet might have entered a new geologic era, the Anthropocene.

All this underscores the urgency, especially for those committed to environmental frames of inquiry, of ensuring that we get our lexical and conceptual bearings straight—and as current as possible—in matters environmental. For the distance between planet Williams and today's discourses of Earth demonstrates that the semantic-conceptual fields within which we operate circa 2015 will continue to change. In such a context the double value of this new compendium is its currency and its amenability to revision as the pertinent knowledge bases, analytical toolkits, fashions, and habits of discourse continue to change, as indeed they will.

FOREWORD LAWRENCE BUELL

Acknowledgments

A book of this scope, exploring topics this complex and urgent, requires the time and effort of a great number of good-willed people if it is to be done well. All the contributors who accepted our invitations to write, and many other experts in their disciplinary fields who, for various reasons, could not accept our invitations, were generously willing to talk at length with us about the project. Each played a significant role in shaping the book by helping us construct an initial list of over 180 possible terms and then identify names of people qualified to write the essays. Later, they helped us decide how to narrow our list to sixty. We thank each of them for their time, expertise, and influence on our thinking, although all decisions on the final list of keywords were ours alone.

The book was incubated in spaces and places that sharpened our thinking about the contents. Bruce Burgett and Glenn Hendler, the coeditors of the first edition of *Keywords for American Cultural Studies*, deserve special thanks for their mentorship and for offering us virtual space on their Keywords Wiki, where our colleagues from around the world were able to meet and reflect on the ways environmental studies disciplines (and interdisciplines) contribute to broader critical conversations surrounding environmentalism, social and environmental justice, sustainability, place, climate change, and other central topics. At actual conference venues, including meetings of the American Studies Association, the Association for the Study of Literature and Environment, and the European Association for the Study of Literature, Culture, and Environment, we benefited from the observations of audience members keen to help us deepen and expand our sense of how the sciences, social sciences, and humanities might translate the terms critical to their disciplines and offer histories of these terms that would increase possibilities for finding transdisciplinary solutions to environmental challenges. We are grateful to all who contributed to these conversations. We are also deeply indebted to the talented staff at New York University Press, and especially to Eric Zinner for his active engagement with the project, from attending our conference sessions to offering greatly appreciated suggestions for potential keywords and authors.

We thank our amazingly dedicated research assistants, Carolina Alvarado at Princeton University and Kyndra P. Turner at Arizona State University. They admirably collated the final manuscript, corresponded with our contributors, and paid incredible attention to detail in compiling, checking, and rechecking bibliographical citations.

Finally, we often hear that humanists and scientists have difficulty communicating with each other, particularly concerning ways to approach ecological crises. We thank each of our contributors for stepping into these gaps and modeling an "intercommunication" among the disciplines represented in this volume. Each author worked with us to write in accessible, jargon-free ways that would make complex information accessible to a wider public. We invite our readers to take up the debates, histories, and information found in each keyword essay and enter into conversations of critical importance to the present and the future.

Introduction

Joni Adamson, William A. Gleason, and
David N. Pellow

This volume creates a new "state of the field" inventory and analysis of the central terms and debates currently structuring the most exciting research in and across environmental studies, including the environmental humanities, environmental social science, sustainability sciences, and the sciences of nature. Inspired in part by Bruce Burgett and Glenn Hendler's *Keywords for American Cultural Studies*, and linked to that volume through Vermonja Alston's essay, "Environment," which she revisits and expands here for *Keywords for Environmental Studies*, we, and each of our contributors, aim to show how, in its broadest sense, the term "environment" enables "a questioning of the relations of power, agency, and responsibility to human and nonhuman environments" (Alston 2007, 103).

The deeper roots of this Keywords project may be found in cultural theorist Raymond Williams's *Keywords: A Vocabulary of Culture and Society* and the iconic "blank pages" at the end of that volume. Williams insisted that his book was not "a dictionary or glossary of a particular academic subject. . . . It is rather, the record of an inquiry into a *vocabulary*" (1976 [1983], 15). He interpreted the blank pages at the end of the book "as a sign that the inquiry remains open, and that the author will welcome all amendments" as contributions "towards the revised edition which it is hoped will be necessary" (1976 [1983], 26). Williams's own revised and expanded *Keywords* edition, coupled with Burgett and

Hendler's first *Keywords* project (recently released in a second edition) and the new Keywords volumes in the New York University Press series, has proved Williams's hunch (that the project would be ongoing) to be correct. From "Ecotourism" to "Eco-terrorism," from "Genome" to "Species," readers can see how dramatically Williams's project is expanding in the fields we group into environmental studies for this volume.

Williams shrewdly observed that "[n]ature is perhaps the most complex word in the language" and that any full history of the uses of the word "would be a history of a large part of human thought" (1976 [1983], 219, 221). Noel Castree, who takes up the challenge of addressing that history in this volume, notes that, today, "the things that certain biophysical scientists are saying (and doing) are pitching one of the Western world's most foundational concepts into a state of crisis. For better or worse, 'nature' appears to have more of a past than a future (hence the scare-quotes)." However, Castree notes, "there's no compelling evidence that 'nature' has lost its semantic importance as a key signifier in both expert and lay discourse. It still performs very important work in various cognitive, moral, and aesthetic registers; a great deal is still said and done in its name." This is why inquiry into the work of "nature" is still crucial these many years after Williams first wrote his own essay. As Castree writes, such an inquiry can help us answer such questions as, "Is the death of nature real or exaggerated, a matter of degree or kind? Which scientists, if any, might we turn to for robust evidential answers to this question?" As Dorion Sagan observes in his essay for this volume ("Evolution"), "massive evidence—from fossils, biology, morphology, organismal behavior, biogeochemistry, ecology, and cosmology"—has made it clear that "life [has been] evolving since its inception over 3.5 billion years ago." An eco-evolutionary perspective has made it even more clear that "for humans to survive in the long run we must become more attentive of, and connected to, the planetmates that support us within the long-lived global ecosystem," a process that will require all the disciplines to be conversant, one with the others.

The essays in *Keywords for Environmental Studies* show how a given term circulates within a particular knowledge community while also circulating across others, including nonacademic publics. Indeed, some of the essays show how terms and problems move from activist and general publics into academic knowledge communities, and in doing so, sometimes reverse the directionality and routes that academics often imagine knowledge follows. Each author is committed to an interdisciplinarity that is accessible to nonacademic audiences while making sure to include the most up-to-date research on a given term.

Readers will discover in the pages to follow that a broad range of approaches—including the quantitative and the qualitative—are critical to addressing the central issues in the field. Readers will also find certain inconsistencies and contradictions within the volume that reveal that some of the most important keywords continue to be used in different ways by writers coming from different scholarly traditions. This is not simply an equivocation or easy relativism but rather a stage in knowledge production that, we would argue, helps develop definitions and genealogies stable enough to

allow the pursuit of broadly collective goals and actions. For example, several essays in this volume use different start dates for the "Anthropocene," a term coined by atmospheric chemist Paul Crutzen and biologist Eugene Stoermer to describe a new epoch in Earth's history. Crutzen and Stoermer intended the term to identify a pivotal transformation in the life of the planet that began some two hundred years ago, around the time of the invention of the steam engine, in which humans became the primary driver of rapid changes across the world's ecosystems. Although this idea has gained considerable traction since the early 1990s, the International Commission on Stratigraphy, the scientific body charged with determining whether or not the planet has moved beyond the Holocene (the geological era that began at the end of the last Ice Age), is still "years away" from a formal decision (Revkin 2015). However, leaders of the Anthropocene Working Group (established in 2009 by the Subcommission on Quaternary Stratigraphy), including Jan Zalasiewicz, Mark Williams, and Colin Waters, who contribute "Anthropocene" to this volume, have already moved substantially beyond asking "*whether* such a transition has occurred to deciding *when*" (Revkin 2015). In a paper published in early 2015 in *Quaternary International*, twenty-six members of the working group point to roughly 1950 as the starting point, indicated by a variety of markers, including "global spread of carbon isotopes from nuclear weapon detonations starting in 1945 and the mass production of and disposal of [approximately six billion tons of] plastics" (Revkin 2015).

The gap between the dates proposed by Crutzen and Stoermer and the working group for the start of the Anthropocene not only suggests why contributors to this volume might place its onset in different eras but also illustrates how ongoing conversations and debates (in this case spanning more than twenty years) help shape

INTRODUCTION   JONI ADAMSON, WILLIAM A. GLEASON, AND DAVID N. PELLOW the production of knowledge itself. In their essay for this volume, Zalasiewicz, Williams, and Waters trace the origin of the concept of anthropogenic change to the earliest days of organized geological study, when the Comte de Buffon prepared "arguably the first evidence-based geological history of the world—*Les Époques de la Nature*, published in 1788." Humanists, familiar with Thomas Jefferson's *Notes on the State of Virginia*, know that Buffon's theories on racial environmentalism have been soundly dismissed, but they may not know that Buffon was the first to describe a "time during which humans dominated and warmed (beneficially, Buffon thought) the Earth" (Zalasiewicz, Williams, and Waters).

The genealogy and history of the keyword "Anthropocene" dramatically illustrate the benefits of bringing humanists, social scientists, and scientists together in one volume at this particular stage of knowledge production in environmental studies. As scientist Daniel P. Schrag, director of the Harvard University Center for the Environment, has observed, it is more important than ever to foster "conversations and interactions between scholars from different fields" (Schrag 2009, vii). Successfully addressing the challenges of accelerating environmental change will require understanding Earth systems, new technologies, economics, and policy. But it will also require understanding that emerges from the humanities about the "cultural components that led us here, the religious and philosophical traditions that affect how people make choices about their interactions with the natural world, and the social norms that are fostered by music, by art, and by literature" (Schrag 2009, viii). Perhaps even more important, as Lawrence Buell, who includes his own influential inventory of key terms in *The Future of Environmental Criticism*, has observed, "we who study in the environmental humanities and sciences are presented with challenges of 'intercommunication' because we have

no 'critical vocabularies' in common" (Buell 2011, 107). More urgent still, writes Noel Castree and colleagues, is the fact that much of the writing about the "human dimensions" of global environmental change in the social and sustainability sciences has offered "little or no sense of humans as diverse, interpretive creatures who frequently disagree about values, means, and ends; and there is nary a mention of power, violence, inequality" (Castree et al. 2014, 765). Castree adds that there now is wide recognition that the natural sciences cannot provide us with all the "knowledge or insight humanity will need to inhabit a post-Holocene environment" (2014, 763). As Andrew Ross writes in his entry for this volume ("Climate Change"), the ongoing task of averting or mitigating drastic environmental change will require a vast collective, creative, interdisciplinary effort to engage in a "social experiment in decision making and democratic action" and "concerted action on the part of political and economic elites as well as comprehensive shifts in the routine behavior of general populations, especially in carbon-rich countries."

Each essay in this volume frames and pursues these conversations and interactions from the perspectives of the humanities, social sciences, and sciences in ways that open access to debates on each keyword to scholars and general readers alike. Building on the work of influential feminists, poststructuralists, sociologists, and science and technology studies (STS) scholars, including Sandra Harding (1986), Donna Haraway (1991), and Val Plumwood (2002), and influenced by the convergent development of narrative, network, complexity, and relationality theories in the social and natural sciences (Latour, 1993, 2004a, 2010), contributors illustrate how the environmental humanities and sciences of nature are, in a sense, "recoding" each other, by blurring the lines separating humans, nonhuman animals, and machines, nature and culture, and the humanities and the sciences. As Dianne Rocheleau and Padini Nirmal observe in their essay on "Culture," scholars who take interdisciplinary approaches are increasingly taking up the "rhizome" (the root structures that supply nutrients to trees and other woody species) as an ideal metaphor for the intellectual, artistic, and scholarly histories and methods that are now profoundly transforming the larger endeavor of "environmental studies." Anthropologist Anna Tsing's study of the matsutake mushroom is often cited as inspiration for this transformation. She and her collaborators, in an open-access digital archive, write compellingly about the global cultural, scientific, and commercial networks surrounding the matsutake, as they explore entangled intellectual, cultural, and natural systems (Tsing 2012; matsutakeworlds.org) that can be "read" or narrated as what environmental literary critics have begun calling "storied matter" (Iovino and Oppermann 2012).

Work such as Tsing's is enriching the environmental humanities and sciences "with a more extensive conceptual vocabulary" that rethinks "the ontological exceptionality of the human" (Rose et al. 2012, 2). New concepts such as "slow violence" (Nixon 2011), "transcorporeality" (Alaimo 2010a), and "queer ecologies" (Mortimer-Sandilands and Erickson 2010), and new modes of research and writing such as "multispecies ethnography" (see Rose, "Ethnography," this volume) and "biosemiotics" (Maran, this volume) reveal the inseparability of the humanities and sciences as they account for humans, nonhuman animals, invasive plants, microbes, and toxins on the move, all of which have material ramifications across place and space, entangling bodies, politics, and ecologies (Kirksey and Helmreich 2010; Rose, "Ethnography"; Maran, "Biosemiotics"). Indeed, each essay in the pages to follow illustrates how research in biology, chemistry, and physics, once on the margins of the humanities and social sciences,

is pressing into the foreground of the fields of anthropology, literary study, history, politics, economics, and geography, among other fields.

These interdisciplinary efforts have implications not just for academics but also for the activists and community members who, as Giovanna Di Chiro notes in her essay on "Environmental Justice," "challenge the disproportionate burden of toxic contamination, waste dumping, and ecological devastation borne by low-income communities, communities of color, and colonized territories." For these groups, the "environment" is located in the places where we live, work, play, and learn. They "advocate for social policies that uphold the right to meaningful, democratic participation of front-line communities in environmental decision making, and they have redefined the core meanings of the 'environment' and the interrelationships between humans and nature, thereby challenging and transforming environmentalism more broadly" (Di Chiro). Likewise, in the fields of urban and political ecology, places that are typically represented as being "outside" of "Nature," including the city, are being reimagined not as "antinature" but as "socionatures"—sites where the human and nonhuman intermingle to reveal a "conjoint constitution" of forces (Swyngedouw and Heynen 2003; Freudenburg, Frickel, and Gramling 1995). Hence nature is no longer a *thing* so much as a cultural "terrain of power" (Moore, Kosek, and Pandian 2003). As Nik Heynen writes in his entry on "Urban Ecology" for this volume, "The metabolic lens offered through Marxist urban political ecology can be marshaled to consider both who and what suffers and who or what benefits from the interrelated and interdependent processes of urban ecological change marching forward into the twenty-first century." In their contribution, Mario Blaser and Arturo Escobar write that political ecology is a field in which scholars engage "reality as an emergent effect of relations and interdependencies that permanently overflow the boundaries based on modern binaries (nature/culture, subject/object, material/immaterial, and so on); and via a revaluation of non-Western knowledges that emerge from conceiving what exists and make up the world in other ways than in terms of those binaries" ("Political Ecology").

These developments have many implications with regard to questions about links between environmental politics and citizenship as well. Environmental justice movements and scholars have recently begun revisiting concepts of environmental citizenship that decenter human beings and expand the category of personhood to other living entities (Adamson and Ruffin 2013; Adamson 2014). As Sheila Jasanoff writes in her essay on "Democracy" for this volume, "the arrival of the 'environment' as a matter of public concern. . . . prompted once-unimaginable questions about the rights and entitlements of human beings and their fellow creatures on this planet." Recognition of rights/citizenship within ecosystems and other-than-humans has the potential to transform the concepts and the institutional power that is traditionally behind them, thus destabilizing state and corporate power and challenging traditional notions of sovereignty (Adamson and Ruffin 2013). Ultimately, then, the essays in this volume point to critical questions that must be asked and answered with regard to what it means to be "human" and what it means to be in multispecies relationships within the "Biosphere," which Tyler Volk in his essay for this volume defines as "the three environmental matrices of atmosphere, soils, and oceans" that "form a closely integrated network" called "Earth."

Finally, the ideas and products of the arts, which make manifest our capacity to be deeply imaginative, creative, and feeling, will be critical to these conversations. As Joni Adamson writes in her essay on "Humanities," in the opening decades of the twenty-first century there is wide recognition of this need among the business and education leaders around the world who are declaring the "environmental humanities"— history, philosophy, religious studies, literary criticism, theater, film, and media studies informed by the most recent research in the social sciences and sciences— crucial to the discovery of solutions to today's entangled environmental and social challenges, from the complex problems associated with extreme weather events to the increasing global inequities associated with rapidly changing environments. In our view, the collaborative work of this volume demonstrates on the one hand that every discipline has a stake in the central environmental questions of our time, and on the other hand that *inter*-disciplinarity and cross-disciplinary conversations not only enhance but are requisite to environmental studies today.

I

Agrarian Ecology

Gary Paul Nabhan

One might wonder whether any twenty-first-century preoccupation with agrarian values, agrarian ecology, and agrarian ideals comes as too little, too late. Less than 2 percent of the North American public lives in rural areas outside towns, cities, and suburbs, and less than half of the world's population now lives outside cities. But the New Agrarianism, which is emerging globally, is not restricted to the rural domain, nor is it necessarily a romantic desire to reenact social behaviors and mores associated with rural populaces in bygone eras. Instead, a New Agrarianism is emerging within urban as well as rural communities, and may indeed be *the* set of values and operating principles that can obliterate the rural-urban divide that, in many ways, characterized and crippled North American and European cultures during the second half of the twentieth century. But what exactly does "agrarian" mean? Why are the concepts associated with it being used once more as rallying cries, decades after most global citizens have become disenfranchised from the land? Finally, why has "agrarian ecology" become a useful focus for anthropologists, biologists, demographers, geographers, historians, and land-tenure lawyers, and why is it being applied to solving problems in at least a dozen countries on four continents?

If we return to its etymological roots, *agri-* can be traced as far back as the proto-Indo-European noun "*h₂éǵros*," meaning field or pasturage, which has cognates not only in Old English but in ancient Greek, Latin, and Sanskrit as well. As used over the centuries in Europe and England, this term refers to a constellation of activities, values, and premises regarding human relationships to cultivated soil or to the land in general. As a prefix in Latin, and then Old, Middle, and modern English, *ager-* and *agri-* relates to soil, fields, farms, land, terrain, landscape, territory, and country. In the related term "agriculture," based on the Latin *ager + colere*, we see the relationship between humans and the land circumscribed by the activities and values of cultivating, tilling, stewarding, tending, and safeguarding. Agrarian ecology, as articulated by agricultural anthropologist Robert McC. Netting in 1974, is the study of both the social and the legal frameworks that guide tenure to and the human uses of cultural "working" landscapes and the interactions between human communities and their agricultural and ecological resources in the landscapes.

Agrarianism, of late, has come to embody a nuanced set of social, political, and ecological values that see rural activities, behaviors, and ethics as functioning on a higher order than urban- or suburban-derived comparables. However, for well over a century, the phrase "agrarian reform" has had broader recognition in Latin America, Europe, and Asia as a movement to keep peasant societies from becoming increasingly landless and in greater servitude to capitalistic institutions by enacting the redistribution of land and other wealth. Agrarian ecologists have paid particular attention to how peasant societies resist such extractive institutions and organize themselves to protect, sustain, and efficiently use the natural resource base and traditional knowledge upon which their members' livelihoods depend.

In a very real sense, agrarian values place heightened importance on the daily human commitment to and daily involvement in rural lifeways as God-given responsibilities. Accordingly, Thomas Jefferson is often designated the best early articulator of American agrarian values, while Henry Wise Wood, Louis Bromfield, Ralph Borsodi, Robert Swann, Wendell Berry, Wes Jackson, Helen Nearing, and Will Allen are granted status as the most elegant contemporary North American defenders of agrarian values in the face of agricultural industrialization and ex-urban growth. However, agrarian values are *not* exclusively Euro-American or even Christian, for Marxist materialists around the world have come to embrace some of the same principles and strategies for valuing the work done by peasant farmers. As eloquent as Jefferson and Berry in the United States and European voices such as Jean Giono and John Berger may be, Japanese farmer Masanobu Fukuoka and Australian permaculturist Bill Mollison exemplify the power and reach of agrarian values outside of the Euro-American geographic and cultural context.

Because twentieth-century agrarian proponents such as Canadian Henry Wise Wood, Japanese Masanobu Fukuoka, and American Wendell Berry have often been diagnosed by urban critics as being afflicted with a nostalgia-emitting dysfunction that has symptoms of being "anti-urban," "luddite," or "retro," some proponents such as Eric Freyfogle and David Walbert call their philosophies "Neo-Agrarian." On his populist website, newagrarian.com, Walbert offers a brilliant articulation of how the New Agrarianism can be distinguished from other forms of agrarianism that may be flawed by romanticism or nostalgia. He argues that New Agrarianism is defined by four elements. First, while it draws heavily on past agrarian practices and thinking, it is not bound by them because New Agrarianism is focused on building the future. Second, New Agrarianism is concerned with creating a new kind of rural community for the twenty-first century—one that is tied neither to traditional models of rural America nor to the dominant large-scale industrial agricultural approach.

Third, New Agrarianism views sustainable community as the ultimate goal, and sustainable agriculture is just one critical part of that vision. Thus, New Agrarian values can be expressed in all sectors of the economy and across all aspects of a sustainable culture and life. Fourth and finally, New Agrarianism recognizes that society is mostly urban and sees this as an opportunity to seed New Agrarian values within and across nations and urban areas, since the core of this philosophy is the desire for sustainable connections among nature, place, and community.

And yet, David Walbert, David Orr, Will Allen, and others concur that an agrarian believes in, if not the primacy, then at least the uniqueness of agriculture among human endeavors. Activists David Hanson and Edwin Marty, coauthors of *Breaking through Concrete: Building an Urban Farm Revival*, believe that agrarian values and practices should and can be expressed in urban, suburban, and ex-urban settings as well as in rural landscapes. Youth groups such as FarmFolk/CityFolk, the National Young Farmer's Coalition, and Greenhorns are moving such an agenda forward as a social movement that now crosses international boundaries. In their view, a foodscape is no longer (and has actually never been) a place beyond the city's limits, and the quest for just, equitable, and sustainable food systems and environmentally healthy foodsheds must engage both rural and urban dwellers of all races, classes, and languages with equal strength. The fact that over twenty-five hundred acres of Metro Detroit's sixty-five hundred acres of formerly built-upon and abandoned urban lots are once again producing food is testament to the survival of agrarian values in an urban setting.

Finally, it is worth noting that agrarian and neo-agrarian advocates link themselves to an unbroken chain of prophetic voices that have critiqued excessively urban, inward-looking, and narcissistic values of those who have become indifferent to the plight of farmers, fishers, ranchers, and foragers, and to the land itself. In theologian Ellen Davis's finely researched book, *Scripture, Culture, and Agriculture: An Agrarian Reading of the Bible,* it becomes historically clear that agrarian voices have risen up as prophets, dissidents, and agents of change whenever urban hierarchical or industrial societies have become too excessive in their consumption, waste, and hegemony over others. Davis deftly links the messages and methods of the Old Testament prophets with modern-day agrarian voices from many countries.

On the academic or scholarly level, it is surprising that biologically trained ecologists are among the least engaged in the documentation and application of agrarian ecology (*sensu* Netting) compared to geographers, anthropologists, agro-ecologists, and rural sociologists. There are, of course, exceptions among broad-based natural scientists such as Mexican Victor Toledo, Chilean Miguel Altieri, and Indian Vandana Shiva, who have trained hundreds of students to apply a broader perspective to ecological issues in food-producing landscapes. Anyone who still believes that agrarianism is something of the past should spend a day with "greenhorns," some of whom are now associated with Via Campesina, Slow Food International, or various young farmers' coalitions and permaculture guilds. Be assured that you will be both tired and fulfilled at the end of one long but fruitful day with them.

2

Animal
Stacy Alaimo

One English word, one Western concept—"animal"—somehow encompasses a vast array of creatures—sponges, spiders, capybara, camels, eels, eagles, ticks, tigers, octopi, orangutans, dinosaurs, and slugs—but it rarely contains humans. Western philosophy and everyday conceptual frameworks define the human against the animal, forcing the multitude of beings other than *Homo sapiens* into one category. Jacques Derrida notes the absurdity and violence of this ostensibly neutral term:

> Whenever "one" says "The Animal," each time a philosopher, or anyone else, says "The Animal" in the singular and without further ado, claiming thus to designate every living thing that is held not to be human (man as rational animal, man as political animal, speaking animal, *zōon logon echon*, man who says "I" and takes himself to be the subject of a statement that he proffers on the subject of the said animal, etc.), well, each time the subject of that statement, this "one," this "I" does that he utters an *asininity* [*bêtise*]. (Derrida 2008, 31)

Uttering the term "animal," in the singular, he argues, is a disavowal that demonstrates one's "complicit, continued, and organized involvement in a veritable war of the species" (Derrida 2008, 31).

"Animal" is an overwrought, overloaded term indeed. The category has been invoked to elevate humans above all other living creatures as well as to denigrate certain groups of people as not-quite-human via racist, sexist, classist, Social Darwinist, and colonialist ideologies that place them "closer" to animals in hierarchies of being. Human exceptionalism, emerging from monotheisms, Enlightenment humanism, capitalist anthropocentrism, and other forces insulates (some) humans from kinship with degraded, brutish beasts. But repressed critters return to bite. Monster movies such as *The Island of Dr. Moreau* feature human/animal hybrids that rouse the viewers' recognition of the animality of the human only to conclude by assuring us that we are certainly not animals after all (Alaimo 2001). Evolution says otherwise. Charles Darwin, who could be considered the first "posthumanist" philosopher, remarked upon the rather obvious correspondences between humans and other animals:

> It is notorious that man is constructed on the same general type or model as other mammals. All the bones in his skeleton can be compared with corresponding bones in a monkey, bat, or seal. So it is with his muscles, nerves, blood-vessels and internal viscera. The brain, the most important of all the organs, follows the same law. (*Descent of Man*, 1936, 395–96)

The word "notorious" unveils the paradox that the fact that "man" is constructed like other mammals is somehow both accepted and unacceptable, both obvious and objectionable. Darwin attempts to make human kinship with other animals more appealing by telling many a charming and humorous tale demonstrating how the animals that humans would denigrate as such actually possess various "human" characteristics, of curiosity, reason, language, affection, tool use, and even religious experiences.

Marc Bekoff's recent work continues this tradition, documenting how various animals think, feel, play, and even behave "morally" (2007; Bekoff and Pierce 2009). These capabilities, Bekoff and many others insist, demonstrate that living beings other than humans deserve ethical consideration. Joy Williams's wry term "animal people" (which is also the title of her scathing essay), denotes humans who advocate on behalf of nonhuman life, but it also suggests that these people are themselves, unflinchingly, "animals" (Williams 2002). Boundary creatures, such as primates, cyborgs, the oncomouse, and dogs populate Donna Haraway's corpus, troubling commonsensical conceptual divisions and definitions that would distance humans from other life forms. One of her most compelling arguments, however, is that what humans are as a species is partly due to canines. In *The Companion Species Manifesto* she argues that humans and dogs do not exist as separate entities but, instead, have co-constituted each other through their significant relations across evolutionary time. "Dogs are about the inescapable, contradictory story of relationships—co-constitutive relationships in which none of the partners pre-exist the relating, and the relating is never done once and for all. Historical specificity and contingent mutability rule all the way down, into nature and culture, into naturecultures" (Haraway 2003, 12). The conceptual abyss between "human" and "animal" that Derrida encounters becomes, in Haraway's work, a fleshy realm of interconnected histories, significant relations within specific "naturecultures."

Whereas Haraway's boundary creatures radically muddle the human/animal divide, strategic boundaries have been drawn at other sites. The Great Ape Project, for example, defends "the rights of the non-human great primates—chimpanzees, gorillas, orangutans, and bonobos, our closest relatives in the animal kingdom" (Great Ape Project). While the appeal for the great apes stresses their kinship with humans, the "Declaration of Rights for Cetaceans: Whales and Dolphins" asserts the "equal treatment of all persons," basing personhood on "scientific research [that] gives us deeper insights into the complexities of cetacean minds, societies and cultures." The list of ten rights includes the right to life, freedom from captivity or servitude, the "right to protection of their natural environment," and "the right not to be subject to the disruption of their cultures" (Helsinki Group 2010). Both the Great Ape Project and the "Declaration of Rights for Cetaceans" release particular species from the "asininity" of "the animal," insisting that certain animals demand separate ethical and legal consideration. Yet, even for those of us who passionately believe in granting the great apes and the cetaceans all the rights outlined in these declarations, these movements do, of course, beg the question of what other living beings should be granted rights and whether the rights framework is adequate or appropriate for addressing the many assaults on the lives and well-being of a multitude of living creatures. Cary Wolfe, commenting on the "Spanish Parliament's approved resolution to grant basic rights to great apes," argues that

> even as it constitutes a monumental step forward
> for our relations with non-human animals within
> the political purview of liberal democracy and its
> legal framework, it might well be seen, within the
> biopolitical context opened up by Esposito and
> others, as essentially a kind of tokenism in which
> non-humans "racially" similar enough to us to
> achieve recognition are protected, while all around
> us a Holocaust against our other fellow creatures
> rages on and indeed accelerates. (Wolfe 2010, 8, 23)

The declaration of the rights of "higher animals" reveals the fault line between animal-oriented philosophies and movements and environmental philosophies and movements. Most notably, biocentrism reverses the established hierarchy of living creatures, asserting that life on Earth could continue without humans, great apes, or cetaceans, but not without bacteria, which are crucial for ecosystems. Animal rights, animal welfare, and animal ethics tend to depart from environmentalism's focus on large-scale, interdependent systems, such as ecosystems and habitats, for example, as well as the threats to those systems—pollution, habitat loss, climate change, overfishing, extinction—as they focus on particular animals. But climate change alone, which threatens to push a million species into extinction by 2050, illustrates why attention to large-scale, interconnected systems is crucial for the continued existence of nonhuman creatures. Conversely, environmental activists, organizations, and scholars rarely consider cruelty to factory-farmed animals, animals used for experiments, or animals exploited for the entertainment of humans. Appeals to wilderness, wildness, conservation, aesthetics, and biodiversity are cordoned off in another discursive universe, far from sterile laboratory cages or filthy feed lots.

While environmentalists examine interconnected systems, "animal people" shift the focus from "objective" views of the big picture to "subjective" standpoints of other living creatures. Scientists, writers, philosophers, and artists have taken on the formidable task of attempting to understand particular animals' worlds and perspectives, from Jacob von Uexküll's *Foray into the Worlds of Animals and Humans* (1934 [2010]) to Anna Sewell's *Black Beauty* (1877), Thomas Nagel's "What Is It Like to Be a Bat?" (1974), artist Sam Easterson's videos from animal perspectives (2001), Les Murray's poems in *Translations from the Natural World* (1992), and the most provocative and perhaps problematic portrayal of elephant culture, kinship, mythology, and even religion: Barbara Gowdy's novel *The White Bone* (2000). Notwithstanding the irresolvable epistemological difficulties with attempting to understand the lives, cultures, and ways of being of other creatures, the scientific, philosophical, literary, and artistic endeavors can spark ethical and political engagements. Despite Joy Williams's justified lament, "We learn more about them, and that does not save them," animal people still somehow hope that such findings will translate into less animal suffering (Williams 2002, 122). Whereas Williams bemoans the fact that the production of knowledge about animals does not diminish cruel and exploitative practices, as the exclusively human, monolithic "we" amasses information *about* animals that rarely translates into movements *for* animals, Bruno Latour proposes a political "collective" that would include the "voices of nonhumans" (2004a, 69) as represented by scientists, with their instruments and their "capacity to record and listen to the swarming of different imperceptible propositions that demand to be taken into account" (2004a, 138). Although Latour does not grant them an official role in his collective, philosophers, transdisciplinary animal studies scholars, writers, filmmakers, and artists seek to shift beliefs, attitudes, and practices toward nonhuman species. Attention to nonhuman creatures within academic disciplines is resulting in new areas, such as "multispecies ethnography" (see Debra Bird Rose's essay in this collection). Whether or not such investigations will benefit diverse living creatures remains to be seen.

Insisting upon the perceptual, cognitive, semiotic, phenomenological, and cultural differences of different species is invaluable for understanding the vast world of living creatures as more than Cartesian machines. But new materialist approaches, which turn their attention to corporeality, substances, and physical agencies, suggest that animals need to be considered within material systems—not only within ecosystems and habitats but also within food systems, big pharma, chemical industries, and other areas of global capitalism. Nicole Shukin's brilliant book *Animal Capital: Rendering Life in Biopolitical Times* urges us to regard animal bodies, as "capital becomes animal, and animals become capital" in a "semiotic and material closed loop, such that the meaning and matter of the one feeds seamlessly back into the meaning and matter of another" (Shukin 2009, 16). Her conception of "rendering" signifies both "the mimetic act of making a copy" and "the industrial boiling down and recycling of animal remains" (Shukin 2009, 20). Drawing on Shukin's formulation allows for consideration of another, disturbing sort of "rendering"—of beings boiled down to data—which begs the question of environmentalism's complicity in animal suffering. In short, to what extent do various environmental arguments, movements, and policy decisions rely upon scientific data on toxicity or radiation that has been produced through laboratory testing on mice, rats, rabbits, and other animals?

New materialisms may interconnect environmental and animal advocacy. In *Bodily Natures: Science, Environment, and the Material Self*, I argue for a conception of transcorporeality that traces the material interchanges across human bodies, animal bodies, and environments. BlueVoice.org, for example, a marine conservation organization that focuses on dolphins and whales, epitomizes a transcorporeal environmental politics by stressing that humans, dolphins, and whales are all vulnerable to the harmful effects of mercury and organochlorines. In the short film entitled *A Shared Fate*, Hardy Jones, a founder of BlueVoice.org, explains how he had dedicated his life to studying and protecting cetaceans. Ironically, his "extraordinary bond" with dolphins

ANIMAL   STACY ALAIMO

becomes undeniably corporeal when he discovers that, like many of the dolphins he studies, he has developed chronic mercury poisoning and multiple myoma, caused by eating the same fish the dolphins eat. Sadly, *A Shared Fate* does not seem to include the fish within its horizons of concern, as it focuses on "higher" creatures. Nonetheless, it does dramatize how human activities produce substances that invade marine ecosystems. Such things as coal-burning power plants, pesticides, and flame retardants result in an ocean riddled with mercury and organochlorides, which threaten marine life. As with most transcorporeal recognitions in risk society, *A Shared Fate* displays both the necessity for scientifically derived data and the need for embedded epistemologies that reconfigure the boundaries between scientific practices, politics, human health, nonhuman animal health, and environmentalism. Even such an anthropocentric concept as Ulrich Beck's "risk society," which stresses the scientific mediation of knowledge, can descend to the bottom of the sea, where benthic worms, sea cucumbers, and krill ingest "toxin-laden microplastics" (Kaiser 2010, 1506). Not unlike ordinary humans navigating the bewildering risks we cannot properly assess, the benthic creatures can no longer rely upon their own sensory organs to detect danger. Animal studies and environmental studies could be enriched by interconnecting investigations of specific animals' modes of being and knowing with theories and data pertaining to ecological systems, risk, and material interactivity. As new materialisms—some of which demonstrate little concern for either nonhuman animals or the environment—develop, it will be crucial to consider the promises and perils of theoretical terminology that supplants the term "animal." "Nonhumans," "things," and "objects" purport to encompass "the animal," but these terms sometimes ontologically and ethically flatten the world in such a way that "plumbers, cotton, bonobos, and DVD players" exist "equally" (Bogost 2009, n.p.). Animal people and environmentalists will need to assess whether such theoretical formulations offer suitable conceptual habitats for diverse living species.

3

Anthropocene

Jan Zalasiewicz, Mark Williams, and
Colin N. Waters

The world today is undergoing rapid environmental change, driven by human population growth and economic development. This change encompasses such diverse phenomena as the clearing of rainforests for agriculture, the eutrophication of lakes and shallow seas by fertilizer run-off, depletion of fish stocks, acid rain, and global warming. These changes are cause for concern—or alarm—among some, and are regrettable if unavoidable side effects of economic growth for others.

How significant are these changes in total? How might they evolve, and what might their ultimate consequences be? One way of studying these changes is to consider them as the latest phase of the many environmental changes that have affected the Earth since its origin, a little over four and a half billion years ago. Humans may be considered as geological agents, and anthropogenic environmental change may be compared with events in Earth's deep history.

Such analysis dates, perhaps surprisingly, from the earliest days of organized geological study. Working before the French Revolution, the Comte de Buffon wrote arguably the first evidence-based geological history of the world—*Les Époques de la Nature*, published in 1788 (Roger 1962): his seventh and final epoch denoted the time during which humans dominated and warmed (beneficially, Buffon thought) the Earth. (While Buffon's theories of racial environmentalism have largely been dismissed, his thoughts on geological time and

process remain important to contemporary discussions of the Anthropocene.) Later, works by scholars such as George Perkins Marsh (1864) and Antonio Stoppani (1873—he proposed the term "Anthropozoic") noted the growing impact of humans on Earth, with the likes of Vladimir Vernadsky (1945) and Robert Sherlock (1922) then developing various aspects of this concept.

Throughout the twentieth century, most geologists (e.g., Berry 1925) dismissed human impact as insignificant when set against the broad canvas of Earth history. This was the case partly because of the time scale. The ten-thousand-year span of human civilization is barely significant on the scale of geological time, while the phase of industrialization is far shorter still. Furthermore, the great natural forces of the Earth—mountain building, volcanic outbursts, meteorite impacts, and so on—were regarded as of far greater long-term significance than the brief human alteration of an ephemeral landscape.

This changed when Paul Crutzen (Crutzen & Stoermer 2000; Crutzen 2002) proposed that we were now living in the Anthropocene, because of the scale of the human-driven chemical, physical, and biological changes to the Earth's atmosphere, land surface, and oceans. The "Anthropocene" (an "Anthrocene" had also been proposed a little earlier by Revkin 1992) caught on, and began to be used almost immediately by a wide range of scientists and made an impact among the general public.

Why the change in opinion? First, it was becoming clear that while human impacts may be geologically brief, they are not trivial. Some "invisible" effects—such as increases in carbon dioxide levels in the atmosphere—can lead to profound physiographic effects by their influence upon global temperature and therefore (through the melting of ice) on sea level. Second, the term was found useful as a means of integrating and conveying the different types of global change. Third—and particularly as regards public impact—the word itself is evocative, overtly placing humans within the same geological narrative (one of almost infinite duration!) that the dinosaurs once occupied.

Where is the concept at present? As an informal term, it now seems firmly established in the sciences—and in the arts too. It has been the focus of studies published by the Royal Society (Williams et al. 2011), of a major initiative of the International Geosphere-Biosphere Programme (IGBP: http://www.igbp.net/), and of a major modern art exhibition at the Haus der Kulturen der Welt in Berlin in 2013–14. The use of this term to denote the current time interval seems set to continue, and probably spread.

Given this widespread use in practice, the Anthropocene is being formally considered as a possible addition to the Geological Time Scale by an Anthropocene Working Group of the Internal Commission on Stratigraphy (http://www.quaternary.stratigraphy.org.uk/working-groups/anthropocene/). It may, therefore, join "Jurassic," "Pleistocene," and other terms that underpin the science of geology (Zalasiewicz et al. 2010). This process will involve much detailed analysis over the coming years, and formalization is not assured. The term has to be shown to be not only technically valid but also of practical value (rather than a hindrance) to working Earth scientists.

One benefit of formalization would be precise definition, for currently there is no accepted date for the beginning of the Anthropocene. It is mostly used in Paul Crutzen's (2002) meaning, as dating from the Industrial Revolution, at around 1800 CE. However, in geology, a geological time boundary must be both synchronous (by definition) and also effectively recognizable in strata. In the case of the Anthropocene, these strata include artificial deposits produced as a consequence of urban development and mineral extraction (Price et al. in Williams et al. 2011; Ford et al. in Waters et al. 2014), damming of rivers, deforestation, and coastal reclamation (Syvitski and Kettner in Williams et al. 2011) and also deposits lacking discernable human influence—those of remote desert dunes, for instance. The "human-made strata" locally date from before the Industrial Revolution, and in general are strongly diachronous (i.e., formed at different times in different places), reflecting the spread of human civilizations around the globe over several millennia.

One potential boundary that is both widely detectable in sediments and approximately synchronous is the spread of radionuclides from atmospheric nuclear weapons tests (Zalasiewicz et al. 2014; Zalasiewicz et al. 2015; Waters et al. 2015) and of fixed nitrogen from fertilizer manufacture (Holtgrieve et al. 2011), both of which date from ca. 950 CE. This date broadly coincides with "The Great Acceleration" (see Steffen et al. in Williams et al. 2011), during which the use of natural resources and the emission of pollutants increased rapidly. The pros and cons of various boundary candidates need to be considered prior to any decision on formalization.

Formally, there is the question, too, of the hierarchical level of the Anthropocene, for the geological time scale consists of smaller units nested within larger ones. Thus, currently (formally), we are living within the Holocene Epoch, which is a part of the Quaternary Period (which began 2.58 million years ago, when the Earth became glaciated at both poles), which lies within the Cenozoic Era (which began at the mass extinction event sixty-six million years ago, when the dinosaurs and much else became extinct), which in turn is part of the Phanerozoic Eon (which began over a half-billion years ago, with the sudden and widespread appearance of complex animal behavior in the fossil record, represented by the burrowing traces of those organisms).

Currently, the Anthropocene is being considered as an epoch (Zalasiewicz et al. 2015). If that is accepted, then the Holocene has formally terminated (though the Quaternary, Cenozoic, and Phanerozoic would continue). There are other possibilities, such as a period on one hand and an age (a subdivision of an epoch) on another. The hierarchical level ultimately chosen should, at least in part, reflect the scale of environmental change (which will in turn determine the distinctiveness of future geological strata). How does the Anthropocene currently seem to be measuring up?

The answer is not straightforward, as the component phenomena of the Anthropocene are evolving rapidly and, for the most part, are in their early stages. Future (imperfectly predictable) trends will develop over many millennia at least.

Geological changes so far include some that are entirely novel in Earth history. Most distinctive are the "urban strata," which may be thought to approximate to (eminently fossilizable) gigantic trace fossil systems currently spreading across and beneath terrestrial surfaces (Zalasiewicz et al. 2014; Zalasiewicz, Waters, and Williams 2014). Of chemical changes to the environment, the human-driven influx of carbon from rock strata into the atmosphere is (as yet) smaller in scale than geologically ancient, natural outbursts such as that of the Paleocene-Eocene Thermal Maximum, some fifty-five million years ago; but, it has taken place more rapidly—over a couple of centuries rather than over several millennia (Ridgwell and Schmidt 2010). That ancient event led to warming of some 5–8° C globally (a process that is now in its early stages) and significant sea level rise (something that, currently, has barely begun).

Ultimately, the most important changes are those to the biosphere, the sum of which are now colossal (Williams et al. 2015). These changes are complex, having begun on the land before affecting the sea. They include species extinctions, though the scale does not yet approach the "big five" mass extinctions of the past five hundred million years. But, with many species now critically endangered, "business-as-usual" will probably see a comparable mass extinction within a few centuries (Barnosky et al. 2011). A related biological change is already on a scale without precedent in Earth history: the species translocated across the globe by humans, consciously or unwittingly. These mass species invasions have led to the proposal of a "Homogenocene" (Samways 1999). Any such biological change has permanent effects, as it determines the course of future evolution. In this sense, the course of Earth history has already been reset.

The Anthropocene is hence multifactorial (and the factors naturally interrelate, for instance as climate warming drives further biological change), complex, evolving, and in its early stages. What does it mean more widely for society? First, it provides an integrated overview of climate change, related to such concepts as planetary boundaries—the limits that should not be exceeded to ensure a functional Earth system. It may be criticized for undermining efforts at conservation (Caro et al. 2011), or invoked to accept (and take responsibility for) human planetary domination (Ellis, Antill, and Kreft 2012). Even its formalization may have significance beyond geological nomenclature, being significant to international environmental law (Vidas, in Williams et al. 2011). Its evolution, as a concept and as observed geological reality, will encompass the future of this planet and of its inhabitants.

4

Biodiversity

Andy Dobson

"Biodiversity" was coined by E. O. Wilson in 1988 to describe all living organisms on the planet (Wilson 1988); it is a condensation of "biological diversity." It is more commonly used to describe all species other than humans; this creates an unfortunate dichotomy between the voting and nonvoting species! Economic, aesthetic, and health benefits that biodiversity contribute to the human economy are termed "ecosystem services" (Daily et al. 1997; Daily et al. 2000). We are only just beginning to quantify these and are rapidly realizing how dependent the quality of human life and economic well-being is upon biodiversity. Ironically, the current increasing rates of species extinction may provide the sharpest way to quantify human dependence upon biodiversity.

Biodiversity's greatest strength is its diversity; it includes huge organisms such as whales, and giant redwoods that live for centuries. At the opposite end of the spectrum are viruses and bacteria that are only visible under electron microscopy, including some of the most dangerous infectious agents on the planet: smallpox, anthrax, and influenza. The abundance of biodiversity is simultaneously its greatest problem; despite several centuries of biological investigation, we still have no exact idea of how many different species share the planet with us. Best estimates range between five and ten million; the number could be as high as one hundred million (May 1988; May 1990). Because we do not know how many species are currently extant, we do not know how rapidly they are going extinct. Our best estimates suggest we are losing biodiversity at around one hundred to a thousand times the normal background rate of extinction (May et al. 1995).

Although the sheer diversity of species inhabiting the planet seems almost overwhelming, taxonomists, ecologists, and evolutionary biologists continue to find ways to organize it from perspectives that reflect the underlying process of speciation and the birth and death processes that both determine abundance and reflect interactions between species in their major roles as consumers of, and as resources for consumption by, other species. New biodiversity is created by speciation when the processes of natural and sexual selection act upon two or more subpopulations of the same species that are partially isolated from each other. If selection operates with sufficient intensity for long enough, then the original species will diverge to form two subspecies that will eventually be incapable of breeding with each other. The amount of biodiversity on the planet at any time always reflects the interaction between speciation and extinction rates. Ultimately, all extinct and extant species evolved from a common ancestor that lived in the primitive organic soup of the world's oceans sometime between two and five billion years ago. The branching processes that are driven by speciation create a natural organization of species into a hierarchical treelike system of branches and twigs that define groups of closely related species with a common evolutionary ancestor (that may be now extinct). The first person to classify all known biological species in this way was Linnaeus (Linnaeus 1735). Subsequent early classifications were based on the morphology of species; more recently this has been replaced by molecular trees that compare the DNA (or RNA) that contains the genetic code that drives the individual physical development of each individual of each species. Mutations in this code are the raw material upon which natural selection operates to modify structure and function and drive speciation. Subdivision of populations through time give rise to new species, which creates an evolutionary treelike structure for all species that have ever lived; the genetic relationships between species are defined by the branch lengths between them, and these define the total time since they shared a common ancestor.

On a broader geographical scale, speciation and dispersal have tended to spread species from their commonest origin in the tropics out into the temperate and Arctic zones. This pattern has in turn been modified by the slow underlying breakup and movement of the continents: thus the oldest plant groups are found in New Caledonia and northeastern Australia, while an arc through Antarctica follows the main trunk and many of the major branches of the evolutionary tree of plants to colonize South America. The ancient diversity of plants then decreases as a subset of the original botanical diversity crosses the Atlantic and passes through Europe and down into Africa, or across the Urals into Asia. The strip of rainforest along the coast of Queensland contains nearly all of the world's most ancient plant families; in contrast, although the forests of Indonesia and the Fynbos of South Africa have a huge number of plant species, they have mainly evolved recently as massive radiations of closely related species from within very few plant families. The insects that feed on these plants show similar patterns of radiation; in contrast, the geographical patterns in the mammals and birds are less distinct, given their much greater ability to disperse more rapidly. Once these patterns of speciation and radiation were recognized, many creation myths were quickly tripped up, particularly the Christian one: it is essentially impossible to reconcile the simultaneous creation of ten million species with the observed underlying genetic and geographical structure of biological diversity that reflects sequential, hierarchical, and intrinsically timed evolutionary radiation from an ancestral ocean-dwelling, single-celled organism (Darwin 1859).

Although the amount of biodiversity seems bewildering, there are curiously beautiful and consistent patterns in the way the abundance and diversity of species seems to be organized: there are many more species in habitats close to the equator than at the poles, suggesting that sunlight, day length, and available moisture are important in determining the number of species that can be supported at any location. Similarly, we tend to find more species on continents and larger oceanic islands than we do on smaller and more isolated oceanic islands (MacArthur and Wilson 1967). Within any location there is always a mixture of a few abundant species and many rare species. This delights bird watchers and plant hunters, who ferociously seek to identify as many rare and uncommon species as possible. Yet, the relative abundance of plants, or birds, or insects, or bacteria at any location seems to follow an underlying statistical pattern that is remarkably consistent and almost predictable. If the numbers of individuals are plotted out as a frequency distribution, boxed into categories that sequentially represent a doubling of abundance, then the pattern is always very close to log-normal (the notorious bell curve), except with logarithmic classes of abundance along the x-axis. Ecologists have long been fascinated with the mechanisms that produce this pattern: once we know the number of species that occupy a region, then we can quickly ascertain their relative abundance (May 1975). The most parsimonious explanation has been provided by Hubbell's Neutral Theory of Biodiversity (Hubbell 2001), which suggests that the observed patterns of relative abundance can be produced by a model that assumes all species are equal and have similar birth, death, and extinction rates (Hubbell 1997). Yet biologists know that these assumptions

BIODIVERSITY ANDY DOBSON

cannot be correct, as different species tend to have different birth and death rates and actively compete with each other. So two of the greatest questions facing ecologists who are interested in diversity are (1) what determines the number of species that can coexist at a particular location? and (2) how does this factor interact with the birth, death, and extinction rates of neutral theory to create the consistently observed patterns of species abundance? Ultimately, we will also need to know whether the observed species abundance relationship creates resilience in the ways in which biodiversity contributes ecosystem services to the human economy.

At any location, biodiversity is organized into a food web: the many plant species will be fed upon by slightly fewer species of herbivorous animals that can digest cellulose and other plant structural material. These in turn will be fed upon by even fewer predatory species. Throughout their lives, all these free-living species will be fed upon by a large diversity of parasitic species that slowly consume them from the inside out, rather than the outside in. These parasitic species may form between 40 and 90 percent of the species networked together within the food web at any location (Dobson et al. 2008; Lafferty et al. 2008). There will also be a large diversity of "decomposer" species that digest individuals that have died, or consume the excretory products that live individuals discard. We have no good estimate of the abundance of decomposer species, but a simple thought experiment tells us that the world would be a much messier place in their absence. We are just beginning to understand the structure of food webs; mathematically, it is one of the deepest and most intractable problems in science and arguably one requiring the most urgent attention. Central to the complexity of this problem is that the mathematical structure of all food webs always becomes increasingly less stable as the number of interacting species increases (May 1974).

Mathematically, there have to be constraints on the intensity with which species interact with each other, constraints that interact with the geometrical structure of the web in ways that allow multiple species to interact, consume, compete, and coexist with each other. Ecologists have begun to identify some of these rules and geometries (McCann 2011). The current rates of extinction emphasize the urgency of fully understanding this problem before there are either no pristine webs left to study, or the reduction in vital services supplied to humans declines dramatically.

5

Biomimicry

Bryony Schwan

Biomimicry is a relatively new design methodology that studies nature's best ideas, abstracts its deep design principles, and then imitates these designs and processes to solve human problems. The term "biomimicry" comes from the Greek words "*bios*," meaning "life," and "*mimesis*," meaning "to imitate." Related to yet also different from terms in earlier use, such as "bionics" and "biomimetics," biomimicry—an approach popularized by Janine Benyus in her 1997 book *Biomimicry: Innovation Inspired by Nature*—entails the "conscious emulation of life's genius" (Benyus 1997), utilizing design strategies that have been fine-tuned through 3.8 billion years of evolution. Whether in the areas of energy, material manufacture, recycling, chemistry, engineering, transportation, or computing, other organisms have managed to do many of the things humans want to do, without depleting fossil fuels, polluting the planet, or mortgaging their future.

The concept of looking to nature for design ideas is not new. Humans have been observing and imitating processes in nature for thousands of years (Merchant 1980). Leonardo da Vinci, the Wright brothers, and Buckminster Fuller, for example, are well known for their observations and innovations inspired by nature, as are many indigenous peoples. The practice of intentionally emulating nature's forms and processes to create more sustainable technologies, however—the principle aim of biomimicry—is a comparatively recent phenomenon.

Innovators in a variety of fields are using biomimicry to develop new technologies to solve a range of human challenges, from health care and water purification to pollution and clean energy to architecture and automation (Lakhtakia and Martín-Palma 2013; Passino 2005; Pawlyn 2011). Inspired by nature's logarithmic spiral, for example, fluid dynamics design firms are engineering fans and propellers that use 30–70 percent less energy. Vaccines that no longer need to be refrigerated because they contain new preservatives that mimic organisms in nature that can survive in a dormant state for decades are now reaching more isolated populations. Colors created from material structure, modeled after the physical manipulations of light that produce the vibrant blue of a Morpho butterfly or the brilliant hues of a peacock feather are replacing harmful pigments in textiles and paints. Nontoxic, waterproof adhesives inspired by the "glue" of Blue mussels are revolutionizing wood composites by eliminating the off-gassing of volatile organic compounds in new buildings. Around the world, the number of biomimicry papers, patents, and products is growing rapidly (Bonser and Vincent 2007; AskNature.org).

Biomimicry requires its practitioners to look beyond the mere imitation of a single function, such as the form or shape of an organism, to the sustainability models provided by more complex natural systems. Where many sustainability frameworks use political or consensus-driven parameters, biomimicry uses ecology to determine appropriate place-based standards. Energy parameters for the design of a building in a desert climate, for example, would be modeled from organisms found in that specific habitat and would thus be quite different from those in an arctic environment. The

Eastgate Center in Harare, Zimbabwe, which mimics the intricate ventilation tunnels of termite mounds to reduce energy costs, is perhaps the best-known contemporary example of such modeling (Grant 2012). Biomimicry can work at even larger scales as well. The international architecture and engineering firm HOK has incorporated the principles of biomimicry into the master plan for the development of the new hill city, Lavasa, in the Mose Valley near Pune, India. By studying and measuring the behavior of local plants and animals, such as the deflection of rainwater by tree canopies, or the mound-building techniques of harvester ants, and then using these observations to shape the city's design, HOK is aiming not only to integrate the project with local ecosystems but also to develop sustainable, site-specific "ecological performance standards" (Peters 2011).

While biomimetic ideas and approaches have gained broad acceptance, they have not done so without critique. Some question whether biomimicry relies too uncritically on a belief in nature as a perfect system rather than an occasionally flawed one, and on a concept of "nature" that is entirely separate from human culture, rather than seeing the natural and the human as already interdependent systems (Armstrong 2013). Although some theorists promote a biomimicry that operates "in the service of a bio-inclusive ethic," others have proposed biomimetic designs that in theory could replace rather than enhance the environments they mimic, giving biomimicry an ethical ambiguity that some environmental philosophers say remains to be resolved (Mathews 2011). Like all energy-saving "green" methodologies, biomimicry may also be subject to the Jevons Paradox, in which savings in production or consumption are often accompanied by overall increased energy usage (given the growth imperative in market economies), potentially negating any ecological improvements unless the broader political, economic, and cultural contexts in which they are applied are also challenged (Alcott 2005). Despite these critiques, it is the hope of practitioners like Benyus that truly sustainable biomimicry will usher in "an era based not on what we can *extract* from nature, but what we can *learn* from [it]" (Benyus 1997).

6

Biopolitics

James J. Hughes

The term "biopolitics" has four distinct but overlapping meanings in modern scholarship. According to Lemke's history of the term (Lemke 2011), political scientists used "biopolitics" in a variety of ways as early as the 1920s, and the Third Reich used it to describe their eugenic plans. But the term really found common usage first among 1960s political scientists interested in the relationship of evolutionary biology and politics (Caldwell 1964). Forming the Association for Politics and the Life Sciences (APLS) in 1981, they defined "biopolitics" as any investigation of the effect of biology on politics, from the primate origins of hierarchy to infectious disease impacts on warfare to the influence of the health of leaders on their decision making (Somit and Peterson 1998).

Biopolitics scholars argued against the prevailing social science model that human beings were a tabula rasa shaped by socialization. During the fights over sociobiology (Wilson 1975) in the 1970s and 1980s, biopolitics scholars were suspected of biological determinism and attempts to condone sexism, racism, and other social problems (Lewontin, Rose, and Kamin 1984; Segerstråle 2001). Thirty years later, the emergence of evolutionary psychology and social neuroscience has validated and depoliticized this form of biopolitical investigation, which has been used in political arguments by both the Left (Singer 2000) and the Right (Arnhart 2005). Now there is growing evidence of the genetic heritability of political attitudes and the neurogenetic variations that govern disgust sensitivity, empathy, egalitarianism, intelligence, and other traits that shape political attitudes (Alford and Hibbing 2008; Hatemi et al. 2015).

The second and still the most influential meaning of "biopolitics," in the humanities, derives from the work of Michel Foucault. Throughout his career, in his work on mental illness (1965), medicine (1973), prisons (1977), and sexuality (1978), formalized in a series of lectures on biopolitics in 1978 and 1979 (2010), Foucault elaborated the idea that institutions develop biopower by creating structures of knowledge about bodies and populations. For Foucault, biopolitics was the effort of states to regulate bodies to ensure their productivity as workers, their obedience as citizens, and their conformity to social norms. This Foucauldian concept of biopolitics has been developed and applied widely, from Giorgio Agamben's writings on death camps as the biopolitical telos of "thanatopolitics" (Agamben 1998) to Nikolas Rose's (2007) explorations of biopower as key to modern governance to Michael Hardt and Antonio Negri's neo-Marxist (2000) interpretation of biopolitics as a counterhegemonic struggle against biopower.

The third use of "biopolitics" has been to encompass all the ways in which public policy shapes medicine, public health, and biotechnology. These topics were always present in the journal of the APLS and acknowledged as part of their scope, although an essay on these topics would rarely have been described as biopolitics. Now, however, there are a growing number of works drawing on the growing biopolitics literature, both ethological and Foucauldian, to illuminate health policy, such as Maren Klawiter's 2008 *Biopolitics of Breast Cancer* and Jan Wright and Valerie Harwood's (2009) *Biopolitics and the "Obesity Epidemic"* collection.

Fourth, the term "biopolitics" has been applied to the emergence of a new biopolitical axis, emerging out of bioethics, comparable to the political-economy and cultural axes that have dominated twentieth-century politics in the liberal democracies (James Hughes 2004, 2009). The origins of contemporary biopolitics can be found in the views of Enlightenment philosophy that advocated for individual autonomy, the use of reason, science, and technology to improve the human condition, the separation of church and state, and the importance of individual subjectivity to personal identity (Hughes 2010). These views formed the basis of liberal Western bioethics after World War II, and shaped the development of policies regulating abortion, animal rights, and death and dying, although at first with little involvement by mass political movements. The politicization of bioethics into biopolitics began with the emergence in the 1970s of the right-to-life movement, which made the rights-bearing status of embryonic life a defining ideological stance. Under the influence of philosophers like Peter Singer (1990), the growing animal rights movement made a parallel move articulating the rights-bearing status of animals. In the 1980s and 1990s these two movements forced an articulation of the biopolitical terrain between advocates of human exceptionalism and advocates of nonanthropocentric rights.

By the mid-2000s, after the appointment of a conservative-dominated President's Council on Bioethics (PCBE) by the Bush administration, American bioethicists had polarized into bioconservative and bioliberal camps with characteristic views on a wide variety of bioethical issues. After polarization, bioliberals drew sharper lines in defense of science, secular liberty, and individual autonomy. While bioconservatives in secular Europe are largely influenced by green politics, bioconservatives in the United States are predominantly religiously motivated. This polarization has reflected and been encouraged by the growing involvement of mass political groups in bioethics policy, from stem cell research to end-of-life care. Since bioconservatives and bioliberals come from both the Right and the Left, however, the biopolitical axis cuts across traditional political ideologies (Rifkin 2002; Sutti 2005; Moreno 2012). As green parties also asserted that they were beyond left and right when they first emerged, and then were assimilated into the broad Left, biopolitics may also succumb to the gravitational forces of the larger left-right political universe. On the other hand, the divisions within the Left and the Right over genetically modified organisms, reproductive technology, vaccination safety, and other issues illustrate the persistent orthogonality of biopolitics (Roache and Clarke 2009).

Biopolitics are just beginning to emerge in the developing world, where they may develop along other lines, especially in Asia. Abrahamic beliefs about ensoulment lead to conflict over abortion, embryonic research, brain death, and animal rights, while Abrahamic beliefs in a divine plan for humanity are hard to square with plans for radical human enhancement. Unshaped by the West's struggle between the Church and the Enlightenment, biopolitical thought in South and East Asia has a different structure. Hinduism, Buddhism, and Confucianism accept a moral and evolutionary continuity between animals, humans, and super-human beings. Surveys of bioethical opinion from India to Japan find far less resistance to genetic engineering and other emerging technologies than in the West (Macer 2014). While Western bioliberals combine individualism with techno-optimism, and Western bioconservatives trump rights claims with communitarian concerns, South and East Asian biopolitical thought often combines techno-optimism with communitarianism and authoritarianism.

With the publication of Francis Fukuyama's *Our Posthuman Future* (2002) and the PCBE's critique of reproductive technology, cognitive enhancement, and life extension (Kass et al. 2003), the debate over human enhancement technologies moved from the fringe to the center of biopolitics. The PCBE focused on the bioconservative belief in an important distinction between the use of biotechnology for therapy as opposed to its use for enhancement, a line that when crossed would lead to hubris and disaster.

While bioliberals like Nicholas Agar (2005), Arthur Caplan, Jonathan Moreno (2012), Julian Savulescu, and John Harris (2007) have defended human enhancement and other biomedical technologies, they do so with many qualifications. "Transhumanists" are the radicals on the biopolitical spectrum (Bostrom 2005; Miller 2006; James Hughes 2009). Ranging politically from anarchocapitalists (Bailey 2005) to leftist "technoprogressives" to millennialist "Singularitarians," the transhumanists embrace radical life extension, cognitive enhancement, and body modification, including the uploading of consciousness to nonorganic platforms.

Biopolitics in the context of the transhumanist-bioliberal-bioconservative debate incorporates competing visions about the desirability of various kinds of posthuman species or transhuman societies. Fukuyama argued that enhancement would fracture the biological similarity that supposedly undergirds democratic solidarity, an argument made more persuasively in Agar's 2010 *Humanity's End.* In response to this alleged threat, the left-leaning bioethicists George Annas and Lori Andrews have campaigned with bioconservative groups for an international treaty to make cloning and germline genetic therapy a "crime against humanity" (Annas, Andrews, and Isasi 2002). Green-leaning groups have broadened their attacks on gene-modified crops to include opposition to gene-modified humans, while left critics believe reproductive technology portends a new racist, classist, and patriarchal eugenics. At these extremes, biopolitics has developed its own revolutionary and apocalyptic wings presaged by the bombing campaign of the eco-primitivist Ted Kaczynski from the 1970s to the 1990s.

Presumably, if the term "biopolitics" continues to be used, it will increasingly refer to ideological schisms around the regulation of biotechnology, and its associations with the Foucauldian or political science academy will decline.

7

Bioregionalism

Keith Pezzoli

Bioregionalism is a social movement and action-oriented field of study focused on enabling human communities to live, work, eat, and play sustainably within Earth's dynamic web of life. At the heart of the matter is this core guiding principle: human beings are social animals; if we are to flourish as a species, we need healthy relationships and secure attachments in our living arrangements with one another and with the land, waters, habitat, plants, and animals upon which we depend. Unfortunately, we have lost our way. Humanity's collective capacity to nurture healthy relationships and secure attachments is not being realized. Thus, bioregionalists argue, we need to establish new, just, ethical, and ecologically resilient ways to reconnect with one another and with the land.

Bioregionalism's core commitments include (1) rebuilding urban and rural communities—at a human scale—to nurture a healthy sense of place and to secure attachments and rootedness among community inhabitants; (2) reintegrating nature and human settlements in ways that holistically instill eco-efficiency, equity, and green cultural values into systems of production, consumption, and daily life; (3) making known (and valuing) the way wildlands, working landscapes, ecosystems, and rural dwellers and resources enable cities to exist; (4) developing authentic community-based participatory processes that empower just and equitable civic engagement in local and regional planning, visualization, and decision making; and (5) building global transbioregional alliances and knowledge networks to support sustainable place making around the world. To meet these commitments, bioregionalists engage in grassroots activism, group mediation, consensus building, not-for-profit and for-profit entrepreneurism, food and fiber production, ecological restoration, art, music, philosophy, spirituality, science, education, and multimedia communications.

The spatial scale of bioregional initiatives varies. Bioregionalists focus on watersheds ("ridge top to ridge top"), multiple watersheds ("landscape scale"), river basins, and even much larger swaths of the Earth's surface, including ocean regions. Kirkpatrick Sale (2000, 55), one of bioregionalism's leading proponents, outlined three nested spatial scales pertinent to bioregionalism. The largest scale is the *ecoregion*—roughly twenty thousand square miles or more (e.g., Ozark Plateau in the United States, Sonoran Desert in Mexico, Chilean Mattoral). The scale beneath this is the *georegion* (e.g., the Central Valley of California nested within the Northern California Ecoregion). At the smallest end of the spectrum is what Sale calls *morphoregions* (several thousand or fewer square miles), identifiable by human settlement patterns, including cities, towns, infrastructure, factories, fields, and farms. These scales are nested one within the other. Globally minded bioregionalists of the twenty-first century will have to forge alliances to address problems that cut across these scales, and across urban-rural divides—for instance, global climate change; globalization of disease vectors; pandemic threats; acidification of the oceans; soil degradation; large-scale disasters; warfare; and injustices associated with race, class, and gender (Evanoff 2011; Hibbard et al. 2011; Pezzoli, Williams, and Kriletich 2010).

Given the emphasis bioregionalism places on reconnecting people to the land in ways familiar to its inhabitants, it is at smaller scales where most bioregional action takes place. In people's lived experience, bioregions are "life-places" where biogeography, ecology, culture, economy, and history interact in a distinct, discernable pattern. Peter Berg and Raymond Dasmann (1977, 400) coauthored one of the first documents in the United States to spell out the meaning of "bioregion": "The term refers both to geographical terrain and a terrain of consciousness—to a place and the ideas that have developed about how to live in that place." Inhabitants define for themselves their bioregion's boundaries. The most relevant factors include climate, topography, flora, fauna, soil, and water together with the territory's sociocultural characteristics, economy, and human settlement patterns.

Robert L. Thayer Jr. (2003, 55), a widely noted bioregional activist-scholar, aptly argues that "the bioregion is emerging as the most logical locus and scale for a sustainable, regenerative community to take root and to *take place*." That a bioregion is a fruitful place-based organizing concept stems from the premise that "a mutually sustainable future for humans, other life-forms, and earthly systems can best be achieved by means of a spatial framework in which people live as rooted, active, participating members of a reasonably scaled, naturally bounded, ecologically defined territory, or *life-place*" (Thayer 2003, 6).

The aims of bioregional initiatives vary. In the United States, the Klamath Basin Coalition focuses on 15,751 square miles of land spanning the U.S. states of Oregon and California and drained by the Klamath River; it is a diverse alliance of local, regional, and national organizations promoting comprehensive social-ecological restoration and sustainable livelihoods in the Klamath River Basin (Doremus and Tarlock 2008; Gosnell and Clover Kelly 2010). The Kansas Area Watershed Bioregion is restoring the ecological health of the prairie and its inhabitants. California's Biodiversity Council (CBC) subdivided the entire state of California into ten bioregions as part of a watershed/landscape scale strategy to protect biological diversity and the economy. In northern California, a range of bioregional groups is focusing on fishing, water quality, timber, and agriculture in ways that address livelihood and conservation. Beyond the United States, the governments of seven Central American countries plus Mexico created a Mesoamerican Biological Corridor, which pays homage to bioregional principles for the protection of biodiversity. New Zealand divided its entire national territory into ecoregions and watershed-based subregions for bioregional policy and planning purposes, backed by an environmental court system (including landscape architects).

Bioregional initiatives conducted at the national and international scale do not necessarily advance the more radical aspects of bioregionalism (e.g., authentic participatory democracy, communitarianism, subsidiarity, mutual aid). Yet national and international efforts may help reframe public discourse, thereby creating new opportunities to advance bioregionalism's core commitments on the ground locally. This may happen, to cite an example, in Bolivia, where the country's legislators passed the Law of Mother Earth in 2012.

The Bolivian Law of Mother Earth is based on Andean philosophy that views "*Pachamama*" (a term used commonly throughout the Andean region, associated with "Mother Earth" but meaning a nongendered "Source of Light, or Source of Life" [cf. de la Cadena 2010]) as a sentient being and sacred home. Article 3 of the law defines Mother Earth as "the dynamic living system formed by the indivisible community of all life systems and living beings who are interrelated, interdependent, and complementary, which share a common destiny." The law provides legal-institutional support for nature itself as a "collective subject" entitled to legal rights, including rights to life, ecosystem health, and regeneration.

Whether or not the law generates significant change is far from certain. However, it may make it easier for indigenous movements to gain leverage for their causes using their cultural heritage and Traditional Ecological Knowledge (TEK)—including sophisticated agroecological knowledge accumulated over thousands of years. Some academics have referred to these Global South movements as "indigenous cosmopolitics" (de la Cadena 2010). Along such lines, the Zapatista movement in Chiapas, Mexico, rebuked neoliberalism in a quest for local autonomy and the preservation of their eco-cultural heritage. These movements have much to offer twenty-first-century understandings of bioregionalism, especially when it comes to ecological governance, sustainable livelihoods, and management of common wealth. Indigenous struggles around food sovereignty are a strong case in point (Adamson 2012b).

Bioregionalists tend to fault foot-loose transnational finance and the reckless growth of the world's resource-intensive, export-oriented, hypermobile, corporate-dominated global economy for many of the twenty-first century's increasingly complex problems (e.g., climate change, peak oil, peak fresh water, ecosystem degradation, runaway economic speculation, food insecurity, poverty, and a widening gap between haves and have-nots). In view of such problems, bioregionalists advocate "localization." Localization is a place-based strategy to create sustainable and resilient communities at a human scale; it encourages local investments in resources, livelihoods, and institutions that build the community's assets (including community power/capabilities) (De Young and Princen 2012).

Peter Berg, Raymond Dasmann, Gary Snyder, Robert G. Bailey, Freeman House, and Ian McHarg, among others, laid the initial foundation for bioregional thought and practice in the United States during the 1960s and 1970s. This early work has roots in utopian and anarchist ideas from the late eighteenth and early nineteenth centuries (Owen; Fourier; Proudhon; Bakunin; Kropotkin) and in the call for ecological regionalism first articulated in the late nineteenth and early twentieth centuries (Geddes; Howard; Odum; Mumford; the Regional Planning Association of America). Bioregionalism in the United States initially developed outside academia. Its identity was countercultural—a back-to-the-land and appropriate-technology movement responding to the perceived failings of North American and Eurocentric conceptions of industrial modernism. Gary Snyder, a civically engaged intellectual and poet, along with Peter Berg, helped get bioregionalism onto a serious academic footing by interweaving counterculturalism with incisive critical analysis and methodology (Snyder 1995; Berg 2009). They accomplished this through civically engaged scholarship and effective science communication enriched with poetry, theater, art, and experiential learning. Doug Aberley (1999), Michael Vincent McGinnis (1999), Robert L. Thayer Jr. (2003), and Tom Lynch et al. (2012), among others, have also added academic rigor to bioregionalism.

Doug Aberley (1999, 13) provides a good historical overview of bioregionalism's development over time, outlining how bioregionalists worldwide have "learned the cultural and biophysical identity of their home territories—their bioregions. They have also worked to share the lessons of this hard-won experience, developing intersecting webs of bioregional connection that now stretch across the planet." Bioregionalism has a track record that includes semiannual congresses (meetings of upwards of two hundred bioregional enthusiasts, from North America mainly but also from other countries around the world). The Planet Drum Foundation (www.planetdrum.org) and the Bioregional Congress (www.bioregionalcongress.net) have been documenting and archiving the congresses. An increasing number of universities offer degrees in bioregionalism, including the University of Idaho, Utah State University, the University of California at Davis, Montana State University, and the University of Pennsylvania. Bioregional programs outside the United States can be found at universities in Mexico, Asia, Europe, and Australia, among other places.

The number of organizations advocating bioregional principles (though not always identifying themselves as bioregionalists) is growing. In the United States, the New Economy Working Group (a progressive, multi-partner, action-oriented think tank) highlights the possibility and potential of localized living economies that support a healthy biosphere (www.neweconomyworkinggroup.org). Ecotrust, based in Portland, Oregon, operates an environmental bank, an ecosystem investment fund, and a range of programs in fisheries, forestry, food, farms, and indigenous affairs (www.ecotrust.org). On a global scale, Bioneers, a group of social and scientific innovators, grows social capital by building bioregional and community-based alliances (www.bioneers.org). The Global Action Research Center (the Global ARC) uses bioregionalism as a framework to help connect the dots among diverse types of globally minded localization (e.g., community gardens, watershed protection, greening of livelihoods, and infrastructure) (www.theglobalarc.org).

At the dawn of the twenty-first century, bioregionalism has an increasingly effective set of tools and concepts from which to draw, including locavore, locavesting, slow foods, food justice, green infrastructure, adapting-in-place, eco-districts, biocapacity, ecological footprint, BALLE (business alliance for local living economies), LOIS (local ownership and import substitution), community-based natural resource management (CBNRM), relocalization, and reinhabitation. The efficacy of bioregionalism as a movement hinges on the application of such tools and concepts through learn-by-doing and "living-in-place." Nurturing bioregional culture and imagination is key. Fortunately, there are an increasing number of groups skilled at doing this—i.e., skilled at cultivating place-based environmental literacy and a new ecological progressivism through the arts and humanities (Lynch et al. 2012). California Northern is a good example (www.calnorthern.net). Their work provides a deeper understanding of bioregions by fostering strong local identity through effective place-based storytelling, art, and communications. Along such lines, Thayer Jr. (2003, 94) expresses hope that as bioregional groups struggle to reinhabit their life-place, "A distinctly regional art, aesthetics, literature, poetics and music can evolve from and support bioregional culture." This hope has deep roots; it resonates with a view expressed three-quarters of a century ago by Lewis Mumford. In his classic work *The Culture of Cities* (1938), Mumford eloquently argued that "[t]he re-animation and re-building of regions, as deliberate works of collective art, is the grand task of politics for the coming generation." This oft-quoted statement rings as true today as it did seventy-five years ago.

8

Biosemiotics

Timo Maran

Biosemiotics, or semiotic biology, is the study of qualitative semiotic processes that are considered to exist in a variety of forms down to the simplest living organisms and to the lowest levels of biological organization. Biosemiotics can be seen as an alternative to the mainstream approaches of contemporary evolutionary biology that use reductionist quantitative methodologies and tend to objectify living processes. Emphasizing the role of sign processes in nature makes it possible to restore the "subjectness" or agency of living organisms that in turn are considered to influence larger ecological and evolutionary processes. Here, a sign process or "semiosis" is defined as a process, in which something—a sign—stands to somebody for something in some respect or capacity (Peirce 1931–1935, 228). A simple example is a bird song that indicates to the singer's species mates that he is guarding his nesting ground. In biosemiotics, processes taking place inside an organism, such as interpretation of DNA for protein synthesis by a cell, are also regarded as sign processes. Although, up until now, biosemiotics has been a paradigm mostly in the biological sciences, this field of study is increasingly referred to in cultural and literary studies.

As a discipline, biosemiotics emerged from comparative studies of animal communication conducted by the Hungarian-American semiotician Thomas A. Sebeok in the 1960s (and referred to, at the time, as "zoosemiotics," Sebeok 1972, 1990). Semiotic approaches to animal communication provided biosemiotics with the concepts of repertoire, code, and context that connect the parties of communication and make mutual understanding possible. Later reconstructions of the history of biosemiotics trace the field back to German biology, mostly to the *Umwelt* theory of Jakob von Uexküll (1982), and to theories of the American philosopher and semiotician Charles S. Peirce. Uexküll's *Umwelt* theory gives biosemiotics its subject-centered perspective. *Umwelt* theory describes an organism's relations with its environment as shaped by its species-specific perceptual and cognitive capacities and organized by meanings that bind the animal to living and nonliving entities in its environment. An important principle for biosemiotics is to consider semiotic and biological processes as they appear to the organism and to treat biological communities as the sum of interconnecting *Umwelten*. Relying on Peirce's semiotics helps biosemioticians to study semiotic processes in other species. Peirce's understanding of sign differs in important respects from the other major semiotic tradition, the semiology of Swiss linguist Ferdinand de Saussure, who proposed a two-part model of the sign. In contrast, the Peircean sign is tripartite (including the "object" that can also be an environmental object) and does not rely on the existence of (human) language. These properties make Peircean semiotics suitable for describing sign processes occurring outside the human species.

Synthesizing different intellectual traditions allows biosemioticians to raise questions about both the general properties of biological communication systems and the special position of human language therein. For instance, is the combination of digital and analogical codes (such as DNA and behavioral codes) necessary for the development of any complex biological system (Hoffmeyer and Emmeche 1991), are there any special rules for communication between the members of the different species (Maran 2012), or what are the similarities and differences between the codes that cells use for interpreting DNA and the codes of human language? Early efforts to organize the field of biosemiotic studies followed largely taxonomical logic by distinguishing zoosemiotics (semiotic study of animals, see Maran, Martinelli, and Turovski 2011) and phytosemiotics (semiotic study of plants, Krampen 1981). Later alternative classifications following hierarchical logic were proposed that distinguished between endosemiotics (study of semiotic processes inside organism), zoosemiotics (study of semiotic processes between animal organisms), and ecosemiotics or ecosystem semiotics (study of semiotic aspects of ecological processes, Nielsen 2007) or were based on the mechanisms of sign processes, distinguishing between the study of vegetative (based on analogical relations and recognition), animal (based on physical linkage and associations), and cultural (based on symbolic relations and conventions) semiosis (Kull 2009).

In the last decade, biosemiotics has advanced significantly, as exemplified by the establishment of the International Society for Biosemiotic Studies, the launch of the journal and book series, both named *Biosemiotics*, and the emergence of biosemiotic courses in curricula of many universities (for historical overviews, see Kull 1999; Favareau 2009). At present the most intriguing topics in biosemiotics seem to include the role of codes and coding in semiotic processes (Barbieri 2010; Markoš and Faltýnek 2011), as well as developing semiotic views on biological evolution (see Schilhab, Stjernfelt, and Deacon 2012). The logic of biosemiotic inquiry can be demonstrated by research questions provided by Kull, Emmeche, and Favareau's 2008 essay, "Biosemiotic Questions": what are the major modes of biosemiosis? How does the world appear to the organism, and what are the methods that allow the study of subjective worlds (*Umwelten*)? What are the general biological functions that are made possible through the phenomenon of semiosis? These questions can be used, for instance, to think about the process of recognition. Recognition is clearly an *Umwelt*-specific phenomenon, as humans do not perceive the same signs (or at least not in the same way) as other animals. Recognition may be a key for important biological functions: for instance, reproduction in most cases is not possible without recognition of the mate. Being a qualitative process, recognition requires special methods of study (*Umwelt* modeling, participatory observation). One example of practical research where biosemiotics has been applied includes the ongoing search for minimal biological entities that show any activity in perceiving and producing signs (e.g., "autocell" in Deacon 2006). Biosemiotic concepts have also been fruitfully applied in landscape ecology to study the engagement of different species with the environment, their use of resources, and their interaction and conflicts both with one another and with human influences (Farina 2008).

Biosemiotics is also contributing to studies of human culture as it is taken up by ecocritics (see Garrard, this volume), ecofeminists (see Gaard, this volume), and multispecies ethnographers (see Rose, this volume), among others. "Biosemiotic criticism," as I have termed it (Maran 2014), is attracting a number of scholars from environmental literary criticism and other environmental humanities (e.g., Wheeler 2008; Coletta 1999; Maran 2010; Tüür 2009; Siewers 2011). Biosemiotics broadens the reach of sign processes into the biological realm as well as into the inner milieu of the human organism, providing the humanities a substantiated expansion beyond human cultural processes and artefacts. There are several possible approaches in biosemiotic criticism. For instance, biosemiotic tools can be applied to cultural phenomena on the basis of analogies between biological organisms and human culture, or through study of different ways that humans and other animals model their surrounding environment (Sebeok 1991), or through study of representations of semiotic processes either from inside or from outside the human body or from the environment. An example of the latter would be to inquire into the ways in which human bodily perceptions and reactions (having a cold, developing an immune reaction, recovering from an injury or a shock) relate to creation and interpretation of literary and artistic works.

Biosemiotic criticism can be invoked to ask questions such as, what are the roles and relations between different modeling processes in cultural representations? Thomas A. Sebeok has distinguished between linguistic and zoosemiotic types of modeling. Linguistic modeling based on human language is specific to the human species, whereas zoosemiotic modeling unites humans with other animals. Zoosemiotic modeling is based on *Umwelt* structure, where signs distinguished by the organism's species-specific perceptual organs are aligned with its behavioral resources (Sebeok 1991). On the basis of this distinction we can ask, where and in what forms does prelinguistic zoosemiotic modeling occur in humans? In literary studies, what is the role of the author and readers as biological semiotic creatures in literature and other cultural artifacts? Are there traces of zoosemiotic modeling in nature essays and in other cultural representations of nature? In which ways do different models and representations of nature loop back to influence the material structures of the world? For instance, in nature essays, authors are often present in the text as bodily creatures perceiving their environment by sights, sounds, smells, and touch. From the perspective of biosemiotic criticism, representing not just the environment but also subtle semiotic ways of relating to the environment (so-called sensory sign, Hornborg

2001) appears to be an important communicative and didactic strategy.

The research program of biosemiotic criticism is still at a very early stage of development and is waiting for further practical applications to different research objects as well as for syntheses with paradigms of both environmental humanities and semiotics of culture. In any case, biosemiotic criticism is coming to be considered a viable alternative to other paradigms, like Darwinian literary studies or posthumanism, that seek to overcome the culture-nature divide.

9

Biosphere

Tyler Volk

All forms of life and the three environmental matrices of atmosphere, soils, and oceans form a closely integrated network that can be called the "biosphere." Thus the biosphere is the system with four main internal, interacting components: air, water, soil, and life. Considering this system, what makes the biosphere dynamically distinct from other layers of Earth, such as lithosphere, mantle, or core, is undoubtedly the presence and influence of life.

Alternative uses of the word "biosphere" in environmental writings, both technical and popular, are (1) the zone that life inhabits or (2) simply, the sum of all of life. Regarding the first, because bacteria are found in deepest ocean sediments, and even floating in air currents, there is little to gain by trying to distinguish it from the definition used in this essay, in terms of where the biosphere begins or ends. Regarding the second, there is an adequate and more precise alternative term for all of life, and that is "global biota."

Even with the definition used here—a system with four components—one might be curious about system boundaries. The biosphere's upper boundary is clearly the top of the atmosphere, which is crossed by incoming solar radiation, outgoing infrared radiation from Earth that exits to space, and minute quantities of space dust. Of course, once in a great while there is also the killer asteroid or two.

The biosphere's lower boundary is admittedly fuzzier. Groundwater reaches kilometers down into pores of rock, and bacteria have been found kilometers down, too. But we can exclude from the definition of the biosphere the minerals in rocks underneath the soil, using the logic that the chemical elements in those rocks have been outside of active circulation in air, water, or soil for millions or hundreds of millions of years. With this logic, which helps clarify the definition of the biosphere, we gain at least two practical results: (1) a focus on the physical fact that creatures live within and exchange gases, liquids, and solids with the environmental matrices of air, soil, and ocean as a system, and (2) an appropriately integrated, bounded system for technical modeling of the impacts of organisms on the chemistry of the global environment on relatively short time scales.

With boundaries, we can inquire into the matter and energy that cross them. The simple upper boundary of the biosphere has been noted. What about its lower boundary? Though compared to the solar and infrared fluxes of energy across the upper boundary, the energy coming up across the lower boundary from the deep Earth is trivial in terms of magnitude; but it is not trivial in terms of effects. The upwelling of deep Earth heat, most visible in volcanoes and deep ocean vents but present everywhere, is responsible for mountain ranges, the movements of continents, the opening and closing of oceans. Across the biosphere's lower boundary, there is matter transfer, too. Chemical weathering of minerals at rock interfaces as well as material and gases that come from volcanoes and deep-sea ocean vents result in inputs of fresh matter upwards into the biosphere. Exiting fluxes that leave the biosphere by going relatively permanently downward include the burial of minerals and organic sediments. These processes remove chemical elements from what had been their times of active circulation among the four components of the biosphere.

A look into its interior time scales sheds light upon the dynamics of the biosphere. There are exchanges of materials among the four main components, and therefore the entire biosphere is circulating (chemically speaking) according to the mixing times of its components. For example, studies of isotopes in the ocean have demonstrated that the huge ocean, from its surface down to depths that average about four kilometers, mixes completely in about a thousand years. Soils can be said to "circulate" on the basis of the times of disintegration of materials that enter soil as detritus from organisms, such as the drop of tree leaves but also including the dying of plant roots. In addition, creatures such as worms that live in the soil literally mix it. The time scales involved here depend on the quality of the materials put into the soil, but we can roughly cite durations of tens to hundreds of years for mixing of most soil (sites such as tundra can be exceptions).

The atmosphere is in constant and rapid motion. Earth's wind-driven, swirling layer of air circulates in merely a year. This is why rising levels of the greenhouse gas CO_2 are essentially the same over Antarctica's ice sheets as they are over the average of North America. Coupled with the fact of the relative inertness of fossil fuel combustion's waste gas CO_2, the rapid mixing of the global atmosphere is the reason why the ever-increasing greenhouse effect is a truly global problem not isolated to source regions of emissions, and is therefore different from many other pollution problems. One can hope that a positive side effect of the CO_2 challenge is to make Earth's citizens more aware of their shared interest in the biosphere.

This mention of CO_2 brings us to the topic of the impacts on life within the biosphere system. Two points need to be introduced. First, there are exchanges of matter among the three environmental matrices (air, soil, ocean) that take place without life. Gases move back and forth between ocean and atmosphere, across the air-sea interface. Rain and wind bring dust with

elements such as iron from the continents to the ocean. Of course, the oceans are the main source of water to the continents, via water vapor in the atmosphere and the hydrological cycle. Rain that enters soil will dissolve minerals from soil particles and carry these as ions to the ocean by rivers.

Second, because all organisms bring materials into their bodies from their environments through their membranes, skins, mouths, and other boundary-pores, whether living as single independent cells or as great blue whales, the creatures act as mechanisms for exchange within the biosphere, too. Organisms must get rid of wastes, which they also do through a variety of membranes and body pores and openings. Green leaves, for example, bring CO_2 into their bodies through tiny pores in their leaves that allow the air to pass in, from which their cells strip some of the CO_2 as a source of carbon to build their bodies. Those same pores, the stomata, release oxygen to the air. Oxygen is a waste byproduct from the plant's photosynthesis, and it just happens to be what oxygen-requiring heterotrophs such as animals and many microbes need to consume organic materials as food, creating waste CO_2 in the process, which they release and the plants can then use.

Actually, this neat linkage doesn't "just happen." One of the drivers of evolutionary change in the metabolisms of organisms is the presence of waste materials from other species circulating in the environmental matrices of the biosphere. These waste materials become available to those creatures who can figure out (evolutionarily speaking) how to use the wastes as, well, nutrients.

So living things flux materials in and out of their bodies: various forms of carbon, nitrogen, phosphorus, sulfur, oxygen, calcium, and other chemical elements. These materials leave and enter the environmental matrices of air, water, and soil. Bring in the nearly four-billion-year-old process of evolution operating in the lives and deaths of organisms. The result is the presence of what are known as global biogeochemical cycles, a great and unique feature of the biosphere.

All the bio-essential elements have unique stories. Sometimes two or more elements travel together for portions of their individual cycles. For example, when you consume a fresh ear of sweet corn, carbon and nitrogen within the corn enter your body together (there is carbon and nitrogen in the amino acids of the corn's proteins, and carbon in its sweet carbohydrates, which you ingest at the same time). In such situations the flux of one element closely corresponds to the flux of another, in proportions required by the living cells of corn, in the case just cited.

Sometimes the pathways of the biogeochemical cycles are decoupled. For example, there are microbes in soils and ocean waters called denitrifiers (many species). These denitrifiers can utilize nitrate ions in the soil's pore water or ocean water as an oxidizer and then some of them release, at the end of a chain of reactions, a waste gas, N_2, the most abundant gas in Earth's atmosphere. In these reactions, nitrogen as a chemical element is traveling its own pathway, separate from the biogeochemical cycle of carbon, to which it is sometimes tied, as the eating of corn showed.

The coupling and decoupling of segments within the biogeochemical cycles of various bio-essential elements means that each element has its own branch of science within the study of the biosphere. However, one overarching fact is that life is a major player in the workings of biosphere. These fluxes and exchanges caused by organisms, summed under the heading of the effects of the global biota, are significant and usually rival or surpass those fluxes and exchanges that are purely physical between ocean, air, and soil noted above.

For example, the carbon that flows annually as CO_2 into plants and algae in global photosynthesis is roughly the same amount as the CO_2 that passes back and forth across the global ocean's entire air-sea interface. As another example, in the conversion of atmospheric nitrogen gas (N_2) into forms of nitrogen usable by plants, both soil bacteria and lightning can do the job. But bacteria dominate the rate of conversion.

What this means is that there is no understanding of Earth's surface system that provides us with habitat—the biosphere—without an understanding of life as an integral part of that system. The giant effects of life make predicting the future of the biosphere particularly challenging. A nontrivial portion of the CO_2 we inject into the atmosphere is being incorporated into tree bodies or soil carbon somewhere, and it has been difficult to discern the details. How will the biological carbon in the peat of the high latitudes respond to climate change? How will marine ecosystems respond to an increase or decrease of nitrogen fertilizer that runs off from croplands in rivers? We humans are all connected to all other creatures, from the mighty to the microbial, and with all of them to the environmental matrices, making such questions and the understandings already won pertinent and also poignant.

IO

Built Environment

William A. Gleason

A term of comparatively recent origin, though a phenomenon of great antiquity, "the built environment" generally refers to those elements of the physical environment that are constructed by and for human activity. The built environment might thus include not only structures and sites such as buildings, roads, bridges, parks, and playgrounds but also (and more broadly) land-use patterns, transportation systems, architecture, and design (Saelens and Handy 2008; Bartuska 2007). Closely identified with cities but not exclusively urban, and often regarded in the modern era as separate from or even opposed to the "natural" environment—epitomized by Lewis Mumford's formulation, "As the pavement spreads, nature is pushed farther away" (Mumford 1938)—the built environment did not figure prominently in early currents of environmentalism either in the United States or globally. In the last decades of the twentieth century, however, theorists and practitioners alike began to question this absence. "Why is it that we tend to think of the built environment of cities as somehow or other not being the environment?" asked geographer David Harvey in the late 1990s. "There is, it seems to me, nothing particularly anti-ecological about cities. Why should we think of them that way?" (Harvey 1997).

Today the built environment is increasingly regarded as vitally involved in many of the central questions of environmental studies, including sustainability, climate change, and environmental justice. Part of this shift can be attributed to the growing realization that the perceived dichotomy between "natural" and "built" environments is a false one. As environmental historian Ari Kelman notes, the field of landscape studies in particular, which rejects the simplistic binary of "human" and "nature" in favor of recognizing (in D. W. Meinig's words) an "intricate intermingling of physical, biological, and cultural features" in every environment has helped expose the artificiality of this divide (Kelman 2003; Meinig 1979).

This shift also gained momentum after the widely publicized announcement by the United Nations in 2007 that for the first time in human history more than half the world's population would be living in cities by 2008. The report also predicted new urban growth on an "unprecedented scale," particularly in the developing world, where the population in towns and cities is expected to double by 2030 (United Nations Population Fund 2007). A more recent report anticipates that this "second wave" of urbanization will place 80 percent of the world's population in cities by 2050, with most of the growth occurring in Africa, Asia, Latin America, and the Caribbean (United Nations Environment Programme 2012). Even acknowledging that the density of cities may make them more energy efficient than the landscape of sprawl, this growth carries substantial implications for global energy consumption and materials use. Given the expectation that this growth may also exacerbate rather than alleviate global inequality, poverty, and vulnerability to climate change, it also carries substantial implications for environmental justice.

"Green" or sustainable design represents one approach (or more accurately, an ensemble of approaches) to the challenges presented by the exponential growth of the built environment. Engineers, architects, landscape architects, planners, and designers are seeking new materials, principles, and techniques to make buildings, infrastructures, and technologies more sustainable and resource efficient. In the building and construction industry, for example, the nonprofit U.S. Green Building Council promotes a certification program known as LEED (Leadership in Energy and Environmental Design) that identifies green buildings according to a set of rating systems and credits. Initially restricted to the United States, LEED now certifies buildings globally. Many other countries also have their own green building certification programs, some of which, such as the U.K.'s BREEAM (Building Research Establishment Environmental Assessment Method) and Germany's DGNB (German Sustainable Building Council) also have an international presence (Galbraith 2012).

These certification programs, particularly LEED, have not been without their detractors. LEED has been criticized, for example, for allowing buildings to accumulate points even for elements that have no practical purpose, such as installing bicycle racks and showers in a building that is difficult to reach by bike. Critics argue that such practices inadvertently enable forms of greenwashing (or "LEED-washing"). A more serious charge sometime levied against performance-based standards for green buildings such as LEED is that by focusing on individual structures, they neglect the context in which those structures are embedded. A highly energy-efficient building, for example, might only be reachable by employees who must drive there in automobiles, rendering the overall environmental gains suspect (Bowen 2014). Global certification programs have also been criticized for ignoring or downplaying regional and local environmental differences in favor of one-size-fits-all technology and design.

Recent academic scholarship on the modern built environment in a broad range of fields—including environmental history, urban studies, civil and environmental engineering, architecture, geography, sociology, literature, postcolonial studies, and critical race theory—is bringing many new questions to the forefront, from the role of waste and environmental management in urban life (Foote and Mazzolini 2012; Douglas 2013) to the prospects for sustainability in the post–Rio Summit era (Roberts, Ravets, and George 2009; Bulkeley 2013) to the impact of race, class, and gender on the social and physical experience of the built environment (Taylor 2009; Gleason 2011; Harris 2012). New studies of antiquity, moreover, are reintroducing the sustainable technologies of ancient architects and planners—from the passive cooling of Indian stepwells to the sewerage systems of Mesopotamia—to the greening of the built environment today (Melosi 1981; Livingston 2002; J. Donald Hughes 2009). In short, there remains no place but the center of environmental studies for the built environment to be.

II

Climate Change

Andrew Ross

Climate change is a significant shift, over a long period of time, in the statistical profile of weather patterns. For most of geologic history, natural factors—solar radiation, continental drift, oceanic circulation, volcanic activity—have forced these shifts. In the period since the late nineteenth century, anthropogenic global warming (AGW) displays the impact of industrial activity, largely through the concentration of greenhouse gases generated by the burning of fossil fuels. AGW can be seen as one component of the "Anthropocene," an unofficial chronological term that acknowledges the significant influence of human behavior on the Earth's ecosystems. Most scientists who favor the naming of this new geological era date its onset to the commencement of the Industrial Revolution, but some backdate it to the rise of agriculture, when humans began to transform land use and biodiversity on a large and global scale.

While many human activities are destructive of local ecosystems, emissions of greenhouse gases—carbon dioxide and methane above all—command most attention today because of their contribution to rising temperatures. Documented consequences of AGW include the retreat of glaciers, permafrost, and sea ice, ocean acidification, soil salinization, desertification, habitat inundation, mega-drought, loss of food security, mass species extinctions, and the more frequent occurrence of extreme weather events. All the available evidence shows that each of these phenomena have been occurring for some time, and that some are proceeding much more rapidly than predicted.

Because these declines are not abrupt, they are not perceived as catastrophic—the Hollywood iconography of natural disaster is spectacular and short-term, and it holds sway over the popular imagination. Nonetheless, large-scale trigger points, such as Arctic methane release from widespread permafrost thaw or the melting of the Greenland ice sheet, have the potential to cause precipitous and irreversible climate change, or what climatologists refer to as Dangerous Anthropogenic Interference (DAI). The threshold for DAI is commonly recognized as a temperature rise of more than 2.0°C (3.6°F) above preindustrial levels, or 450 parts per million (ppm) of atmospheric carbon dioxide, although not a few climate scientists have put the critical limit around the 400 ppm mark.

In 1824 Joseph Fourier was the first scientist to suggest that greenhouse gases helped to warm the Earth's lower atmosphere by absorbing and emitting infrared radiation. This "greenhouse effect" was investigated quantitatively by Svante Arrhenius in 1896. Contemporary estimates show that the atmospheric volume of anthropogenic carbon dioxide emissions from the combustion of carbon-based fuels has risen from preindustrial levels of 275 ppm to 397 ppm. Thermometer data records show that the Earth's mean surface temperature has increased by about 0.8°C (1.4°F) since the early twentieth century. About two-thirds of the increase has occurred since 1980. The detection of this AGW signal has been further corroborated by climate profiles of much longer historical span generated from analyses of ice cores and sediment, as well as floral and faunal and sea-level records.

The most authoritative scientific surveys of human-induced climate change are produced by the Intergovernmental Panel on Climate Change (IPCC). These comprehensive assessments (the last issued in 2007, the next in 2014) are supposed to provide policy advice in the implementation of the United Nations Framework Convention on Climate Change (UNFCCC). Progress in the international treaty process of the UNFCCC has been hampered by the reluctance of the major carbon powers—the United States and China especially—to enter into binding agreements about emission-reduction targets. The race to decarbonize has many willing contestants, but their number does not include those whose contribution would have the biggest impact. Among the obstacles standing in the way of international policy making is the powerful lobbying pressure of the fossil fuel industries, widely perceived to be responsible for encouraging "climate change denial" (Washington and Cook 2011; Norgaard 2011).

The prospect that inadequate political action will guarantee the continuing rise in emissions has generated widespread concern on the part of scientists, a community not readily distinguished by a rush to judgment, let alone by public alarmism. Their increasingly activist standpoint can be contrasted with that of oceanographer Roger Revelle's milestone 1956 testimony to Congress about the rise of CO_2 emissions. "From the standpoint of meteorologists and oceanographers," he submitted, "we are carrying out a tremendous geophysical experiment of a kind that could not have happened in the past or be reproduced in the future" (Revelle 1956, 467). His choice of words has long been criticized for suggesting that the increase of atmospheric CO_2 presented merely a rare opportunity for scientists to study its impact. That clinical mentality no longer prevails. Natural scientists are in the forefront of lobbying for radical policy measures to address the climate crisis, issuing periodic warnings, signed by the leading figures in a variety of fields, beginning with the 1992 "World Scientists' Warning to Humanity."

Today, it is the task of averting drastic climate change that might be described as an experiment—a vast social experiment in decision making and democratic action. Stabilizing greenhouse gas concentrations will require concerted action on the part of political and economic elites as well as comprehensive shifts in the routine behavior of general populations, especially in carbon-rich countries. Success in that endeavor will not be determined primarily by large-scale technological fixes, though some—such as building an adequate infrastructure for sustainable energy sources—will be needed. Just as decisive to the outcome is whether our social relationships, cultural beliefs, and political customs will allow for the kind of radical changes that are necessary. That is why the climate crisis is as much a social as a biophysical challenge. Moreover, if this social experiment is to avoid an authoritarian turn, then its mechanisms of public consent cannot be bypassed by geo-engineering schemes, such as seeding the oceans with iron or reflecting sunlight though orbiting mirrors and brightened clouds (Kintisch 2010; Fleming 2010). Such schemes for intervening in complex and unpredictable natural systems are likely to generate a variety of unintended consequences, both geophysical and sociological.

In addition, as part of a new phase of green capitalist development, most technical green innovations are being targeted toward the affluent populations in the market segment known as LOHAS, or Lifestyles of Health and Sustainability, covering almost 20 percent of adult consumers in Organisation for Economic Co-operation and Development (OECD) countries. LOHAS evangelists assume that reforming the high carbon consumption of this population may be enough to stave off the scenarios of eco-collapse. The problem is that the carbon savings to be gotten out of this 20 percent cannot possibly outweigh the commercial neglect of the other 80 percent (Jones 2008). The sociology of climate change has made it quite clear that no one can opt out, or be left behind. Either the green wave has to lift all boats, or else we may find ourselves with the dismal "lifeboat ethics" first proposed by ecologist Garret Hardin—where those not already on board are left behind (Hardin 1974).

Advocates of climate justice, who draw attention to the uneven geographic distribution of environmental risks, argue that sustainable technologies need to be developed as remedies for social inequalities. Otherwise, affluent communities in the Global North and in the global cities of the South will turn into eco-enclaves, hoarding resources and knowledge about resilience from others. If resources tighten rapidly, a more ominous future beckons in the form of triage crisis management, where populations are selected for protection inside these eco-enclaves, while those outside or across the borders are abandoned. In the "climate wars" to come, the threat of global warming will increasingly be used to shape immigration policies around a vision of affluent nations or regions as heavily fortified resource islands.

Anti-immigrant sentiment in the North has been stoked by fears about population pressure on resources, and by the manipulation, on the part of right-wing nativists, of the new carbon consciousness—immigrants, it is imputed, increase their carbon footprints when they move to the North. Among the broad social spectrum of border crossers, climate migrants (alternately termed "environmental migrants" or "climate refugees") pose a particular ethical dilemma. This fast-rising population comprises those forced off their land or deprived of their livelihoods by the impact of climate change. A much greater population is being severely impacted by climate change but cannot, for one reason or another, leave their homes. Migrants who cross borders, especially from poor to rich countries, attract the most attention. Alarmist estimates of their numbers feed into neo-Malthusian sentiment about overpopulation, on the rise again since the mid-1990s despite the overall slowing demographic trend in population growth. Those who migrate within the borders of poorer countries do not generally register as a concern among overpopulationists. China, for example, has tens of millions of internal climate migrants—set in motion by increased droughts, flooding, coastal erosion, saltwater inundation, glacial melt in the Himalayas, and shifting agricultural zones—but they are not recognized, identified, or counted as a distinct population, nor do they attract the anxious ire of groups like the UK's Population Matters (formerly the Optimum Population Trust) and the United States' Population Connection (formerly Zero Population Growth).

Those who are counted as climate refugees numbered in the tens of millions globally by 2000, and estimates from the IPCC, the 2006 Stern Report, and other sources predict that climate change will generate upwards of two hundred million such refugees by 2050 (Myers 2001; Stern 2006). To date, no international convention has recognized the needs and rights of climate migrants, even though, by 2010, according to the Red Cross's estimate, they had outnumbered the population of refugees from war and violence (International Federation 2001). The International Committee on Migration has been pushing for multilateral legal recognition, though the consequences are not clear. Some fear that legal refugee classification would only create another level of second- or third-class immigrant status that would hold migrants in limbo. More promising are the moral demands that issue from the climate justice movement, which has embraced climate migrants as a living embodiment of the debt owed by carbon-rich beneficiaries of industrialization in the North to populations in the South most affected by the impact of climate change.

Under these terms, what is owed to climate migrants? What rights and resources can they demand in compensation from the beneficiaries of carbon-rich industrialization whose emissions are responsible, however indirectly, for setting them in motion (Ross 2011)? They should be offered sanctuary and legal protection at the very least, but other forms of reparations may gravitate to the center of international treaty attention in the years to come.

A similar understanding of repayment applies to this central principle of the climate justice movement that rich nations owe a carbon debt to poor ones. Advocates argue that reparations are appropriate since the industrial development of the North was achieved at the environmental expense of the South. Negotiations over the terms of this debt have been a sticking point in the biennial UNFCCC climate summits. In addition, representatives of southern nations insist that the industrialized countries transfer adequate resources, both financial and technological, to enable them to build renewable energy systems as part of a low-carbon pathway of development.

Currently, the favored mode of negotiating climate debt is through carbon trading markets and mitigation swaps of the kind favored by the United Nations program Reducing Emissions from Deforestation and Forest Degradation (REDD). Large industrial polluters buy pollution credits openly through the carbon markets, or they bargain for the right to continue polluting by purchasing a chunk of rainforest, most often without the consent of the indigenous communities who live there. Another way for them to evade direct climate debt repayment is to transfer accountability onto the consciences of individuals. This is an ongoing ideological shift, whereby individuals are encouraged to assess the carbon footprint of every personal act or material that they consume. The guilt and sense of responsibility is thereby reassigned, or outsourced, away from the large polluters who have the most effective power to lower emissions. This "carbon calculus" introduces a new kind of moral tyranny. Hydrocarbon becomes the loud scourge of our civilization, rather than an outlawed by-product of it.

For real evidence that progress is being made in building responses to the climate crisis, we can look to cities. Today, there is a thriving "sustainable cities" movement all over the developed world, and in many developing countries. Mayors toot their horns whenever their cities move up in the national sustainability rankings. Unlike state and federal lawmakers who rule on energy regulation, city managers and politicians are not subject to the fierce lobbying power of the fossil industries, so they are generally freer to push green policies. The Large Cities Climate Leadership Group (C40) is comprised of forty of the world's largest cities collaborating on extensive decarbonization programs. C40 issued its own Climate Communiqué at the UNFCCC summit in Copenhagen, in which mayors pleaded with the national representatives of the carbon powers to "recognize that the future of our globe will be won or lost in the cities of the world" ("C40" 2009). The mayors' statement reflected a growing consensus that only in dense urban environments could efficient, low-carbon living be achieved on a mass scale. Humans are fast becoming an urban species, and our survival will depend on how we live in cities, which already consume 75 percent of the world's energy and emit 80 percent of the greenhouse gases (Newman, Beatley, and Boyer 2009). Even without a decisive shift in energy supply away from fossil fuel, more compact patterns of urban growth are delivering a sizable boost to efforts at decarbonization.

Advocates of urban sustainability push on a number of fronts: water conservation, decentralization of energy production and distribution, transformation of transit and transport, redesign of buildings and infrastructure, establishment of closed-loop waste systems, growth of a bioregional food supply, and municipal planning along carbon-neutral lines (Kahn 2006; Birch and Wachter 2008; Portney 2003). But if these initiatives do not take shape as remedies for social and geographic inequality, then they are likely to end up reinforcing existing patterns of environmental injustice. A "green gap" has already opened up between the eco-oases of affluent carbon-conscious communities and the human and natural sacrifice zones on the other side of the tracks, where populations have to fight to breathe clean air and drink uncontaminated water. To prevent this gap from widening, municipal administrations will have to break with the habit of making environmental policy solely at the behest of moneyed voters, and craft their sustainability plans as vehicles for civil rights, primed to deliver justice to those who have been sidelined, ghettoized, and stranded by a century and a half of unsustainable growth.

The future of green governance hangs in the balance, and not only in those cities that have taken the lead in low-carbon policy making. What if the key to combating the climate crisis lies in innovating healthy pathways out of poverty for populations at risk, rather than marketing green gizmos to those who already have many eco-options to choose from? These are not mutually exclusive options. After all, the high-carbon lifestyles of affluent communities need to be targeted. But building a low-carbon economy by catering only to the LOHAS demographic will end up doing little more than adding a green gloss to patterns of chronic inequality. Likewise, placing all of our faith in clean-tech fixes will cede too much decision making to a closed circle of experts who, regardless of their technical prowess, will have no power to prevent the uneven application of their solutions.

I2

Conservation-Preservation

William G. Moseley

The concepts of conservation and preservation typically are used to differentiate between forms of human management and use of renewable resources (e.g., forests, fisheries, many sources of water). While these terms are sometimes used interchangeably in public discussions, environment-related fields carefully deploy these words to refer to particular management regimes for renewable resources. The difference between conservation and preservation is said to be important because these are the "major ideological camps . . . which still dominate debates over natural resources today" (Cutter and Renwick 2004, 41).

"Conservation" (sometimes also described as the "utilitarian approach" in environmental history, or as "resource conservation" in the UK) typically refers to use within certain biological limits, or within the annual growth increment of a particular resource (McManus 2000). In the case of forests or fisheries, this annual growth increment is also referred to as the "sustainable yield" or "maximum sustainable yield" (Dana and Fairfax 1980). In most national contexts, the government natural resource management agency most closely associated with the conservation approach is the forest service, agency, or ministry (Clarke and McCool 1985). Many forest agencies around the world manage their forests under the principle of maximum sustainable yield. Typically the formula (forest area/age to maturity) is used to determine the percentage of the total forest area that may be harvested and replanted each year. For example, in an even-aged monoculture of white pine that is 150 hectares in size, and for which the age to maturity is 30 years, it would be harvested and replanted at the rate of five hectares per year (150 hectares/30 years).

In contrast to "conservation," "preservation" (also known as "nature conservation" in the UK) generally refers to the nonuse or nonconsumptive use of natural resources in an area (McManus 2000). The practical expression of the preservationist approach is often the wilderness area or park. In some instances, an area is completely off limits to humans. More frequently, nonconsumptive uses are allowed (e.g., hiking, camping). The rationale for preservation is that certain areas must be set aside for compelling aesthetic or biodiversity reasons. This approach to preservation has been described as the "Yellowstone model," a model that emphasizes national parks as places where people may visit as tourists but may not reside or exploit to support a resource-based livelihood (Millington 2005). The preservationist approach also was introduced to developing countries during the colonial era, when many parks and wilderness areas were established (Adams 2001). Parks and preserves in the world's tropical regions have received considerable attention since the 1992 World Summit on Environment and Development in Rio de Janeiro (Bryant and Bailey 1997). Neumann (1997), for example, notes that over 10 percent of the territory in some African countries is now managed by state and international organizations for preservation purposes.

Attention to scale (or the size and shape of the area being analyzed), land-use patterns, and the connections between places and regions complicates conventional understandings of preservation and conservation (Moseley 2009). In practical terms, the scale at which the preservationist approach may be implemented is limited by the need for humans to use natural resources. As a result, unless an area is lightly populated,

it is challenging to set aside extremely large tracts of land as preserves because people need to use some land to sustain themselves. With the possible exception of Antarctica (a continental example), most lands set aside for preservation are modest in scale. While parks appear as preservation, if the unit of analysis stops at the park boundary, this perception quickly changes at broader scales of analysis if surrounding areas are overexploited. For example, national parks in Costa Rica have been referred to as diamonds in a sea of devastation (Sanchez-Azofeifa et al. 2002).

In contrast to preservation, conservation could (at least in theory) be implemented at a much broader scale because it allows for human use of resources within biological limits (acknowledging that biological limits are also contested). In such a situation, people in all places would be allowed to tend to their needs, yet would be required to operate within the biological limits of the environment. This is a very integrated and spatially broad vision of conservation that shares commonalities with certain (i.e., the strong or radical green) conceptions of sustainable development (e.g., Goodland et al. 1991; Redclift 1992).

In the real world, conservation (like preservation) is often implemented at a more local scale. As described earlier, foresters managing a wood lot under the principles of sustainable yield carve it up into equal-sized plots (derived via the formula forest area/age to maturity) and then harvest and replant one such plot per year until eventually they return to the first plot that was cut and replanted. Here again, examining the situation at a variety of scale frames allows one to recognize that the management of the forest as a whole might be labeled as conservation. Conversely, when examined at the scale of the individual plot being harvested (often several hectares in size), the situation might more aptly be described as exploitation (or use without regard to

CONSERVATION-PRESERVATION   WILLIAM G. MOSELEY

long-term consequences). "Exploitation" could be the more appropriate term at this scale, as such plots are often clear-cut and lie barren for some time before they are replanted. While such a process incrementally impacts biodiversity, soil stability, and infiltration at the scale of the forest management unit, all three of these factors decline dramatically at the plot scale after a clear-cut.

If one pulls back from the land management unit and begins to analyze the situation at broader scales, at least two other issues begin to become apparent: (1) patterns of land use (i.e., how the landscape is divided up into different land-use units); and (2) the economic and ecological connections between different areal units—and the politics of these linkages. In the first instance, for example, preservation at a limited scale means that humans must divide up the landscape into areas designated for preservation and those designated for other types of land use (ranging in use from overexploitation to conservation). Thus, when the situation is viewed from a broader scale perspective, one sees a patchwork landscape of exploitation, conservation, and preservation, which (by definition) could not be considered preservation. Such patchwork landscapes, with preservation in some areas and different uses in others, may represent conservation at best (use within biological limits) and (more likely) overexploitation in many instances. In other words, preservation at the local scale could violate conservation at a broader scale if it leads to overexploitation on other parcels.

Underpinning the land-use mosaic are a variety of economic and ecological linkages between preservation areas and other points on the landscape. At a very basic level, the nonuse or nonconsumptive use of resources in certain areas implies that uses that could have occurred in these areas have probably shifted elsewhere. While it is acknowledged that national parks often are established on economically marginal lands, it is difficult to deny that these could have been sites of resource extraction. In other words, it is the "subsidy" provided by intensive use of "normal use areas" (both as sources of resources and as sites of human habitation) that allows people to set aside areas for preservation (a point that has been made by other authors, including Zimmerer 2000). The subsidy provided to preserves from normal land-use areas may come from inside or outside the country. For example, a reduction in U.S. timber harvests since 1990 because of increasing requirements to sustain biodiversity and other ecosystems functions was possible in the face of growing U.S. demand for wood products because of increasing imports from Canada (Martin and Darr 1997) and several tropical countries (Tucker 2002).

Beyond shifts in resource extraction, a second connection between preserves and other points on the landscape may be the dislocation of peoples. A large literature on parks and peoples has documented how the creation of parks in developing countries often implies the relocation of peoples to other areas (e.g., Neumann 1998; Colchester 2004). For example, Guha (1997) describes an ongoing controversy in Nagarhole National Park in Karnataka Province, India, where the Forest Department has been trying to relocate six thousand local people. Relocated peoples, while (arguably) lessening impacts within the preserve, often augment impacts elsewhere (another dimension of outside areas "subsidizing" preserves). Furthermore, displaced peoples often bear considerable costs for the creation of such preserves in terms of compromised livelihoods. While it is a lesser-known phenomenon, the establishment of national parks in North America also involved the displacement of native peoples (Burnham 2000; Braun 2002). Furthermore, the construction of public municipal parks often accompanied the displacement of marginalized populations who lived there first, many of whom were later excluded from those spaces once they were constructed, for example, Central Park in New York City (Taylor 2009).

Not only is preservation a problematic concept once it is inflected with a scale perspective (because it is highly contingent upon the scale at which it is examined), but Cronon (1995a) suggests that an overemphasis on the preservation of wilderness areas creates dualistic thinking that allows people to be blind to the intensive use of normal landscapes. In other words, preservationist approaches may lead people to view wilderness or nature as separate and apart from humans, rather than all around us in our daily lives. Such dualism may lead people to behave one way in wilderness (e.g., leaving nothing behind except footprints) and another way in their everyday lives.

What the above examples suggest is that preservation is only preservation (nonuse or nonconsumptive use) at the scale of the preserve. When such a preserve is viewed at broader scale frames, we see that this is not really preservation because nonuse/nonconsumptive use in one area is almost always subsidized by use in surrounding areas (or even distant centers of production and consumption). In either instance (conservation or preservation), the distinctions among exploitation, conservation, and preservation begin to become blurred when they are analyzed at multiple scales. Key to understanding this multiscalar analysis is attention to how humans divide up space to apply either principle (conservation or preservation) and to socioeconomic and ecological connections that exist between regions and places.

13

Consumption
Andrew Szasz

In discussions of environmental conditions or problems, the concept of consumption appears, typically, in two ways. First, consumption is said to be the *problem*. We are hurting the environment because we consume too much or we consume the wrong kind of things. Second, changing consumption is said to be an important part of the *solution*. If we wish to have a more sustainable relationship with our planet, we must consume less or, at least, smarter and better.

Is consumption the problem? Environmentalist rhetoric asserts that it is. We hear statements such as, "The United States has 4.5 percent of the world's population but consumes about 25 percent of the world's resources" and, "If everyone on Earth lived like Americans, that would require four Earths' worth of natural resources."

Is changing consumption (at least part of) the solution? Important strains of environmentalist rhetoric say it is. Concerned citizens are exhorted to consume less, to choose the green, natural, organic, sustainable alternative, to reduce their carbon footprint. We will live more sustainably, have fewer adverse impacts, if each consumer "walks lighter on the planet," uses fewer resources, generates less waste. In the longer term, if enough people make the greener choice, that will send a market signal, motivating manufacturers to bring more such goods to market. (We might think of this as the work of a Green Invisible Hand.)

Sociological analysis does support some version of such claims. Consider the well-known "IPAT" formula:

(Environmental) Impact is a function of a society's Population, its level of Affluence (i.e., the level of its consumption), and its Technology (Commoner 1972; Ehrlich and Holdren 1971). Empirical studies using the IPAT formula confirm that the higher the level of consumption, the higher the environmental impact (York, Rosa, and Dietz 2003). Other, prominent social scientists argue that getting people to consume less or to consume differently can significantly reduce their ecological footprint. O'Rourke (2011) has developed an app for smart phones that helps consumers make ethical choices when they shop. Willis and Schor (2012) argue that green consuming can *politicize* the consumer. Being mindful about one's choices, choosing the greener alternative, and doing all that repeatedly, can change a person's consciousness, they say. It can sensitize one, make one more aware. And that can lead a consumer to increasingly identify himself or herself as someone who "cares about the environment," increasing the likelihood that he or she will then engage in other forms of environmental activism.

Sociological analysis acknowledges some aspects of such claims, but the preponderance of sociological thought has to be seen, overall, as expressing a certain skepticism toward them. Consumption, though certainly not unimportant, is only one, and probably not the main, cause of environmental problems. A sociological understanding of consumption suggests that changing consumption is difficult and may not be the best strategic choice if one is looking for the most effective way to address contemporary society's environmental crisis.

The first point to emphasize is that the relationship between consumption and environmental impacts is more complex than the overconsumption condemned by environmentalists. Sociology has always emphasized the centrality of inequality when analyzing societal phenomena. Today, inequality is as important as ever. Eighty percent of the world's peoples live in nations in which inequality is rising (Shah 2013). In the United States, income and wealth inequality have been increasing for thirty years and are now up to levels not seen since just before the Great Depression (Saez 2015). Globally, the top 1 percent has 48 percent of all wealth (Oxfam 2015), while three billion people, almost half of humankind, live on $2.50 a day or less (Shah 2013). For the idea that "consumption is the problem," extreme inequality means that consumption is really two problems: at the top, overconsumption, as the usual critique says; at the bottom, desperate takings of basic materials needed for bare survival, even when that threatens to extinguish species or terminally exhaust local ecosystems.

To focus just on consumption also tends to ignore other drivers of environmental harm, drivers that are certainly linked to consumption in some way, but would not be improved or ameliorated with a politics aimed *only* at changing consumption patterns. First, as the work on IPAT shows, population—sheer human numbers—is an important driver of environmental impact. Every one of the world's seven billion peoples has minimum physiological needs, for water, protein and calories, warmth, and shelter. Feeding all those people, alone, requires vast quantities of land and water—degrading those resources, transforming ecosystems, depriving other species of their use—plus vast outlays of energy, chemicals, etc. Second, consumption is just one phase of a cycle that runs from resource extraction to production, distribution, and marketing to consumption and, finally, to the disposal of things when they are no longer of service. Every phase of this cycle—including the damage associated with mining; waste discharges to air, water, and ground during manufacturing; and the accumulation of postconsumer wastes—is a significant cause of environmental degradation.

I turn now to what sociological analysis suggests about an environmental politics focused on changing consumer behavior. The most important point to make is that a politics that aims at consumer behavior does not address the other causes listed above. Such a politics will have little, if any, impact on the destructive ways resources are extracted, on the toxic methods used in manufacturing goods, on the impacts of transporting goods globally, and on disposal. A politics aimed at changing consumption ignores the question of population and social inequality, hence the environmental consequences of mass poverty.

Furthermore, a politics that aims at changing consumer behavior may not accomplish much even on the terrain on which it aims to operate. When activists talk of "changing consumption," they are talking about *individual* consumption, when many forms of consumption are *collective* and/or *fixed*. Many cities and suburbs, especially those that grew to their present size in recent years, have a social geography—a spatial organization—that is built around, and hence presumes and requires, the use of the private automobile; these "facts on the ground" are fixed, *given*, largely beyond the influence of the individual consumer. In important ways they limit or *constrain* individual consuming choices. For example, when it comes to buying or renting a home, only a few privileged, wealthy individuals get to design and build their own home, deciding what kind of house, how big, how energy efficient, etc. Most individuals are presented with preexisting choices: homes in an existing, given social geography that are products of many earlier decisions, made by others, about what kind of housing will be built; decisions to put residential neighborhoods here, shopping there, and workplaces someplace else; other decisions to build roads, parking lots, and gas stations rather than light rail; and so forth. The social geography, already there, to be *found*, not within one's power to *make*, constrains choices about what home to buy or rent, how much to drive, and many other mundane aspects of everyday life.

For the fraction of consuming that is properly the realm of individual consumer choice, sociology has emphasized the degree to which such choices are far from free or unconstrained. Just over one hundred years ago Thorsten Veblen coined the phrase "conspicuous consumption" to describe how consumption had become a form of status display, way beyond any notion of consumption as necessary to satisfy basic needs (Veblen 1899). Much later, Herbert Marcuse argued in *One-Dimensional Man* (1968) that advanced capitalist societies create new needs, needs that promise to be fulfilled through the consumption of specific commodities. Consuming the right stuff promises to make one "happy." One consumes to construct, lay claim to, and display a desirable social identity. Consumption is no longer just the satisfaction of basic needs; consumption is driven by powerful societal/cultural motives. Not many persons are so deeply committed to the environment that they are willing to forgo all that and voluntarily choose less attractive green alternatives.

The point is that changing individual consumer behavior can have a material impact only if it is a mass phenomenon, only if millions of consumers make such choices and do so consistently. Sociological analyses of contemporary consumer society imply, however, that there are powerful cultural forces arrayed against such mass shifts in consumer choice; hence green consuming is likely to remain a fringe phenomenon, with limited material impact on environmental conditions. Marketing data appear to support this pessimistic assessment of the promise of targeting consumer behavior as the way to achieve positive environmental change. Hybrid cars continue to account for only a tiny percent of cars sold in the United States. The average home sold in the

United States is many square feet bigger than the typical home sold in past decades. Although some forms of green consuming are thriving, there is scant evidence of a truly mass shift away from conventional consumer items.

Finally, some sociologists are skeptical of the argument that green consuming can politicize people and encourage civic engagement. It is just as plausible, they say, that doing a bit of green shopping may well lead, instead, to feeling less urgency, because "I have done something"; "I have done what I can"; "I have done enough" (Szasz 2007, 2011).

Is consumption part of the problem? Yes, but one has to "unpack" the concept if one is to understand exactly how. Is it—can it be—part of the solution? Possibly, if one avoids the seemingly obvious "let's persuade consumers to go green." True, if enough consumers look to buy fruits and vegetables free of pesticide residues, that can foster the growth of organic agriculture. It is more powerful, though, to advocate for policies that can change consumer behavior "from above," policies that result in design of smarter cities, that incentivize the building of energy-efficient homes, that direct automakers to increase mileage. New, innovative movements, such as the Transition Town Movement, which advocate more holistic, community-based, collective action "to build community resilience in the face of such challenges as peak oil, climate change," are also promising (Transition 2013). Sociological analysis suggests, however, that a politics that focuses on consumption will not, by itself, be a solution to our environmental crisis unless we also address, at the same time, inequality/poverty, population, the whole cycle of extraction, production, distribution, and waste disposal.

14

Cosmos

Laura Dassow Walls

"Cosmos" is one of the most important and most deeply misunderstood words in our vocabulary. In common use, it designates the stars and planets beyond Earth, realms accessible only by telescopes or the most futuristic of technologies. But the complex history of this ancient word suggests that it has much more to teach us—indeed, that we need it now more than ever, for popular usage masks its long history as humanity's oldest ecological vision of our planet.

In ancient Greece, "*kosmos*" meant not "the universe"—for this the Greeks used "τὸ ϖᾶν," "the all"—but rather, the universe comprehended as a unified system that was both ordered and beautiful. How, they asked, did this this system come into being? Of what did it consist, what was its fate? Their answers elaborated a plethora of possibilities that initiated what we now call "science." *Kosmoi* were variously imagined as finite or infinite; designed and fated, or bubbling up by chance; single and identical with the universe, or many coexisting in a pluriverse, or perhaps rising, flourishing, and dying in a succession of rebirths—but in all cases, the ancient Greeks imagined their *kosmoi* arising from some prior state of "chaos," not, as in the Christian thought that supplanted them, *ex nihilo*, from nothing, commanded by a Creator God. Instead, Greek *kosmoi* arose by virtue of some form of ordering principle that was natural, invariant, and consistent over time, hence not subject to interference by the caprice of the gods—and hence, subject to empirical investigation (Gregory 2007, 1–24).

By the nineteenth century, all this had devolved into, in Henry David Thoreau's words, "only a curious philological fact," including the further curiosity that the English word "cosmetics" carries forward the earliest meaning of "*kosmos*" as ornament or adornment. "Cosmos" in its modern usage originated with Alexander von Humboldt's resurrection of the archaic Greek word for the title of his best-selling work *Cosmos* (1850 [1997], 184–62), a pathbreaking multivolume hybrid of science and intellectual, cultural, and environmental history. To find his keyword, Humboldt ransacked the available lexicon of his day and came up empty. All the available choices—"universe," "*monde*," "Earth," "world"—failed his primary test: did it recognize the "reciprocal incorporation" of mind and nature, form and history, intellect and awe? None did. So Humboldt turned to the ancient Greeks, reviving "Κόσμος" to name "a Cosmos, or harmoniously ordered whole. . . . one great whole (τὸ πᾶν) animated by the breath of life" (*Cosmos* I:72, 69, 24–25). Only science and poetry combined could be worthy of such a word, which signified "*universe, order of the world*, and *adornment* of this universal order" (I:79). Or, in more modern English: while the physical, material universe exists without us, the cosmos exists only as the historical entangling of nature and global human intelligence, brought into being through philosophy and song, literature and art, sciences, technologies, explorations and migrations, gardens and farms, preservation and ecological understanding.

Humboldt spent over five decades and dozens of volumes elaborating his holistic concept of cosmos, becoming the popular icon of science, the Einstein of the nineteenth century: Emerson remarked that he lived in "the Age of Humboldt." He founded several branches of natural science, inspired a fleet of artists and writers, agitated for the end of slavery and for recognition of the dignity of indigenous peoples; but even in his own lifetime, many in science rejected his view that "nature, human nature, and human history were locked in a close embrace" as "outmoded, naïve, or impossible to implement" (Reill 2005, 23, 238). The word "cosmos" devolved from an ecological vision unifying cosmic, geological, organic, and human history on "*the planetary world*" amidst the perpetually evolving "great garden of the universe" (*Cosmos* I:68, 84) to a term for the heavens removed from our sublunary concerns.

This philological curiosity is a symptom of today's crisis, precipitated by our profound inability to think the cosmos—which we now call the Anthropocene, acknowledging at last our historical role as agents reciprocally intertwined into a geo-planetary system. The philosopher Charles Taylor observes that Euro-American modernity was marked by the "transformation in outlook from a limited, fixed cosmos to a vast, evolving universe," an overwhelming opening "onto unencompassable space" and incomprehensible deep time. As Taylor remarks to his readers, "Our present sense of things fails to touch bottom anywhere" (2007, 325–27). Faced with this abyss, the solution that prevailed was to delink human from natural, humanities from sciences, and to "treat the environment simply as a silent and passive backdrop" to historical narrative (Chakrabarty 2009, 204). This was precisely the path Humboldt swerved to avoid by devising a new style of thinking: treating nature as a fully historical agent (the insight developed by Darwin); science as "the labor of mind applied to nature"—in effect, the cosmos studying itself (*Cosmos* I:76); and humanity as simultaneously a unified natural species and a diversified cultural innovator.

Humboldt's inceptive vision was not entirely lost. It inspired many who pursued his insights into literature, the arts, and environmental thinking, including such founding figures of American environmentalism as Emerson, Thoreau, Whitman, John Muir, Susan Fenimore

Cooper, and George Perkins Marsh, who all credited Humboldt as their inspiration. Highlights include Thoreau's troubled warning that "[w]e have to be told that the Greeks called the world Κοσμος Beauty—or Order, but we do not see clearly why they did so" (1862 [2007], 217), part of his own project to write a new natural/poetic cosmos into being; Whitman's self-introduction in "Song of Myself" as "an American, one of the roughs, a kosmos"; artists, including Fredric Church, Martin Johnson Heade, and Ferdinand Bellermann, who heeded Humboldt's call to paint nature's full diversity and followed him to the South American tropics, inaugurating a new school of landscape painting; Muir's vow "to be a Humboldt" and enter into "a plain, simple relationship with the Cosmos"; Cooper's history of the human feeling for nature, which paraphrases Humboldt's *Cosmos*; and Marsh's use of Humboldt's conceptual framework to document the environmental damage humans had done to the landscape (see Walls 2009, 251–301). Franz Boas declared himself a Humboldtian "Cosmographer," and in taking up the Humboldt brothers' unfinished work on American Indian languages and cultures, founded the modern science of anthropology (Walls 2009, 210–14). Thus the threads linking today's environmental thought to Humboldt's original vision are many and strong; as Carl Sagan said in his 1980 book and TV series, *Cosmos*, "There are a million threads from the past intertwined to make the ropes and cables of the modern world" (336).

Most recently, cosmos-thinking has emerged again in critical theories of "cosmopolitanism," an inquiry related to Humboldt's assertion that "true cosmical views" are not the work of a single nation or civilization, but a collective composition made possible "by a great, if not general, intercourse between different nations" and begun sometime before the beginning of recorded history, where already "we see many luminous points, or centers of civilization, simultaneously blending their

rays"—including the ancient civilizations of Egypt, India, Iran, and China, to name only a few of the largest stars still visible today among so many forever lost in time (*Cosmos* II:116, 114). This concept of cosmos as a *collective* collaboration originating well before written records, and as a *work* sustained through millennia by the multitude of the Earth's peoples toward a cosmopolitan and ecologically sustainable future, has recently reemerged in Bruno Latour's call for a political order "which brings together stars, prions, cows, heavens, and people, the task being to turn this collective into a 'cosmos' instead of an 'unruly shambles'" (1999, 261). As Latour details, "The cosmos," a word intended to bring forward the ancient Greek meaning of "arrangement" and "harmony" as well as the better-known meaning of "world," "is thus synonymous with *the good common world*"—that is, a world that brings together, by "cosmopolitics" or "due process," the pluriverse of peoples and natures into a commons resilient enough to embrace the future (2004a, 237–40; emphasis added).

The challenge posed by Latour's "due process" is explored by his colleague Isabelle Stengers, who in her pathbreaking essay "The Cosmopolitical Proposal" cautions us to slow down, lest we impose our particular cosmos or tradition upon all: "There is no representative of the cosmos as such," she reminds us. "In the term cosmopolitical, cosmos refers to the unknown constituted by these multiple, divergent worlds and to the articulations of which they could eventually be capable" (2005, 995). The problem of our time, then, is to think the cosmos without merely thinking our own society—even our own species. Who, Stengers asks, can speak for the cosmos? No one: any decision about its creation must be made in the full presence of those who will be its victims. For, as she writes in *Cosmopolitics*, "Ecology . . . doesn't understand consensus but, at most, symbiosis" (2010–2011, I:35).

Never, in the millennia-long history of this vast collective project, have the stakes been so high: now that humans have become a geological agent committed to altering the very planet itself to a state never experienced by human beings; now that the looming realities of climate change, discerned by Earth systems scientists from Humboldt forward, are pressing politics into novel and unknown forms; now that recorded history is no longer a reliable guide to the future, or even to the assurance that there will be a future; now, more than ever, we need to think with a concept so deep that its protocols outwit our current stalemate. That concept is cosmos, for it ties together our future hopes with our deepest past in a symbiotic web whose resources we have barely begun to explore.

15

Culture

Dianne Rocheleau and Padini Nirmal

The popular understanding of culture in mid-twentieth-century America and Europe was arguably the symphony orchestra, the ballet, the art museum, and a "national" white, elite etiquette dictating, and explaining, how people should behave. Since the mid-1960s, we have witnessed the production and recognition of a proliferation of "cultures" and "multiculturalisms" within popular culture and informal political and economic institutions. Current usage is replete with compound cultures: counterculture; pop culture; office culture; indigenous culture; urban culture; peasant culture; global culture; "mainstream" culture; and "other-cultures."

The biological and agricultural roots of the word have also expanded, as illustrated by frequent references to "lab culture" (a living bacterial or fungal assemblage produced by humans in a laboratory); and living "micro-cultures" purposely fostered in yogurt, sourdough, tofu, and other living foods. The latter sense of the word is by no means irrelevant to the prior "social" sense; they are all about the terms of relationship among various elements in complex assemblages of humans, other beings, the Earth, and things. In contemporary vernacular understanding, culture is the ongoing collective sense making of how we *be* in relation with each other, other living beings, and the living world, and may include everything from microbes to artificial intelligence, virtual worlds, and the "viral" phenomena that sweep across the worldwide web.

In academic contexts, "culture" refers to a process, or a code of practice rooted in a shared understanding of the world(s). It can be an artifact, or a performance that simultaneously creates and conveys meaning through symbolic representations. Social scientists often define it as the behaviors and beliefs characteristic of a particular social or national group. Most natural scientists place culture outside their domain of species, spaces, patterns, and processes of the biophysical world, notwithstanding the convergence of social, biological, and physical categories in popular culture and in theoretical breakthroughs across disciplines (complexity, network theory).

Raymond Williams (1973) noted that "culture" described persons, and whole peoples and their practices, whose civilized, aesthetic sensibilities and intellectual focus lifted them above the gritty physicality of farming, manufacturing, and artisanal work. The cultured person transcended nature, Earth, the body, and "animal instincts" through the life of the mind and the "fine arts," reflecting class overtones and delineating clear boundaries between types of people (Williams 1976 [1983]; Mitchell 2004; Yudice 2007). He cited the origins of the word in Europe as related to "cultivate," meaning "to work with or upon nature," a perspective that informed cultural-landscape and cultural-ecology approaches to the crafting of cultural landscapes and the appropriation of nature by humans (Mitchell 2000). This framing of nature-culture connections was dualistic, hierarchical, nonreciprocal, and partial, yet asserted universality. Whether they celebrated or mourned the fact, scholars generally agreed that culture stood outside of nature, transforming it. Contemporary scholars have suggested that the rift between the humanities and sciences, described by C. P. Snow (1964 in Williams 1976 [1983]) as the "two cultures" phenomenon, as well as the current "culture wars" and "science wars," all rest on a deeper shared assumption of active human cultures working upon the passive raw material of a separate and opposite nature (Ingold 2004; de Sousa Santos 2010).

Three key points have emerged since Williams's keyword essay that warrant special mention here. The first is the breakdown of the nature-culture binary by feminist, poststructuralist, and science and technology studies (STS) scholars. The second is the convergent development of network, complexity, and relationality theories in the social and natural sciences and in social movements. The third is the new wave of decolonial thought and action that derives from and supports cultures *of* resistance and culture *as* resistance in relation to power struggles over land/territory/ecologies and the nature of the world. We present each of these separately and a brief summary of their convergence in current theory and practice.

Since the 1980s, feminist and poststructuralist theorists across disciplines have challenged the nature-culture binary and have called out the power relations inherent in this construct. Poststructural feminist questioning of culture and nature-culture binaries is best epitomized by three critiques of science and culture. Sandra Harding (1986) advanced a critique of scientific notions of objectivity, based on assumptions of rational scientific actors standing outside of culture, creating knowledge about nature. She proposed an epistemology of situated knowledge(s), taking into account the culturally embedded positionality of all knowers. Val Plumwood (1993) criticized the culture of science that purports to govern nature, and directly challenged the dualistic constructs that lie beneath the ecological crisis of reason. Building on Harding and Marilyn Strathern (1987), Donna Haraway (1988, 1991a, 1991b, 1994, 2007) exhorted feminists to blur the lines separating human, animal, and machine, and to see and speak from an acknowledged, situated position with the "power of partial perspective" in the nature-culture borderlands that we all inhabit.

With the exception of Plumwood, most ecofeminism critiques maintain elements of essentialized dualistic categories of nature and culture, identified with woman and man, although they criticize the ways in which these are deployed by powerful patriarchal institutions, including science, to the detriment of nature and women (Merchant 1989; Shiva 1988 [1989]). Marxist and socialist feminists have maintained dualistic constructs of humans and nature, focusing on the gendered distribution of labor and property, though several works have complicated the use of nature and culture and note how they often enable hegemonic political and economic power (Katz 1998; Jackson 1993).

Feminist political ecology (FPE) scholars have countered the nature-culture binary through culturally and ecologically situated feminist critiques of science, sustainable development, and environmental injustices, in place and across places (Rocheleau, Thomas-Slayter, and Wangari 1996; Rocheleau 2008; Nightingale 2006; Hawkins and Ojeda 2011; Sundberg 2004; Harcourt 2009; Harcourt and Escobar 2005; Escobar 2008). Several works have moved beyond men and women, applying insights from feminist theory and practice to analyze the intersecting fields of power in specific contested ecologies (Mollett 2010).

Critical and poststructural nature-culture theorists across disciplines have also questioned nature-culture dualities on the basis of ethnographic research (Nash 2001), archival and oral accounts of environmental history (Braun 2008), social constructivism (Whatmore 2006; Braun and Castree 1998), and case studies of power in political ecologies. While much of the early work in political ecology left culture largely unexamined, feminist, poststructural, postcolonial, and decolonial political ecologies have embraced culture

and challenged nature-culture dualisms (Rocheleau, Thomas-Slayter, and Wangari 1996; Blaser and Escobar, this volume). Proponents of STS and actor network theory (ANT) have also explored the cultures of science and its social construction(s) of nature (see for instance Haraway 1991a, 1991b; Latour 2005; Law 1992; Law and Mol 2002; Stengers 2000). Several ethnographic scholars have concluded that wild and domesticated are not fundamental and universal categories across all peoples, nor are nature and culture to be found always and everywhere (Strathern 1987; Croll and Parkin 1992; Descola and Palsson 1996). Similarly, Tim Ingold (2011) concludes that ways of life, of being, seeing, and doing, are not about fixed repertoires contained within symbolic systems of representation. He eschews culture and nature for "the dwelt-in-world," which we know through immersion, encounter, process, sensibility, and becoming (Ingold 2000, 42). He notes that what we call cultural transmission occurs by cultivating habits and possibilities of attention and of "being alive to the world" (Ingold 2000, 2011).

What is made visible through these various polemics is the artificial separation of nature and culture by hegemonic, and often colonial, thought. What remains are some key questions about the different connections and relationships that make dwelling in the world possible. For instance, how are people connected to each other and to other beings/elements of the world? Social connections that have long been posited within social theory as hierarchical structures are increasingly characterized as "flat" social networks devoid of power. Labor-based understandings of human relationships to nature, in terms of raw material to be transformed, are giving way to metaphors of humans rooted in place(s). Visions of rhizomes, mushrooms, rooted networks, networked places, and relational territories are increasingly invoked to reconcile these dichotomous formulations with the complexities of the more-than-human world (Whatmore 2002, 2004; Rocheleau and Roth 2007; Escobar 2008; Massey in Harcourt et al. 2013).

The network, the web, and the rhizome are rapidly gaining over the tree as metaphor in the study of culture, nature, and technology. The transgression of the nature-culture binary requires a vision that does not see a focal, self-contained organism, human or otherwise, planted upon nature as substrate. Rhizomatic metaphors have become especially fashionable in social theory, as well as in technology studies and communications (Deleuze and Guattari 1987). The humble mushroom is perhaps ideal for purposes of thinking about culture(s) and the relationships of humans to other beings, whether in the case of rooted social movements (Rocheleau and Roth 2007) or explicit associations of people and other species (Tsing 2012; Rocheleau 2015). The actual organism, the thing called mushroom, is the living network below ground, and the thing we usually refer to as mushroom is only the fruiting body, which is an ephemeral "organ" of reproduction. It presents a visible expression that allows us to read the presence of the organism, though we then tend to conflate that fruiting body with the thing itself. That subterranean web of relations is what "powers" the ongoing development, performance, and result of visible, legible connections among humans, and among the Earth, humans, other beings, their artifacts, and their technologies—hence, culture.

Rooted networks (Rocheleau and Roth 2007) bring the material into assemblages of actors, while recognizing the importance of place and the rooting of all networks (that include humans) in the Earth. All networks go to ground somewhere, even if that ground is dispersed and networked in a web of sources. There is a need to ground network and assemblage-based visions to accommodate actually existing and emergent nature-cultures (see also Ingold 2011). These range from indigenous peoples and farming communities in specific ancestral lands to the contingent territorial roots of pastoralists, agropastoralists, and urban nomads who are at home, on the move, over vast regional homelands. Likewise, this encompasses the various rhizomic modes of "rooting" in the emergent urban, rural, and regional nature-cultures of the pluriverse.

"Pluriverse" has been invoked by many indigenous and decolonial thinkers, activists, and scholars to express the "coexistence of multiple interconnected worlds" (Escobar 1995 [2011], viii). While the Western academy credits William James (see Ferguson 2007) with the concept, we suggest that he codified, rather than invented, a phenomenon already alive in many cultures across place and time. The dictionary provides a pluralistic definition of "universe," though James's "pluralism" went beyond the common usage today. It evokes a sense of coexistence, simultaneity, and copresent realities, going beyond multiple epistemologies to multiple ontologies. Writing about such a pluriverse, decolonial thinkers Arturo Escobar, Marisol de la Cadena, and Mario Blaser suggest that we are actually talking about modernism as a reduction, collapsing the pluriverse into a single world where real politics is impossible and what passes for politics polices culture, while a rationalist, modernist science polices nature. The attempted hegemony of modernist "economy" seeks to govern both. It is a world of the pluriverse denied, of many worlds occluded to serve the interests of an aspiring (but always incomplete) dominant power structure. This "mainstream" globalized culture is built on one lie of universality, a second lie of invented and enforced duality between Nature and Culture, and a third lie of multiple domesticated versions of culture(s) as proof of difference to deflect recognition of (attempted) hegemony as well as the real diversity of multiple worlds.

Pluriversal thinking stands in opposition to Euro/capitalocentric constructs that reify difference and mask hegemony. It also contrasts with denials of difference such as those by David Harvey (1990), encapsulated by Don Mitchell (1995, 111) as "'[c]ulture' makes 'others.' 'Others' do not make 'culture.'" Instead, the inversion of that same "othering" gaze has repeatedly made visible, and politically powerful, the multiple and overlapping cultures, including oppressed and oppositional groups within "the West" and globally, among "the rest" (Hall 1992). The reemergence of decolonial thought and practice and the rise of explicit cultures *of* resistance/culture(s) *as* resistance (Yudice 2007; Nash 2001; Simpson 2011) has roots in the work of Franz Fanon (1963), as well as in several waves of "Third World" and other liberation struggles throughout the world and the United States, including the Black Power movement, the American Indian Movement (AIM), and Chicano/Latino movements. Nationalist and identity-based politics of liberation derived in part from shared cultures of resistance and to some extent from cultures as resistance, though the language invoked peoples, races, and classes, in relation to land and freedom, rather than nature-cultures in more-than-human worlds.

New social movements, scholars, artists, and activists have brought decolonial, indigenous, and subaltern thinking into the emergence of a new liberation culture and politics, based on "decolonizing knowledge and reinventing power" (de Sousa Santos 2010; Hopkinson 1998, 2000, 2012), much of it linked to Earth, land, territory, and the more-than-human world. These demonstrate that "others" *do* make cultures, and choose to defend certain ways of being, in many (but not all) cases against capital, empires, authoritarian/totalitarian states, and other repressive patriarchies. La Via Campesina makes culture(s) and knowledge(s), along with the World Social Forum, the Zapatistas, the Indignados, Occupy, and a profusion of indigenous movements worldwide. While violent racial supremacists, militant patriarchal religious movements, and international corporations also make cultures, the presence and practices of liberatory social movements clearly demonstrate that cultures do not, of necessity, emanate from coercive power. They can also be powerful agents and advocates of emerging nature-cultures based on principles of social justice and ecological viability.

As Arturo Escobar notes, "Culture is political because meanings are constitutive of processes that, implicitly or explicitly, seek to redefine social power. When movements deploy alternative conceptions of woman, nature, development, economy, democracy, or citizenship that unsettle dominant cultural meanings, they enact a cultural politics" (Escobar 1998, 64). In fact, to some indigenous scholars like Brendan Hokowhitu (2009), such a cultural politics is enacted on and by indigenous bodies serving as sites of resistance. Various indigenous movements, too, conceptualize culture as the enactment of a radical politics of change. The "Idle No More" movement, in Canada, to protect lands and waters against tar sand extraction and mining is a contemporary decolonial indigenous movement that sees culture as political, and cultural practice as the enactment of political thought. For scholar-activists like Leanne Simpson, the revisioning of political thought from an indigenous way of being-in-relation-in-the-world implies a continual struggle against a culture of extraction, and for a creative culture of "continuous rebirth" (cited in Klein 2013). Here, a decolonial feminist understanding of culture emphasizes not only a reintegration of nature and culture but the decoupling of modernist and colonial interpretations of culture from its actual enactment as a site of political struggle to legitimate different ontologies/epistemologies of being in the world.

The reemergence of decolonial thinking is converging with other radical currents that question basic categories of knowledge and language as deeply steeped in Eurocentric, racist, sexist, classist, and heteronormative binaries, hierarchical assumptions, and patriarchal values and practices. Decolonial theorists are in conversation with pluriverse and feminist thinkers, and are coalescing with resurgent indigenous and alternative movements, under the umbrellas of world anthropology, decolonial theory, indigenous theories of liberation, and alter-globalization. Together, they treat nature-cultures as particular responses to colonial acts, on the one hand, and as existing also within acolonial space-times, on the other. The decolonial turn validates both the precolonial existence of multiple nature-cultures and the possibilities that particular existing nature-cultures provide for decolonization. Building on these decolonial visions, we see culture as the habit-forming practices and politics of connection and disconnection that shape, and are shaped by, the dynamic experiences of being-in-relation-in-the-world(s).

16

Degradation
Stephanie Foote

"The global economy," Wendell Berry writes, "institutionalizes a global ignorance, in which producers and consumers cannot know or care about one another, and in which the histories of all products will be lost. In such a circumstance, the degradation of products and places, producers and consumers, is inevitable" (Wirzba 2002, 244). For Berry and other critics, the new global order, oriented toward achieving unprecedented material abundance for a small number of people, thrives when it contaminates natural resources like land, water, air, and species habitat, and prospers by degrading less tangible, though equally crucial, aspects of human life: a sense of place, human dignity, a belief in the interconnectedness of whole ecological systems not limited to those that support only humans (Harvey 2006; Nixon 2011; Tsing 2005). But it is also clear that for writers, scholars, and activists, among the most insidious effects of a market system designed to exceed the carrying capacity of the planet is its uncanny ability to degrade a sense of the richness of local histories and ways of knowing in the name of progress, and thus paradoxically to erode the sense that the future can be otherwise.

"Degradation," like other key terms in environmental discourses, has both technical and cultural meanings. It can describe a biological process (all organic matter decays and breaks down); a chemical process (all inorganic material deteriorates under the right conditions); as well as an unwelcome mechanical effect (technological and mechanical systems will ultimately lose their efficiency). But "degradation" does not merely refer to a predictable decay or deterioration. It is a politically and morally charged term. It can, for example, describe the deliberate debasement of human social actors. To degrade people is to rob them of their full humanity, to turn them into objects. Work in environmental justice has revealed the powerful connections between the degradation of habitats and ecosystems and the subsequent sustained degradation of the people who worked and lived in them, loved and fought for them (Bullard 2000; McGurty 2009; Pellow 2004; Pezzullo 2009). In environmental terms, the moral charge of degradation is critical for understanding its ecological impact; degradation indexes the loss of richness and resilience of a place or a system, a loss resulting from the indifference of global economies to the very people and places that sustain them.

But can environmental critics harness the multiple implications of degradation to move our work forward? For some critics, degradation does not merely measure deliberate ecological damage, nor does it only describe the human cost of that damage. Rather, it is both a process and a narrative, for degradation happens—often unpredictably—in different temporal and spatial registers (Adamson, Evans, and Stein 2002). Thus degradation, far from simply erasing objects and subjects, can produce knowledge about them in the global context of capitalism—indeed, can reveal capitalism's faultlines and errors. In such a narrative, degradation coordinates different orders of knowledge and experience, different scales of place and time in a global order that can often seem as though it seeks to standardize and homogenize them.

Perhaps the most immediate example of an object that narrates and embodies material and human degradation is the most common tool of the new global economy: a computer of the sort I have used to write this entry. This laptop was made and marketed by a multinational corporation. It is constructed of component parts that were assembled all over the globe. Those parts were sourced from raw materials in still different nations, and when it finally fails me—when the computer chips that store my information degrade, or when the monitor cracks—it will make its way to yet another place I cannot yet know, and its parts will be broken down. Some will be harvested and used again, some will be thrown into a landfill. Every element of the life cycle of this tool will degrade—the political climate in the nations that supply rare Earth metals may well continue to worsen, the materials themselves will decay, the health of the workers who assemble it and those who recapture it will suffer (Braungart and McDonough 2002). The landfill in which it is stored will fail and contaminate groundwater, further degrading the bodies and the lives of the people around it. And yet recognizing different scales of degradation and decay—the "slow violence" of a long view of resources extraction, to use Rob Nixon's term—makes visible new narratives about the way in which human and ecological costs are managed by a global economy, and allows us to imagine them differently, to create counternarratives that might mitigate the systemic degradation on which they were founded and that they perpetuate (Morton 2010). That is to say that degradation—its different registers, temporalities, and scales, and the human and ecological costs embodied therein—can be the beginning points, not the final decayed moments, of our ability to hear and produce narratives about the deliberate damage to ecological systems wrought by both large systems and daily practices.

17

Democracy

Sheila Jasanoff

Democracy, like religion, offers a powerful vision of collective order, and like religion, democracy retains its vitality through constant adaptation and transformation. No recent social phenomenon has so comprehensively tested the plasticity and regenerative capacity of democracy as environmentalism in the late twentieth century. The arrival of the "environment" as a matter of public concern changed the playing field for democratic thought and action in fundamental ways. It framed new issues for political conflict, opened up novel possibilities for action and alliances, expanded the time scale of responsibility toward nature and future generations, created unprecedented tensions between expert and lay knowledge, limited to some degree the sovereignty of nations, gave rise to new demands for international cooperation, and prompted once-unimaginable questions about the rights and entitlements of human beings and their fellow creatures on this planet. The resulting explosion in environmental controversies, laws and treaties, court decisions, forms of expertise, and modes of civic expression disrupted the conventional meanings of both components of the word "democracy": who is the "demos," or populace, whom governments should represent and protect, and what does it mean to govern in relation to environmental problems? Indeed, to repeat Robert Dahl's famous question of the 1960s, who governs (Dahl 1961)—when the thing being governed is not a polity but the environment? Less transparently as

well, responses to environmental problems by states and societies revealed unexpected divergences in national commitments to the idea of democracy itself.

Salient transformations in the relationship between environment and democracy can be grouped under the headings of *risk*, *property*, and *nature*. In each category, it makes sense to think of democracy as a set of culturally inflected practices rather than as a globally homogeneous norm that applies universally across all times and places.

The early years of environmentalism increased awareness that technologically achieved improvements in the human condition—longer lives, richer harvests, greater control over nature's caprices—had come at a cost. Rachel Carson's elegiac *Silent Spring* drew U.S. national attention to soil, air, and water pollution that had created wastelands in place of thriving ecosystems and driven songbirds to the edge of extinction (Carson 1962). Hers was a starkly realist, though some claimed overdrawn, account of the dangers posed by chemicals carelessly dispersed into the environment and persisting there to nature's detriment. Carson's focus on chemical risks framed U.S. environmental politics for a decade and more (Brickman, Jasanoff, and Ilgen 1985), giving rise to a host of new laws and rules, as well as a form of citizen mobilization centered on individual and community rights and distinctive enough to merit its own acronym: the NIMBY ("not in my backyard") syndrome.

Risk consciousness, defined mainly in terms of hazards to health, safety, and property, shaped American environmental democracy in specific ways, exacerbating the tendencies toward fracture and polarization inherent in pluralism and translating political debates into scientific ones. Interest groups allied with science as a means to achieve political ends (Ezrahi 1984) and vigorously deconstructed technical arguments unfavorable to their preferred policy outcomes (Jasanoff 1995). Resource-rich companies funded research to destabilize expert consensus in a process that some termed "manufacturing uncertainty" (Michaels 2007). These moves and countermoves placed intense pressure on regulatory agencies responsible for making decisions on issues marked by high levels of scientific ignorance and indeterminacy. A bureaucratically popular response was to insist that risk assessment rested on, and could be conducted as, "sound science"; political values would enter decision making only at the later phase of risk management (U.S. NRC 1983). The discursive boundary between "assessment" and "management"—or facts and values—offered welcome relief to beleaguered regulators, who could claim legitimacy on the ground that they were doing good science (Jasanoff 1990). But by channeling debate toward technical arguments, risk discourse directed democratic politics away from the political power structures (Winner 1986) and "sociotechnical imaginaries" (Jasanoff and Kim 2009) that underpin risk creation in modern political economies.

More insidiously, insistence on the scientific character of risk assessment went hand in hand with a deskilling of the demos. Influential analysts embraced studies in social and behavioral psychology purporting to show that lay people suffer from biases that prevent a fully rational (i.e., scientifically sound) appraisal of risks (Breyer 1993; Sunstein 2005). People, in other words, are born irrational. This characterization of lay capabilities contradicted the more optimistic Jeffersonian and Deweyan assumptions of a "knowledge-able" public that had animated an earlier generation of environmentalists (Jasanoff 2011). It reinforced the notion that environmental policy is a domain of expert governance, in which lay people can have standing only if their views align comfortably with scientific and economic notions of rationality.

The politics of risk unfolded along quite different lines in other countries, eliciting correspondingly different responses from citizens and governments. In Europe, it was a work of social analysis rather than an eloquent scientist's lament for nature that captured the public imagination: *Risk Society* by the noted German sociologist Ulrich Beck (1986 [1992]). Beck, like Carson, took risk for granted as a byproduct of the contemporary world. But whereas U.S. experts emphasized the need to calculate and control risk, Beck and his followers called attention to the inescapable uncertainties and gaps in responsibility opened up by advances in science and technology. Not surprisingly, Beck's ideas initially found greatest resonance in (then) West Germany, where social mobilization against nuclear power (Nelkin and Pollak 1981) and, later, biotechnology (Gottweis 2000) expressed a troubled national sense that risks are fundamentally ungovernable, rendering democracy itself problematic. In sharp contrast to U.S. practice, German policy tried to secure through legislation certainties that science could not deliver, establishing clear lines of authority and banning or stringently regulating technologies that present moral and political, as well as physical and biological, hazards (Jasanoff 2005).

The rise of environmental thinking also called into question notions of property rights derived from a Lockean view of land as having value only through economically productive uses. Three significant questions arose. First, who is entitled to speak for undeveloped land, or wilderness? Second, to what extent can governments restrict property rights by controlling, or even prohibiting, development in the name of environmental protection? Third, how can present-day decision makers secure the rights of future generations? In the United States, the first two questions characteristically received their most extended hearing and analysis in the courtroom. During the expansive 1970s, U.S. federal courts granted standing to environmental groups to litigate against harmful development, holding that members' aesthetic or recreational interests were sufficient to allow such groups to speak for nature (*Sierra Club v. Morton*, 1972; *United States v. SCRAP*, 1973). Later Supreme Court decisions cut back on standing for environmental groups (*Friends of the Earth v. Laidlaw*, 2000) and used the doctrine of "regulatory takings" to restrict government's power to limit development for environmental purposes (*Dolan v. Tigard*, 1994). The question of representing future generations, by contrast, devolved into a struggle between disciplinary experts, with ethicists and legal scholars (Kysar 2010) contesting the economists' practice of discounting, and thereby rendering moot, the claims of those not in a position to claim environmental rights in the present or near future.

From the standpoint of democracy's effectiveness, one may set these developments, grounded in judicial and scientific understandings of economic value, against the direct-action strategies adopted in many other countries. Indian environmentalism provides especially striking contrasts. There, the Chipko movement, featuring village women tying themselves to trees to prevent felling, served as a model for nonviolent ecofeminist action to protect local livelihoods. In a process of governmental transformation, the Chipko protests helped redraw political relationships and even state boundaries within India (Guha 1989). The Narmada Andolan, an Indian social movement against large dams, caused the World Bank to stop funding the project and to rethink its environment and development policies (Roy 1999; Goldman 2005).

Still unfolding at the turn of the twenty-first century were questions about whether human representation of nature provides adequate safeguards for the environment or whether natural objects and entities should enjoy rights of their own. These issues were broached in the early years of American environmental litigation, most notably by Justice William O. Douglas in his

dissent in *Sierra Club v. Morton*: "The voice of the inanimate object, therefore, should not be stilled." Displaying similar faith in a rights-centered approach, the legal scholar Christopher Stone argued that trees should have standing to represent themselves in court (Stone 1974). European thought approached the same issues from somewhat different angles. Whereas Stone was concerned with nature's right to advocate for itself, constitutional protection for primates in Spain and for ecosystems in countries such as Ecuador and Bolivia granted present and future existential rights to nature, implicitly drawing nonhumans and natural objects into the human moral community. Meanwhile, some science studies scholars queried whether maintaining a strict ontological separation between humans and nonhumans provides a broad enough foundation for democratic politics, or whether nature must be allowed to represent itself in "parliaments of things" (Latour 2004a). The way these as yet scattered and uncoordinated moves to give legal and political voice to nature play out in practice will profoundly affect what environmental democracy means, nationally and globally, in the unfolding future.

18

Eco-Art
Basia Irland

Eco-art involves a transdisciplinary, multimedia, activist-oriented process, which addresses environmental and sustainability issues. There is a shift away from art as commodity and toward new creative possibilities of art in service to communities and ecosystems. Eco-art includes artists who consider it their role to help raise awareness and create actions about important issues and natural processes; invite participation and devise innovative strategies to engage diverse communities; work directly with others to augment the knowledge associated with particular fields; and produce works that inspire people to reassess the notion of commons. Eco-artists emphasize collaboration. Many of these artists work with indigenous and local community members, as well as specialists from a range of disciplines, including media, education, architecture, performance arts, sociology, engineering, and the gamut of sciences.

Ethics is at the core of this practice and has been explored from multiple perspectives by writers in systems analysis, deep ecology, and complexity theory. Eco-art is flexible, adaptive, far reaching—theoretically, philosophically, methodologically, and geographically—with extensive legacies, abundant intentions, and widely (wildly) variable parameters.

Works might be sited, or unsited, anywhere in the world—within old mines, under water, on top of utility poles, in forests, throughout digital media, in the air, at a massive slagheap, in the produce aisle of a grocery store, at an urban dump, on frozen tundra, or under the lens of a microscope. The investigative probe of the eco-artist might include a minute flea, an oil spill, cyclical processes, a bird song, water-borne diseases, weather phenomena, sewage discharge, an entire watershed, or even a continent. The process of creation is often as important as, or more important than, the end "product," which might be ephemeral, existing only through documentation. The materials and techniques used for the creation of the art, if there is a physical object at all, are as different as chunks of coal, seeds, computer-based social media, gardens, Geiger counters, malformed frogs, bees, soil, fungus, ice.

The word "ecology" (from the Greek "*oikos*," meaning "house, habitat for living," and "*logos*," meaning "discourse"), is defined by the *American Heritage Dictionary* as the "science of the relationships between organisms and their environments," and better describes a more deeply engaged and research-oriented approach than does the term "environment." Eco-artists see themselves as devoted to compassionate positive change, and often embed solutions to ecological problems into their art.

Eco-art is about extending the rather narrow definitions of site, context, and relationship that have been central to much contemporary art. It is holistic and integrated work, which addresses the impacts humans have on ecosystems, the places where we live, and the other species with whom we share these places; as a result, it can be said to be art in which the works have purpose. Some eco-art projects focus on the idea of resilience—a system's capacity to be buoyant, endure, overcome, or adapt to changes—which might be enhanced by using three other "r's": restoration, remediation, and reclamation, in addition to the traditional reduce, reuse, and recycle.

The ethos for this group of artists encourages the long-term flourishing of social and natural environments in which we reside, so questions about appropriateness are endemic. Does a practitioner take an airplane to a place to create work or attend a conference, thereby increasing the carbon footprint? What are the most "green" materials that can be used for a project, which will leave no objectionable trace? Even apparently irreverent questions concerning the validity of categorization and the defining of terms emerge as both challenge and possibility for future generations of artists.

19

Ecocriticism
Greg Garrard

In the beginning, a small group of scholars at a 1992 meeting of the Western Literature Association at the University of Nevada–Reno founded the Association for the Study of Literature and the Environment (ASLE). One, Cheryll Glotfelty, America's first Professor of Literature and the Environment, went on to edit, with Harold Fromm, *The Ecocriticism Reader*, which defined ecocriticism as "the study of the relationship between literature and the physical environment" (Glotfelty and Fromm 1996, xviii). It would be, like feminist and Marxist criticism, a politicized reading practice that would challenge ecocidal attitudes and promote "an earth-centred approach to literary studies." Its twenty-five essays examined American nature writing, Native American literature, ecofeminist "herstory," and environmental philosophy. The characteristics of ecocriticism risked metropolitan condescension due to its focus on the literature of western America as opposed to the civilized East; its admiration for nonfictional writing about wild places rather than self-referential postmodern concoctions; some ecofeminists' emphasis on the intrinsic environmental virtues of women rather than the social construction of gender; ecocriticism's frank engagement with spirituality; and—worst of all—its seeming hostility to human-centered literary theories emanating from "old Europe." A common response to early ecocriticism was sniffy dismissal of "backpacker critics" entranced by "poems with trees in them."

Pioneer ecocritics resented having to choose between defiant advocacy of poems with trees in them and embarrassed disavowal of such naïve interests. They were sustained by environmentalist convictions, which by the early 1990s included escalating concerns about climate change, ozone depletion, and the global biodiversity crisis. Ecocritics saw themselves, in sanguine moments at least, as defending, with their writing and teaching, those beings environmental philosopher Val Plumwood calls "Earth Others," much as social movements like Earth First! might try to defend them with direct action. Yet the tanned, healthy, Caucasian hiker-professors depicted by Jay Parini in the *New York Times* magazine in 1995 seemed unlikely revolutionaries in what one of them, Lawrence Buell, later called "the ecocritical insurgency" (Buell 2005, 12).

Buell's *The Future of Environmental Criticism* characterizes the early years (in which his 1995 book *The Environmental Imagination* was prominent) as "first wave," by analogy with popular histories of feminism. The "second wave" might be dated to the 1999 ASLE conference in Kalamazoo, where a Diversity Caucus was formed as a challenge to the demographic homogeneity of the plenary speakers and delegates. Since then, the caucus has continued to promote diversity of both personnel and perspective, emphasizing the "intersection" of the domination of nature by humans with various forms of intraspecific power relations organized around gender, class, sexuality, race, and disability. While Buell cautions against schematizing history, he nevertheless discerns a decisive—and largely welcome—shift towards a "sociocentric" rather than a "biocentric" approach in this second wave.

As with all origin stories, the tale of how ecocriticism began is partial and somewhat distorted: Glotfelty and Fromm included critical material on John Updike and Don DeLillo as well as Edward Abbey, and several essays engaged constructively with continental literary theory. The importance of urban as well as ex-urban environments and the limitations of the "wilderness" as an aesthetic phenomenon were emphasized from the beginning, albeit by a minority. While ASLE and its growing list of affiliates in Europe and Asia organized a vital community for scholars, the reappraisal of diverse ethnic and national traditions it sponsored revealed numerous antecedents, such as Puerto Rican poet Juan Antonio Corretjer, German critic Jost Hermand, and British Marxist Raymond Williams (DeLoughrey and Handley 2011; Goodbody 2015; Head 2002). Moreover, as Greta Gaard has recalled, since ecofeminism and feminisms of color arguably predate both the ecocritical origin and the putative first wave of feminism without featuring prominently in either history, "feminists and ecocritics utilizing feminism's 'wave' metaphor will inadvertently erase the history of ecological feminism and feminisms of color from both feminism and ecocriticism alike" (Gaard 2010a, 646). Without questioning the importance of canonical literature and criticism, Elizabeth DeLoughrey and George Handley "call attention to an implicit production of a singular American ecocritical genealogy that, like all histories, might be reconfigured in broader, more rhizomatic, terms" (DeLoughrey and Handley 2011, 15). Such breadth involves both intellectual possibilities and risks.

Where ecocritics might once have allowed Muir's ecstasy on Yosemite's Half Dome to prod them into awe and wonder, it is now the ambition of the ecocritical enterprise itself that is sublime. The centripetal, territorializing impulse in ecocriticism, registered in pedagogies of outdoor experience and political philosophies of bioregionalism popular in the United States, now seems both insufficient as a response to planetary environmental crisis and absurdly disingenuous for globe-trotting academics, yet Heise's "environmental imagination of the global" appears to be reflected in the work of environmental-justice, feminist, and postcolonial critics who, following Latour and Vivieros de Castro, are exploring "cosmopolitical" or "micro-minority" environmentalisms based on concepts of "multinaturalism" (Adamson 2001, 2011; DeLoughrey and Handley 2011; Nixon 2011; Pellow and Brulle 2005).

As the geographical horizon has receded, so has the historical: there are substantial ecocritical studies of medieval, Renaissance, and eighteenth-century literature, as well as British Romanticism and American Transcendentalism. Furthermore, the encounter of ecocriticism with identity-based theories familiar in mainstream literary study—organized primarily around race, disability, sexuality, and postcolonial subalternity—has multiplied, far beyond gender, the cultural permutations it must comprehend. Finally, now that ecology is being addressed as a science rather than, in Dana Phillips's terms, "a slush fund of fact, value, and metaphor" (Phillips 2003, 45), it has become distressingly apparent how deficient it is in generalizations (Kricher 2009, 101)—let alone the kind of "laws of ecology" proposed by Barry Commoner in 1971—suitable for metaphoric redeployment, still less political corroboration for ecocritical theory. Not only is culture multicultural; nature is multinatural (De Castro 1998; Latour 2004a), a configuration beyond the grasp of any one human mind. The herd of largely genial cats that is the ecocritical community will need to develop new forms of deep, sustained, and systematic collaboration to allow it to grapple successfully with the vast field of inquiry it has adumbrated.

From poems with trees in them to a Theory of Everything in just two decades is an impressive trajectory. The student or scholar making her own beginning at this point, though, will rightly quail at the prospect. So how might one map the naturecultures postulated by ecocritical theory? *Ecocriticism* (Garrard 2004, 2011 2nd ed.) proposes "master metaphors" such as pollution, pastoral, and the Earth as organizing concepts in criticism, an approach that places texts, rather than literary theories or human social identities, at the center of the analysis. In these, at least, we are expert. Additional tropes worthy of exploration might include health, climate (Hulme 2009), and—why not?—the most moribund metaphor of them all, "humanity." As yet, the explicitly Anglo-American and "aborescent" configuration of ecocritical history Garrard presents has not been supplemented by an accessible "rhizomatic" history of the kind proposed by DeLoughrey and Handley.

The most perceptive interpreter of ecocriticism as a Theory of Everything, Timothy Clark, builds his *Cambridge Introduction to Literature and the Environment* around thirteen "quandaries," such as, "What isn't an environmental issue?" The size and structure of the family can no longer be regarded as a merely personal matter given the ecological ramifications of additional infant consumers, while even "[d]ivorce . . . becomes an environmental issue if it creates two households instead of one" (Clark 2011, 86). The risk is, as Clark observes, that "environmentalist concerns, refusing either to fit given understandings of the political or yet transform them, threaten to become only a new kind of personal moralism in the smallest affairs of day-to-day life" (86). Whereas second wave ecocriticism tends to assume too tidy an alignment between the interests of oppressed humans and those of Earth Others, Clark's deconstructionist approach acknowledges that thorny dilemmas and painful paradoxes are, and will be, the norm.

Here is a paradox, inspired by Clark, of political subjectivity in the era of climate change: the unbearable lightness of green. Drawing on Milan Kundera's *Unbearable Lightness of Being*, we might contrast the weight of existence experienced by Beethoven, during whose lifetime human population passed one billion, with the painful sense of insignificance that assails us now that there are over seven times as many. I may "count" as roughly ten Nicaraguans or one hundred Tanzanians in terms of carbon emissions, but I "count" as only 1/62 million Britons politically. At the same time, in ecology the reassuring notion of the "balance of nature"—a biological myth of Eternal Return—has been replaced by the disorienting idea of perpetual flux within broad geographical limits (Kricher 2009). On one hand, the cumulative impact of our species is so great that some scientists argue that a new geological era has begun: the Anthropocene. On the other, the individual actions purported to contribute to this dramatic change seem ridiculously slight: leaving a light on or buying a steak. The language of climate change activism, says Clark, "enacts a bizarre derangement of scales, collapsing the trivial and the catastrophic into each other" (2011, 136). To want to be "green," then, is to experience at once a terrifying collective weight—enormous carbon footprints trampling the planet—and the unbearable lightness of eroded agency and illegible ecological signposts. Since we dwell among such quandaries henceforth, it should be the business of ecocriticism to bring critical intelligence and humanistic scholarship to bear, and seek either to resolve or, more likely, to help us endure them.

20

Ecofascism

Michael E. Zimmerman and Teresa A. Toulouse

Most literally, ecofascism (ecological fascism) names a collectivist political regime that uses authoritarian measures to achieve its major goal, protecting nature. No such regime has yet existed, although German National Socialism (in)famously included two components that later, both explicitly or implicitly, became viewed as ecofascist. The first was an organic-corporatist authoritarianism that overrides all individual liberties and the second was a nativist racism that justified protecting German blood and land (*Blut und Boden*) from the polluting presences of non-Germans. Beginning in the early 1980s, claims of "ecofascism" have been used by philosophers of varying kinds to criticize a range of views that could be regarded as "eco-authoritarian." Some have also used "ecofascism" to attack arguments that environmental wholes trump the interests of individual organisms (human and nonhuman). In the 1990s a number of historians, literary critics, cultural studies scholars, and philosophers, turning attention to the intersection of nineteenth-century nature Romanticism and racism, have argued, without explicitly using the term "ecofascism," that the mainstream American environmentalist movement could usefully analyze the darker origins and consequences of some of its own assumptions and aims.

German National Socialism has been retroactively described as having ecofascist dimensions. Calling on Social Darwinism and the monism of Ernst Haeckel (who coined the term "ecology" in 1866), the Nazis conceived of humans as organisms struggling for survival in a hostile world. To succeed against (supposedly) inferior and degenerate races, so the Nazis surmised, would require not only "purifying" mystically intertwined German blood and German land but also demanding that individuals sacrifice themselves and their property for the good of the corporate-organic whole. Condemning Judaism and Christianity for their otherworldliness and alleged contempt for nature, some Nazis developed what amounted to a racist religion of nature (Pois 1986). While Nazi politicians drew on American eugenics to justify racial hygiene programs that later culminated in mass murder, many German nature-protection advocates borrowed ideas from American environmentalists to protect the German "homeland" from industrial exploitation. In 1934 the Nazi *Reich* adopted what were at the time the world's most far-reaching environmental protection laws. Although some nature-protection advocates opportunistically supported the Nazi Party, others willingly aligned themselves with the racist, antimodernist, and militaristic Nazi program. Even the opportunists tacitly consented to purging the homeland of "rootless" (landless) peoples such as Jews and Gypsies (Zimmerman 1995; Uekotter 2006). Despite the role played by environmental protection in Nazi ideology, Germany's decision in 1935 openly to re-arm revealed that in practice National Socialism was primarily interested in developing the human biopower required to achieve its expansive political aims.

Memory of the twisted "green" side of National Socialism ruled out environmental-protection discourse in West German politics for an entire generation. When in the 1970s the new German Green Party made its slogan "neither right nor left, but out in front," the party implicitly criticized anthropocentric, industrialist modernity's goals of individual liberty and collective human self-realization (Biehl and Staudenmaier 2011).

Many European and American corporate executives, labor union leaders, and politicians suspected that the Greens and some other environmentalists adhered to a pro-nature, anti-industrial stance that would not only undermine material well-being but also, in the process, curb important political freedoms.

The idea of an eco-authoritarian polity did emerge in the 1970s, when American neo-Malthusian environmentalists Paul Ehrlich and Garrett Hardin, among others, concluded that *only* authoritarian regimes could prevent human overpopulation—made possible by modern food production, public health measures, and industrialization—from causing a global "tragedy of the commons" (Ehrlich 1968; Hardin 1972, 1977). Some radical environmentalists called for draconian cuts in human population, or even human *extinction*, an aim to be achieved either voluntarily or otherwise.

In the 1980s, some eco-philosophers interpreted the influential German philosopher Martin Heidegger as a forerunner of the deep ecology movement, because he condemned techno-industrial modernity's assault on nature and called instead for humans to "let things be" (Zimmerman 1983; Foltz 1995). Viewing Heidegger as a proto–deep ecologist became problematic, however, after new scholarship revealed that he had joined the Nazi Party precisely because of its antimodernism and its emphasis on purifying the German homeland. In a famous 1987 diatribe, social ecologist Murray Bookchin specifically called deep ecology "ecofascist" for aligning itself with a far right-wing thinker like Heidegger (Bookchin 1987). While Bookchin's angry critique was overstated, he did remind radical environmentalists that nature-protection ideologies could have problematic sources and political applications that philosophers should investigate.

Still other philosophers in the 1980s used the term "ecofascism" to disparage the view, shared by many environmentalists and ecological scientists, that individual organisms are less important than species and the biospheric whole. Philosopher and animal-rights activist Tom Regan maintained that such environmental holism parallels fascist holism, in which the social collective trumps the rights of individuals (Regan 1983). A possible target of Regan's complaint was J. Baird Callicott, an important eco-philosopher who once promoted an authoritarian ecological holism, recommending that society sharply limit individual freedom in order to promote the overall well-being of nature (Callicott 1980). (Later repudiating this stance, Callicott explored how to *reconcile* environmental holism with rights for individual organisms.)

Dimensions of the broader collectivist vs. individualist debate also appear in the history of mainstream postwar environmentalism, which gained influence after Rachel Carson's best-selling *Silent Spring* aroused widespread anxiety about the dangers of chemical pesticides for people as well as for birds (Carson 1962). The growing perception that industrialism was wreaking havoc on nature and threatening public health led the U.S. Congress and the European Common Market to support environmental legislation that required private industry to clean up polluted air and water. U.S. environmentalism in some ways traded on the consensus of Franklin D. Roosevelt's New Deal, according to which large-scale government programs were needed to preserve the common or *collective* good.

In 1980, however, the *individualist* version of modernity enjoyed resurgence in the United States with the election of Ronald Reagan. Thereafter, some American libertarian property rights activists began using the term "ecofascism" to condemn allegedly authoritarian government interference in individual and corporate property rights, and to criticize radical environmentalists who wanted not merely to halt industrial pollution

but to bring down industrialism and the modernity associated with it. In reply, some environmentalists accused organizations like the libertarian Wise Use Movement of promoting an *anti*-environmental fascism, in league with the right-wing industrial-military complex that used political influence and wealth to undermine environmental policies and to ruthlessly exploit the land.

Whereas accusations of ecofascism first arose in the context of 1970s–1980s debates about the ethics of environmental holism vs. individualism, and the politics of collectivist and authoritarian vs. libertarian approaches to environmental issues, in the 1990s, attention shifted to another facet of the nascent ecofascism of National Socialism, its racism. The race issue was raised prominently by Robert Bullard in *Dumping in Dixie* (1990 [2000]), a study that helped to shape the environmental justice movement's criticism of mainstream environmental organizations for their primarily white membership and leadership, and for their focus on wilderness preservation rather than on the industrial pollution affecting the poor and people of color, in the United States as well as in the Global South.

Academics had long explored the roles played by Social Darwinism and eugenics in the Nazi doctrine of *Blut und Boden*, but now they began to examine the role played by racist ideologies in the intertwined history of nationalism and an emerging environmentalism in late- nineteenth- and early-twentieth-century America. Arising in the context of a white supremacist ideology, late-nineteenth-century American environmentalism was shaped in part by a version of the Romantic nature aesthetic that would also later influence National Socialism. Historian William Cronon's lead essay in his anthology *Uncommon Ground* (1995), which maintains that the nationalistic American "wilderness ideal" depends on historically specific and often problematic understandings of "nature," helped to open up the question of how different cultures interpret and value nature according to their own perspectives.

According to Cronon, the wilderness ideal long popular in mainstream environmentalism arose in part from the desire of late-nineteenth-century white, male, elite, and urban Americans to preserve a reminder of the nation's settler past, and to define a sense of American male identity expressed in encounters with a "wild" and untamed American nature. This ideal had both salutary and deeply tragic aspects (Cronon 1995). On the one hand, the wilderness ideal helped to inspire a growing moral conviction that nature is not merely a resource but also an end in itself worthy of respect, admiration, and protection. The neo-Romantic assumption that "wild" land unsullied by human civilization is worth preserving played a crucial role in establishing the U.S. national park system and in passing the American Wilderness Act (1964). On the other hand, ahistorical and racist assumptions that the land was or should be "unsullied"—this, after many centuries of Native American occupancy and management—had dire consequences throughout the nineteenth century. Armed with an ideology of "manifest destiny" and an overt or covert sense of racial superiority, a modern democracy ordered its soldiers either to kill or forcibly evict from their ancient homelands countless Native Americans and relocated them in reservations, thereby opening the land for white settlers.

In the view of American cultural and environmental studies scholars such as Joni Adamson (2002), Robert Nixon (2011), Lisa Sun-Hee Park and David Naguib Pellow (2011), and Andrew Ross (2011), similar racialist assumptions surprisingly persist in new instances of displacement and enclosure. These critics have variously argued that the ideal of an untrammeled, pure nature has been used by some Americans to justify resistance to immigration by Mexicans, who presumably do not share the "wilderness ideal." That ideal has also been exported by both American and European environmental organizations to Third World countries, in which acts of enclosure have established game preserves and national parks. Funding for such parks has come in part from wealthy eco-tourists who prefer to view wild animals in landscapes unsullied by local people (Guha 1989).

Finally, such critics have pointed to how uninterrogated assumptions about environmental activities in the First World have had unintended consequences in the age of globalization. For years, because of the tireless efforts of environmental advocates, First World consumers have enjoyed a decreasing burden of pollution. Overlooked in the process, however, have been the environmental consequences of outsourcing First World industrialization and its pollution to the largely nonwhite Global South.

Most mainstream environmental organizations have taken steps to address how their assumptions and activities might reflect possibly racialist or even racist cultural assumptions. In 2014 the Sierra Club even created a new national award named after Robert Bullard. Deep ecologists and some other radical environmentalists, however, have criticized scholarly claims that nature is "merely" a human construct insofar as this claim assumes that nature has no independent standing outside of human sociocultural frames. Even legitimate concerns about environmental racism, so they argue, must not be allowed to once again efface the claim that *all* forms of life and even the biosphere itself deserve respect and care.

Although charges of ecofascism have often been overstated, given the different positions of varying actors in these debates, such charges have importantly prompted philosophical and historical inquiry into questionable assumptions influencing a variety of different forms of environmentalism. To the extent that environmentalism as such can affirm certain positive achievements of modernity, even as it simultaneously draws attention to modernity's patent environmental failures, environmentalism can surely be seen as a moral *development* beyond an anthropocentric modernity. Presented in this way, environmentalism could forestall implicit or explicit charges of ecofascism.

Still, this much is clear: the eruption of a murderous, antimodernist, and racist regime in the educated and industrially advanced Germany of the 1930s and the authoritarian and possibly racist strands at work in the historical formation of even apparently progressive and mainstream environmental ideologies and practices indicate how daunting is the task of sustaining modernity's professed goals of universal liberty and equality, while at the same time promulgating environmentalism's goal of expanding the domain of moral considerability to include nonhuman beings. Environmentalists who call for authoritarian measures to "save the planet," who engage in blanket critiques of modernity as a whole, and who ignore racialist dimensions of their own past history and current views undermine the credibility of the entire environmental movement at a time when conceptualizing a balance between the well-being of humankind and that of nature is more important than ever.

21

Ecofeminism
Greta Gaard

Was Rachel Carson an ecofeminist? Technically, no—the term "ecofeminist" did not appear until a decade after her death—yet Carson's work exemplifies ecofeminist praxis (the inseparability of theory and practice). Observing an unusual and alarming phenomenon of environmental health (massive bird deaths), Carson used scientific methods to trace the various avenues for sustenance in the songbirds' lives—air, water, food, habitat—and discovered that overexposure to pesticides was the lethal agent. Making interspecies connections among the environmental health of avians, humans, and ecosystem flows (air, water, food), Carson (1962) argued for an end to pesticides, publishing her findings in a voice both literary and scientific. In the final months of her life, Carson faced down strong corporate assaults on her work while battling breast cancer, an illness later found to have significant links to the synthetic chemicals she studied.

Carson's intersectional approach, linking environmental issues with social concerns, anticipates ecofeminist perspectives. As an evolving praxis, ecofeminism grew out of many women's interconnected sense of self-identity—a deep recognition of interbeing that bridges socially constructed boundaries of class, race, species, sexuality, gender, age, ability, nation, and more—and an "entangled empathy" (Gruen 2012) that brings both compassion and action to the task of alleviating conditions of eco-social injustice. Ecofeminism is one of many feminist environmentalisms that have taken shape across the academic and applied fields of art, literary studies, geography, agriculture, energy, forestry, labor, indigenous rights, animal studies, and environmental health. What makes an action or analysis ecofeminist is its intersectional approach to understanding and acting to correct eco-social and environmental problems. Specifically, ecofeminism is unique for bridging human justice, interspecies justice, and human-environmental justice, while other feminist environmental perspectives ignore the species question, or subordinate it (they also fail to challenge the culturally produced links among gender/race/sexuality/nation/nature, and often reinscribe them as a result). For example, feminist peace camps of the 1980s and 1990s—the Seneca Falls Women's Peace Camp of 1983 (United States), Greenham Common Women's Peace Camp (1982–2002, UK)—and the eco/feminist defense of British Columbia's Clayoquot Sound (1993) moved ecofeminism beyond maternalist peace politics to challenge the gendered politics of environmental activism and the contested linkages between gender and nature that were variously manipulated among the environmental activists as well as the logging companies (Moore 2008; Mortimer-Sandilands 2008a). Documents such as the "Unity Statement" from the Women's Pentagon Action and Clayoquot Sound's "Code of Non-Violent Action" make explicit these activists' awareness of the linkages among human injustices, economic and political structures, and the exploitation of species and environment. From Chile's Con-spirando collective and its response to economic neoliberalism to Buddhist women's environmental activism in Taiwan, the strategies for change are shaped by specific contexts, cultures, and religious/spiritual beliefs (Eaton and Lorentzen 2003). Across diverse contexts, ecofeminist praxis is unique for bridging human justice, interspecies justice, and human-environmental justice.

In other instances, ecofeminists have persisted in developing critiques and actions around milk, beginning with the introduction of rBGH in the United States in 1993: at that time, ecofeminists framed the issue as harming female bovines and their offspring, consumers, and small farmers, while increasing profits for one large biotechnology corporation, Monsanto (Gaard 1994). The foundational feminist critique can be seen in the fact that, in both humans and cattle, motherhood is seen as a proper and necessary site for scientific intervention. In the twenty-first century, environmental justice advocates and environmental feminists together emphasize the associations among environmental toxins, breastmilk, and the body burden transferred to nursing infants (LaDuke 1999; Steingraber 2003; Boswell-Penc 2006). An ecofeminist perspective underscores these approaches, expanding their analysis to examine the manipulation of motherhood across species, for in the theft of cows' nursing milk and the enslavement of "veal" calves, female bovines and their offspring suffer from Western industrialized culture's social construction of "motherhood" as self-sacrifice and perpetual nurturance of others, even unto death (Gaard 1994).

Approaching ethics as a narrative about right relationships among humans, animals, and nature, ecofeminists have critiqued the romantic and heroic narratives underlying traditional environmental ethics. In her essay "From Heroic to Holistic Ethics," Marti Kheel (1993, 256) develops her theory of the truncated narrative, a theory that foregrounds the rhetorical strategy of omission: "Currently, ethics is conceived as a tool for making dramatic decisions at the point at which a crisis has occurred. Little if any thought is given to why the crisis or conflict arose to begin with." Creating "ethics-as-crisis" conveniently creates an identity for the ethical actor as hero, an identity well suited to what Val Plumwood

(1993, 258–59) defines as the Master Model. "Western heroic ethics is designed to treat problems at an advanced stage of their history," Kheel argues, and "run counter to one of the most basic principles in ecology—namely, that everything is interconnected" (258–59). As an alternative to truncated ethical narratives and heroic ethics, Kheel proposes retrieving "the whole story behind ethical dilemmas," uncovering the interconnections of social and environmental perspectives, policies, economics, and decision making, including all those affected by the ethical "crisis" (258–59).

Kheel's inclusive strategy is helpful in retrieving the missing pieces of the climate-change story—or, more specifically, the gendered human-justice and interspecies elements of the story. For example, while the Food and Agriculture Organization of the United Nations' report, "Livestock's Long Shadow" (2006), Campbell and Campbell's *China Study* (2006), and other nutritional research (not funded by the agricultural industries!) make the climate-change/human-health/industrial-animal-agriculture links visible, they do not offer a gendered analysis. Along with First World overconsumption and wastes from unsustainable energy and transportation technologies, animal-based agriculture contributes 18 percent of the worldwide annual production of greenhouse gases—only energy production (21 percent) produces more than this (Gaard 2011). And while climate change is produced largely by the world's elites (the United States produces 25 percent of global greenhouse gases annually), it most severely affects those nations and communities least able to make adaptations for survival. Within those communities around the world, women are the ones hit hardest by climate change and natural disasters due to social roles, discrimination, and poverty: women's gender roles restrict women's mobility, impose tasks associated with food production and caregiving, and simultaneously obstruct women from participating in decision making about climate change, greenhouse gas emissions, and strategies for adaptation and mitigation. An ecofeminist ethical narrative listens to a larger story of climate change by attending to the quality of human-environment, human-human, and First World/two-thirds-world relations and interactions. Instead of heroic ethics and the heroes they feature, ecofeminists argue for detecting and correcting injustices as they occur, and advocate policies and practices of economic, environmental, interspecies, and gender justice that offer more effective strategies for intervention and prevention of social and ecological crises.

Empathy, care, and connection figure strongly in ecofeminist discussions of animal defense and vegetarianism. Deane Curtin's (1991, 70) concept of "contextual moral vegetarianism" affirms that our dietary choices have cultural, ecological, economic, and relational contexts and consequences. Acknowledging that every act of eating—from hunting wild caribou to caging "broiler" hens to growing a vegetable garden—requires the death and suffering of some life form, Curtin maintains that vegetarianism moves us in a *moral direction*. "If there is any context . . . in which moral vegetarianism is completely compelling as an expression of an ethic of care," writes Curtin, "it is for economically well-off persons in technologically advanced countries." Industrialized animal agriculture, world hunger, climate change, and gender politics all suggest that First World feminists and environmentalists may be most compelled to choose the path of moral vegetarianism as a strategy for reducing suffering.

Taking a material approach to the treatment of animal bodies and women's bodies, ecofeminists have examined the politics of sexuality and reproductive justice (Gaard 2010b). Both Maria Mies and Vandana Shiva (1993; Shiva and Moser 1995) have explored questions of genetic engineering, reproductive technologies, colonialism, and population, highlighting the ways in which reproductive technologies are used to enforce economic hierarchies, dividing elite First World women from disadvantaged women through the availability, affordability, safety, and reliability of reproductive control. While "two-thirds-world" women and women of color are depicted as "breeders" whose rampant sexuality poses a threat to the environment and must be controlled, pharmaceutical and biotechnology companies court elite women with fertility-enhancing procedures and reproductive choice (Roberts 1998). Women's right to bodily self-determination, the foundation of feminism, becomes an ecological issue as well in a context of economic globalization that continues social, political, economic, and gendered forms of colonization that appropriate women's bodies and environmental bodies alike. Biotechnology asserts the scientist's right to manipulate human, plant, and animal genetics for the benefit of "humanity," but it is primarily the elite and male sectors of humanity who benefit from these manipulations of subordinated others—women, indigenous communities, animal others, and environment (Shiva 1997).

The associated devaluation of women and environments matters more than mere cultural ideology or gendered stereotype; it means economic profit in a system of global accounting that describes the reproductive and subsistence work of women in the home—birthing, breastfeeding, caregiving, cooking, cleaning, water gathering, waste removing—and the work of environments (sustainability) as "externalities" that count for nothing (Waring 1988). For our future survival, an ecological and feminist ethic is needed to guide our social, economic, and political relations.

22

Ecology
Reinmar Seidler and Kamaljit S. Bawa

Since the nineteenth century, ecology has been defined as the study of the functional interrelationships of living organisms, played out on the stage of their inanimate surroundings. Ecology has developed through an ongoing dialogue between two distinct ways of seeing the world. We might call these "synthetic" and "analytic" tendencies. Throughout the history of ecology, they have competed for attention, at times replacing one another sequentially, at others coexisting uneasily.

Aristotle laid the philosophical foundations for ecology by presenting natural history as a distinct area of inquiry, but the power of ecology to explain the origin and diversity of life on Earth was not fully demonstrated until the work of Charles Darwin (1859) and Alfred Russel Wallace (1860). Their research—especially that of Wallace—revealed how interactions between organisms and environments have powerfully shaped every species' distributional pattern and history. Darwin's concept of natural selection emphasized mechanisms of competition and sexual selection. In contrast, Wallace's years of biogeographical field work in the tropics led him to focus on cases where related species occupy neighboring, but ecologically distinct, regions. Wallace recognized that environmental conditions can drive speciation by molding population traits through time. Along with the concept of evolution itself, this vision of mutable species responding to environmental constraints made possible the later elaboration of ecology as a formal discipline for organizing knowledge about a vast array of relationships among species and their environments.

The term "ecology," however, was not coined until 1866 by German zoologist Ernst Haeckel. Ecology as a discipline separated itself only gradually, over the next several decades, from related focal areas within contemporary biology. The first book to use the term "ecology" in its title was the seminal *Oecology of Plants* (1895) by Danish botanist Eugenius Warming. Warming drew inspiration from the work of Alexander von Humboldt (1805), an early pioneer of systematic, large-scale, "scientific" natural history. Humboldt's wide-ranging interests and field experience spurred him (nearly a century before Warming) to combine observations from biology, geology, and meteorology to show that plants occur in predictable associations along geographical gradients.

The idea that plants form well-defined, stable communities inspired Warming's work and dominated the still-youthful field of ecology well into the twentieth century. Frederic Clements (1916) developed the idea that plant communities change over time along a predictable pathway called "succession"—a process he compared with the development of an organism through its life stages. Clements even suggested that plant communities could be considered *superorganisms.* His erstwhile student Henry Gleason (1926) came to vehemently oppose Clements's ideas, maintaining that species in plant communities have *individualistic* properties, and are brought together at particular places by essentially random processes.

The split between Clementsian and Gleasonian understandings of plant associations signaled a fundamental rift in the field of ecology—one that is still operative today. One might even call it the split between two ecological *paradigms* (Kuhn 1962). A *synthetic* branch of ecology highlights systemic properties, emergent patterns, and functional connections among ecological elements. An *analytic* branch views organized, goal-oriented activity as restricted to the levels of individual and population, finding talk of higher-order or "holistic" properties overly teleological. The synthetic branch focuses on the functional and organizational characteristics of a system; the analytic, on developmental and evolutionary pathways. Dialogue between these two orientations has animated many of the important advances in ecological theory and understanding.

During the 1920s, animal ecology developed a more theoretical, quantitative, and predictive orientation partly as a result of work by mathematicians Alfred Lotka (1925) and Vito Volterra (1926). The Lotka-Volterra differential equations predicted the behavior of interacting predator and prey species, laying the foundations of an analytical, mathematical ecology focused on individuals, populations, and species rather than on communities.

About the same time, however, Arthur Tansley at Oxford (1935) was articulating his concept of the *ecosystem* as a dynamical unit, at first relying heavily on Clementsian concepts. After the war, the development of ecosystem theory was taken in new directions by the Odum brothers, Eugene and Howard (1953). Where Lotka's work had explored the evolutionary-ecological implications of the physical laws governing the transformation of energy—portraying competition among individuals basically as a struggle for available energy—the Odums' *systems ecology* approach focused on stocks and flows of energy and materials within highly idealized food webs and large-scale communities. This systems approach has found a more recent echo in the concept of *ecosystem services* (Daily 1997), broadly defined as the "benefits people derive from natural ecosystems" (MEA 2005). As with systems ecology, the stocks and flows involved have proved difficult to measure and quantify empirically (Norgaard 2010; Seppelt et al. 2011).

During the 1950s and early 1960s, partly as a counterbalance to the emphasis of systems ecology on the physical parameters of ecosystems, a new *evolutionary ecology* centered on biotic interactions began to take shape. Robert MacArthur (1958) and G. E. Hutchinson (1959) first explored patterns of resource partitioning and niche segregation among species. By incorporating processes of predation (Janzen 1970; Connell 1978) and coevolution (Ehrlich and Raven 1964), evolutionary ecologists highlighted the central role of biotic interactions in shaping community structure. Their models strengthened the sense that ecological communities are infused with an orderly functionality rooted in evolutionary processes tending to divide up the available resources among well-adapted species. John Connell recalled, "At the time I had a mind-set that ecological communities were all neatly regulated around some equilibrium level. I was sure that the component species had been evolutionarily selected until they were coadjusted in their niches so as to fit with each other. This was the prevailing view of most ecologists at that time" (Connell 1987). Expanded with new inputs from statistics and economics, the language of mathematics seemed to hold ever-greater promise of adequately expressing the complexity of ecological relationships in concise and elegant terms.

Meanwhile, systems ecology had led to a period of experimentation with "Big Ecology" in the 1960s and 1970s. "Big Ecology" organized landscape-scale, "whole-systems" modeling and experimental programs such as those of the International Biological Program and Hubbard Brook (New Hampshire, United States). An early outcome of the latter program was the demonstration that industrial emissions were exerting measurable impacts on natural ecosystems through acid rain. The first Earth Day in 1970 channeled the energy of antiwar and civil rights movements together with growing popular awareness of U.S. environmental degradation. A new popular environmental movement was coalescing, inspired by the work of Rachel Carson (1962), Barry Commoner (1971), and others. Concern about forest die-back and lake acidification in New England and *Waldsterben* in Europe joined with a growing awareness of the risks of nuclear energy and the rapid disappearance of old-growth forests across western North America. Soon, the terms "ecology" and "ecological"—vocabulary drawn from what had been until then a rather esoteric scientific discipline—exploded into popular use, almost as terms of approbation. "Ecological thinking" was considered holistic, inclusive, all-embracing, as opposed to the conventional narrow, analytical, reductionist scientific approach. This uptake of the term into popular parlance was correlated with a resurgence of the organicist concept of ecological systems. The "Gaia theory" of the 1970s represented perhaps the apotheosis of this radical modern organicism, analogizing the functioning of the entire *biosphere* to that of a single organism.

In the academy, growing interest in tropical ecology brought the richness of species interactions in the tropics to center stage in debates about community diversity and assembly—i.e., how *biodiversity* (Wilson 1988) originates and is maintained. Tropical plant ecologist Steve Hubbell (2001) attempted to reconcile theories of ecological association based on niche-assembly with those based on dispersal dynamics. The resulting "neutral theory" predicts species diversity within trophic levels solely from estimated rates of species dispersal, extinction, and speciation. The theory is "neutral" in the sense that in explaining how species coexist (especially in hyper-diverse tropical forests), it avoids the need to elucidate complex details of ecological interactions. It does this by assuming "*per capita* ecological equivalence"—meaning that each individual of each species in a trophically defined community has what amounts to equivalent chances of survival and dispersal. The theory goes sharply counter to the instincts of many field ecologists, but it "unifies" systems theory with evolutionary ecology in the sense that it shows how, mathematically, knowing and incorporating all species traits and relationships into a complex community model produces essentially the same result as treating them all as equivalent in a simpler model. As Hubbell points out, arguments about community structure are partly the result of ecologists working at different scales.

> The perspective to which a person adheres can largely be predicted by the scale on which the person works. Most proponents of niche assembly come out of a strong neo-Darwinian tradition, which focuses on the lives of interacting individuals and their fitness consequences. . . . Proponents of dispersal assembly, on the other hand, typically work on much larger spatial and temporal scales, using biogeographic or paleoecological frames of reference. (Hubbell 2001)

To what degree neutral theory corresponds to real-world data, however, remains in dispute (Nee 2005; McGill, Maurer, and Weiser 2006; Leigh 2007).

Neutral theory is built explicitly on the old equilibrium theory of island biogeography (MacArthur and Wilson 1967), but over larger time scales it portrays natural systems in a never-ending flux driven largely by random influences. It is therefore akin to the explicitly nonequilibrium models of the so-called New Ecology of the 1980s and '90s (Botkin 1990). These models deeply question the privileging of particular associations or community profiles. They emphasize stochastic processes over deterministic patterns; unpredictable disturbance over regular succession; resilience over stability. They were developed partly from close observation over time of disturbed and regenerating communities that increasingly dominate modern landscapes. These models demand a set of newly available analytical and statistical methods—including Bayesian and Boolean approaches—which, it is argued, may be better suited than conventional methods to interpreting and articulating the levels of uncertainty inherent in these new models.

This "New Ecology" was also influenced by the rapid rise to prominence in the 1980s of *conservation biology*. The roots of conservation biology can be found in a variety of domains, the common denominator being an increasing awareness of large-scale loss of biodiversity. Accelerating rates of deforestation in the tropics, and the run away loss of natural habitats elsewhere, made ecologists aware of the dire need for a new, formal discipline oriented toward conservation. From the start, conservation biology was concerned with the problems of small and threatened populations, and with the drivers of extinction risk. Soon it became apparent that conservation biologists could not avoid participating in wider societal discourses on the management of natural resources, and even of economic development planning. Such themes quickly drew conservationists into new realms of knowledge making: ethics, economics, ethnography, historical ecology, political ecology, and others. Perhaps more than any other field of science, conservation biology has had to wrestle with the paradoxical requirement of finding tractable ways to measure and represent the world, even while incorporating an ever-increasing array of potentially relevant influences on system outcomes.

The urgent need to understand interactions between nature and human society is by no means new. Agriculture and animal husbandry have always demanded sophisticated awareness of the dynamics of social-ecological systems. In the 1930s and 1940s, the principles of demography and population ecology were first systematically applied to fisheries and forestry, marking the beginnings of a formal *applied ecology* oriented toward the management of renewable natural resources. What is new is the understanding that the interrelationships of organisms, ecosystems, and human economy are now shaping not only the natural world but also the future of humanity at large. These trends have produced a great expansion of ecology's field of view, amounting to a redefinition of the discipline.

Perhaps history will look back on the early twenty-first century as a transitional period when the environmental and the social sciences—with the help of the humanities—matured enough to be able to work together on global-scale problems. Underlying these efforts must be the will to redeploy the tools of science in imagining, defining, and moving toward sustainable ways for the human enterprise to persist—both *with* and *within* natural systems. Given such a will, we can imagine a future in which the discipline of ecology is folded into a larger, more encompassing enterprise known (not too grandly, we hope!) as "sustainability science."

23

Ecomedia
Michael Ziser

In its most basic sense, "ecomedia" (a contraction of "ecological media") is shorthand for representations of and communication about the human and natural environment in media beyond traditional print. A more general category than eco-art, with which it shares many concerns, ecomedia includes environmentally engaged film, television, music, visual arts, installation and conceptual art, as well as work in new-media venues like web pages, video games, and mobile operating systems. Because the term is currently used more often by scholars than by cultural producers, its most immediate value is to facilitate discussion of environmental representation both within new media studies and across the traditional disciplines constructed around single media (literature, art history, film studies, etc.). Under the ecomedia umbrella, a diverse array of creative and critical projects that may have struggled for recognition within their home disciplines can be presented together in exhibitions, book series, special journal issues, and even graduate programs. The advent of an ecomedia blogosphere where scholars can make connections with one another and with a broader readership is already one happy outcome of this relatively recent coinage.

Ecomedia has a deeper claim upon environmental studies than this thematic and pragmatic gloss might at first indicate, however. Because they arise from an inaugural estrangement and denaturalization of the dominant print environment, ecomedia projects are natively critical about the materiality of their objects and methods in a way traditional single-medium environmental disciplines often are not. This reflexivity is in turn a good fit with ecology and ecocultural studies, which in their own ways pay close attention to the histories, meanings, and consequences of specific forms of mediation, embodiment, and material connectedness in natural and social systems. The complementarity of subject matter and critical approach in the field gives it an excellent vantage on many serious contemporary environmental problems in which the "objective" scientific issues cannot easily be separated from the aesthetic and social ones. Such problems lend themselves to two basic ecomedia approaches: one that traces the physical footprint of our media practices to mount an environmental critique around resource depletion and other environmental concerns (media as environment), another that follows environmental ideas and images as they are disseminated and remixed across various media (environment through media).

The first approach proceeds through extended engagement with the full environmental context of the production, distribution, consumption, and cultural afterlife of particular media genres and works, borrowing the network-tracing methodologies pioneered by science and technology studies and actor-network theory. For example, there is a story worth telling about the environmental history of twentieth-century representational technologies—the toxic legacy of still and moving film processing, the production of cathode-ray televisions and transistor radios, the rapid obsolescence of computing technology, etc.—and the degree to which this material history surfaces at the diegetic level of films, cartoons, video games, and other artworks of the period. Looking forward, what are the likely environmental impacts of undersea fiber-optic cables, server farms, and wireless networks? To what degree can these be represented in ecomedia projects and clarified in critical accounts of existing work? On the consumption side, how do sites ranging from traditional movie theaters to home video systems to streamed audio and video on handheld devices differently enable our tacit sense of the environment and our ability to represent it? Ecomedia artists and scholars with insight into the hardware, software, and social infrastructure undergirding both old and new media will play key roles in answering such questions.

Ecomedia's broadest impact is likely to come by virtue of the simple fact that nonprint media increasingly provide the primary platform for environmental engagement of all kinds (animal welfare, energy production, agricultural policy, etc.) for the majority of citizens in the developed world—and perhaps even more so for those in the developing world. As a result, ecomedia will eventually supplant text-based approaches on basic questions of environmental representation. It is also likely to offer the best critical access to key questions about environmental affect and other population-level phenomena that play an immense but poorly understood role in the emergence of environmental narratives and the governmental policies that sometimes emerge from them. The larger and more abstract the environmental issue, the more significant become the nuances of the verbal and imagistic discourses in circulation. (The media profile of the current global climate change debates is the best current example.) Only critics savvy about the production and circulation of media images and narratives will be able to supply the kinds of analysis required by these technologically and socially complex mass-mediated environmental problems.

24

Economy

Robert Costanza

The words "economy" and "economics" derive from the Greek roots "*oikos*," meaning "house," and "*nomia*," meaning "management." Literally the word refers to how we manage our "house" at scales from individual households to communities, nations, and the whole world. In modern usage, "the economy" is often equated with "the market economy"—the sum of all the transactions of goods and services for money. But this is far too narrow a definition. A broader definition often used is "the allocation of scarce resources among alternative desirable ends" (Farley and Costanza 2002). This definition implies the analysis of what ends are desirable, what resources are scarce, and how best to allocate the resources to achieve the ends. Certainly the market and the goods and services exchanged in it are a part of this, but many desirable ends exist outside the market, as do many scarce resources. So a better way to think about "the economy" is as everything that is scarce and that contributes to desirable ends—broadly, human well-being and its sustainability. In our current "full world" (Daly and Farley 2004), almost everything is scarce, including natural and social capital assets, most of which are outside the market. In addition, human well-being is a function of much more than the consumption of marketed goods and services (Costanza et al. 2007). The emerging "science of happiness" (Layard 2005) and "positive psychology" (Seligman 2012) attempt to better understand what influences people's subjective sense of well-being. It is clear from

these and many other studies that nonmarket factors are extremely important.

Also, to fully understand the economy, one must understand its interdependence with and embeddedness in society, culture, and the rest of nature. This kind of whole system or ecological economics is what is needed in a world rapidly filling up with humans and their artifacts if we hope to better define and fulfill desirable ends. These include the recognition that (1) our material economy is embedded in society, which is embedded in our ecological life-support system, and we cannot understand or manage our economy without understanding the whole, interconnected system; (2) growth and development are not always linked and true development must be defined in terms of the improvement of sustainable well-being (SWB), not merely improvement in material consumption; and (3) we need a healthy balance among thriving natural, human, social, and cultural assets, and adequate and well-functioning produced or built assets. One can refer to these assets as "capital" in the sense of a stock or accumulation or heritage—a patrimony received from the past and contributing to the welfare of the present and future.

These assets, which overlap and interact in complex ways to produce all human benefits, are defined as follows:

- *Natural capital*: The natural environment and its biodiversity, which, in combination with the other three types of capital, provide ecosystem goods and services—the benefits humans derive from ecosystems. These goods and services are essential to basic needs such as survival, climate regulation, and habitat for other species, water supply, food, fiber, fuel, recreation, cultural amenities, and the raw materials required for all economic production.

- *Social and cultural capital*: The web of interpersonal connections, social networks, cultural heritage, traditional knowledge, trust, and the institutional arrangements, rules, norms, and values that facilitate human interactions and cooperation between people. These contribute to social cohesion, strong, vibrant, and secure communities, and good governance, and help fulfill basic human needs such as the needs for participation, affection, and a sense of belonging.
- *Human capital*: Human beings and their attributes, including physical and mental health, knowledge, and other capacities that enable people to be productive members of society. This involves the balanced use of time to fulfill basic human needs such as fulfilling employment, spirituality, understanding, skills development, creativity, and freedom.
- *Built capital*: Buildings, machinery, transportation infrastructure, and all other human artifacts and services that fulfill basic human needs such as shelter, subsistence, mobility, and communications.

Recognizing the embeddedness of the economy in society and the rest of nature implies that we balance these assets in a way that achieves the three broad subgoals of (1) a *sustainable scale* of human economic activities that recognizes basic planetary boundaries (Rockström et al. 2009a, 2009b); (2) a *fair distribution* of resources and opportunities among groups within the present generation of humans, between present and future generations, and between humans and other species; and (3) an *economically efficient allocation* of resources that adequately accounts for protecting the stocks of all assets, whether marketed or not, including stocks of natural and social capital.

Returning to our original etymology of "economy," there is another word that also stems from "*oikos*"—"ecology." Ecology is literally the study of the house. What we need to manage our multiscale house in the new full world context we now find ourselves in is an adequate understanding of ecology in its broadest sense—an ecological economics.

Ecological economics is a transdisciplinary field that integrates the study of humans and the rest of nature (Costanza 1991; Costanza et al. 1997; Daly and Farley 2004). It is not a subdiscipline of conventional economics. It is an ongoing effort to bring together the study of what is desirable with what is sustainable on our finite planet. Ultimately, it is what "economics" should be all about.

25

Ecopoetics
Kate Rigby

"Ecopoetics" is an ecocritical neologism referring to the incorporation of an ecological or environmental perspective into the study of poetics, and into the reading and writing of (mainly) literary works.

"Poetics" and "poetry" derive from the classical Greek word "*poiesis*," meaning "making," and as Bate observes in his landmark theorization of *ecopoiesis*, this is an activity that might in principle be practiced in any medium (Bate 2000, 45). Making is by no means an exclusively human practice. Many other species make things, some of which display not only high levels of craftsmanship but also an aesthetic sensibility, as Darwin observed of the highly decorated structures created by the Australian male bowerbird, the sole purpose of which is evidently to charm the female. The natural systems that have enabled the emergence of these diverse creative practices might also be seen as *poietic* or, rather, *autopoietic*, continuously generating new forms and patterns, and dissolving old ones, in a dynamic process of open-ended becoming. Human "poesy" is thus both continuous with that of other species and sustained by what the early German Romantics referred to as the "unconscious poesy" of the Earth (Rigby 2004a, 102–3). One of the core concerns of ecopoetics is to consider how what we make—especially, but not exclusively, with words—might in turn help sustain these other-than-human poietic practices and autopoietic processes.

Within literary studies, "poetics" generally refers to the theory of poetry, which "attempts to define the nature of poetry, its kinds and forms, its resources of device and structure, the principles that govern it, the functions that distinguish it from other arts, the conditions under which it can exist, and its effects on readers or auditors" (Brogan 1993, 930). Much work in ecopoetics follows this usage in privileging the verbal art form of poetry. Bate, e.g., argues that "the rhythmic, syntactic and linguistic intensifications that are characteristic of verse-writing frequently give a peculiar force to the *poiesis*," and he suggests that poetic meter might be perceived as "answering to nature's own rhythm, an echoing of the song of the earth itself" (Bate 2000, 75–76). Following Martin Heidegger, Bate locates the value of ecopoetic writing, as a "making" of the "dwelling place" (*oikos*), in its capacity to counter the technocratic reduction of all things (including fellow humans) to "standing reserve" by disclosing the Earth and its diverse denizens in words that present them as irreducible to human ends and understandings, as does, e.g., Elizabeth Bishop in her poem "The Moose." Bate nonetheless acknowledges that the material process of disseminating poetic writing takes its own toll on the Earth, since it "can only occur through technology," while stressing that "the poetic articulates both presence and absence," bearing no more than a trace of the earthly experience to which it responds (2000, 281).

Within ecological literary studies, this element of absence is sometimes emphasized even more strongly. Ecocritic Leonard Skigaj, whose approach, like Bate's, is phenomenological (although drawing on Maurice Merleau-Ponty rather than Heidegger), proposes a theory of poetic "*référence*," incorporating an acknowledgment of the limits of language, in order to refer the reader's perception "beyond the printed page to nature, to the referential origin of all language," in search of practices of "atonement or at-one-ment" with "nature's rhythms and cycles" (Skigaj 1999, 38). Similarly, David

Gilcrest (2002) favors a poetics of "restraint," such as that practiced by Marianne Moore, arguing that the domination and destruction of other species is facilitated by forms of symbolic appropriation that translate the nonhuman world into words tied to human desires. As Jonathon Skinner observes in the inaugural edition of his journal *ecopoetics*, any writer "who wants to engage poetry with more-than-human life has no choice but to resist simply, and instrumentally, stepping over language" (Skinner 2001, 105). Contrary to Heidegger's view, this does not necessarily imply that human speech and writing is radically discontinuous with, or sets its users apart from, and over and above, the diverse communicative processes of the rest of the biosphere. In my account of the "ecopoetics of negativity" (Rigby 2004b), it is nonetheless crucial to value the ways in which poetic writing discloses its own inadequacy as a form of representation, or mode of response, in order to resist the logic of substitution: the illusion, that is, that the text can stand in for something else, whether embodied experience, empirical knowledge, or ethico-political action. Pointing beyond itself by foregrounding its own status as textual artifact, and marking its own limitations as such, poetic writing can nonetheless indicate why such experience, knowledge, and action in support of the flourishing of more-than-human life might be called for.

Ecopoetics, however, is by no means restricted to the consideration of ecologically oriented, or environmentalist, texts. Bate actually excludes the latter from his definition of the "ecopoem" on the grounds of their instrumental use of language, while John Felstiner's "fieldguide to nature poems," encompassing a huge historical sweep from the Hebrew Bible to Gary Snyder, omits other contemporary "ecologic" writers as deserving a "chronicle of their own" (2009, xiii). For those ecocritics who have begun to attempt such a chronicle, a key question is the relationship between earlier nature poetry and contemporary "ecopoetry." Terry Gifford (1995), in identifying a body of "green poetry" that engages directly with environmental issues, shows how much of this writing extends the "anti" and "post-pastoral" tendencies that he traces back to the late-eighteenth- and early-nineteenth-century English poetry of Crabbe, Blake, Wordsworth, and Clare. By contrast, Scott Bryson (2002), assuming a more reductive view of Romanticism, makes a sharp distinction between earlier nature poetry and current "ecopoetry," identifying the latter as characterized by "an ecocentric perspective . . . humility in relationship with both human and nonhuman nature . . . [and] intense skepticism concerning hyperrationality . . . that usually leads to an indictment of an over-technologized modern world as a warning concerning the very real potential for ecological catastrophe" (2002, 5). He nonetheless recognizes that "any definition of the term *ecopoetry* should probably remain fluid at this point" (5).

Theorists and practitioners of ecopoetry disagree about both the meaning of the prefix "eco" and the "poetic" forms best suited to advancing the preferred ecological orientation. To some extent, this follows differences in the wider field of ecocritical debate, with earlier understandings of ecology, premised on order and stability, such as inform John Elder's *Imagining the Earth* (1985), giving way to models of "discordant harmony" and disorderly change, while concerns about preserving the wild jostle with the quest to overcome environmental injustice. Because of the cultural assumptions, social relations, and historical experiences that shape human understandings of the natural world, "there can be no 'innocent' reference to nature in a poem" (Gifford 1995, 15). This can be seen in a comparison between the meanings and values attributed to various natural phenomena, and the places and situations in which they are encountered, in the white American poetry favored by Skigaj and that anthologized by Camille Dungy (2009) in her landmark volume of African American nature poetry. Differences also surround formal considerations, with some ecocritics privileging more traditional lyric poetry, while others argue that experimental forms, such as those that feature in Skinner's *ecopoetics*, are more conducive to engaging with contemporary eco-social contingencies, especially anthropogenic climate change, which renders any attempted realignment with "nature's rhythms and cycles" anachronistic. Harriet Tarlo (2009), e.g., makes a strong case for the use of found text (also referred to as "cut ups" or "recycles") to create open textual structures incorporating multiple voices and perspectives (including that of the reader), which are both imitative of, and continuous with, ecological networks of open-ended interdependency (see also Fletcher 2004). Juxtaposing and defamiliarizing diverse public discourses of nature, such works can also highlight our desensitization to environmental bad news stories, reminding us of our failure to act on what we know (Kerridge 2007). In refusing to occupy the cultural niche conventionally accorded to poetry in the lyric mode, avant-garde ecopoetics seeks to avoid becoming a "tool for placation" (Kinsella 2009, 146). It is important to recall, however, that since the impact of any type of ecological art is ultimately decided on the level of reception, a variety of ecopoetic forms, as well as foci, are required in order to engage and inspire a range of differently situated recipients.

Discussions of ecopoetics also extend beyond poetry *per se* to include nonfiction prose "nature writing" (e.g., Tredinnick 2005), novels (e.g., Hogue 2011), mixed-media work, involving collaborations between poets and visual or sound artists, and public art installations that put poetry in the landscape (e.g., Morley 2008). At its most encompassing, ecopoetics entails nothing less than the art of living genuinely sustainably. One writer who understands it as such is Australian poet Patrick Jones (2010), whose utopian practice of *permapoesis*, conjoining digital communication, permaculture, and postindustrial foraging, is committed equally to ecocentric bioregionalism and transnational social justice (http://permapoesis.blogspot.com.au/). While few environmental writers or ecocritics would be prepared to join Jones and his family in this radical experiment, most would agree that unless our words, however artfully crafted, emotionally compelling, or intellectually challenging, get linked to deeds, ecopoetics might amount to little more than "fiddling while Rome burns."

26

Eco-terrorism

David N. Pellow

The term "eco-terrorism" invites and courts confusion, misinterpretation, and misuse. It is a fine example of doublespeak, and is probably best thought of broadly as a terrain of power or, in a narrower vein, as one scholar writes, "nothing less than one vast attempt at control" (Gibbs 1989, 339). The term is believed to have been coined by anti-environmental activist Ron Arnold (Arnold 1983, 1997 [2010]), whose writings caught the attention of conservative media and political leaders who injected it into national and international discourses to exert greater control over a critical public policy issue, leading to hearings in the U.S. Congress and the passage of laws targeting eco-terrorism in most U.S. states and increasingly in other nations. Arnold famously defined "eco-terrorism" as a "crime committed to save nature" and is just one of many public voices that generally characterize "eco-terrorism" as any violent act against property or persons in the defense of a pro-environmental or animal-rights ideology. Many activists and scholars who are critical of this use of the term counter that rather than affixing this label to nonviolent activist movements seeking ecological sustainability and animal liberation, states and corporations that routinely harm ecosystems and nonhuman animals should be branded "eco-terrorists." As animal liberation activist-scholar Steven Best (2004, 309) puts it, "It speaks volumes about capitalist society and its dominionist mindset that actions to 'save nature' are classified as criminal actions while those that destroy

nature are sanctified by God and Flag." Adding to the confusion, many scholars use the term "environmental terrorism" to describe unlawful actions taken by groups (presumably nonstate and noncorporate actors) to deliberately target and harm an ecological resource base needed to sustain a human population—a discourse with clear global applicability and implications (Chalecki 2002; Miller, Rivera, and Yelin 2008, 113).

While radical movements for Earth and animal liberation have been most visible in Global North communities, in Global South communities there are numerous movements for environmental justice that may also be subject to such labeling, and many activist groups in the South have vocally used the term "eco-terrorism" to critique the actions of companies, militaries, and states harming ecosystems in places like Indonesia and elsewhere ("Bukit Lawang" 2003).

The majority of scholarship on "eco-terrorism" falls into two basic camps: works written by authors who accept the term and support the pursuit and prosecution of activists so labeled, versus those who challenge it and question its social, political, legal, and philosophical genealogies and implications. Interestingly, even scholars and activists who are highly critical of the discourse of "eco-terrorism" generally accept the idea that there are groups that can be unproblematically labeled "genuine terrorists" or "real terrorists" because, unlike committed environmentalists, they are apparently guilty of "actual terrorism" (Vanderheiden 2005, 425, 426, 427; see also Best 2004 and Lovitz 2010; for an exemplary exception see Hadley 2009), which involves harming or threatening people. For example, as Miller, Rivera, and Yelin (2008, 119) write, one should distinguish between legitimate social protest and civil disobedience versus "terrorism" so that "people will be more secure in their rights to free speech and political opposition." The problem here is that these authors assume that there can ever be an unambiguous and universal method of defining "terrorism" that can be reflected in some broad common sense. Moreover, such a position explicitly supports an uncritical acceptance of the dominant narrative and law enforcement practices of "anti-terrorism" as long as they are not directed at environmentalists and animal-liberation activists.

Here we should step back and consider the weighty and controversial term "terrorism" and then explore its discursive and legal relationship to "eco-terrorism." "Terrorism" has conventionally been defined as "the calculated use of violence or threat of violence to attain goals that are political, religious, or ideological in nature . . . through intimidation, coercion, or instilling fear" (Chomsky 2003, 69). Under the USA PATRIOT Act (passed in the wake of the attacks of September 11, 2001), this earlier definition was expanded to include attacks on inanimate objects, via arson or explosives, particularly objects or property used in interstate or foreign commerce. One result of this broadened scope was that many activities that radical environmental and animal-liberation activists had been practicing (vandalism, sabotage, ecotage, etc.) were now well within the parameters of U.S. federal terrorist law. Thus "eco-terrorism" became a federal offense with additional sentencing enhancements and penalties. While many scholars believe that this shift to include inanimate objects was new, in fact, property destruction was often included in early immigration exclusion laws that sought to criminalize and exclude foreigners who were politically radical—what I might call antiterrorist laws of the early twentieth century. The Immigration Acts of 1917 and 1918 made it a deportable offense if an immigrant/noncitizen in the United States was convicted of sabotage or property destruction. So in that sense, the PATRIOT Act is not actually a shift or expansion, but rather a reassertion and recuperation of a longstanding

practice that was always about the protection of property in a capitalist state.

Fortunately, there is some critical scholarship on the topic that is of use here. Sociologist Jack Gibbs (1989, 330) offers a definition of "terrorism" that acknowledges nonhuman objects as possible targets: "Terrorism is illegal violence or threatened violence directed against human or nonhuman objects." Gibbs continues with a related observation:

> Despite consensus about violence as a necessary feature of terrorism, there is a related issue. Writers often suggest that only humans can be targets of violence, but many journalists, officials, and historians have identified instances of destruction or damage of nonhuman objects (e.g., buildings, domesticated animals, crops) as terrorism. Moreover, terrorists pursue their ultimate goal through inculcation of fear and humans do fear damage or destruction of particular nonhuman objects. (Gibbs 1989, 331)

Even taking into account Gibbs's important intervention, there is an unexamined logic of anthropocentrism inherent in the discourses of "terrorism" and "eco-terrorism" in that they are both variously defined as violence or threatened violence against humans and/or property owned by humans. The significance here is that neither discourse allows for the consideration of the view of "eco-terrorism" as violence against nonhuman animals and/or ecosystems in their own right. Even the term "environmental terrorism" is focused on attacks on ecosystem elements reserved for humans. And while some scholars refuse the charge of "eco-terrorism" directed at groups like the Earth Liberation Front, they do so largely on the grounds that the ELF does not target human beings, but that position implicitly accepts the dominant definition of "terrorism"—one that enshrines the primacy of humans as the ultimate subject of state protection and value (Best 2004; Vanderheiden 2005). Thus, by defending the ELF within that framework, scholars such as these unfortunately also support dominant political and legal discourses that produce a social terrain that undermines the ELF. This contradiction is particularly pronounced given the fact that the ELF's primary goal is to prevent violence directed at nonhuman populations and ecosystems. Ultimately, I would argue that this is the kind of humanist and dominionist thinking that reinforces ecological destruction.

One of the compounding problems with the lack of definitional clarity regarding the concept of "eco-terrorism" is that, more and more, groups that operate with fully legal nonprofit status and only practice non-violent protest are frequently labeled "eco-terrorists." For example, Greenpeace and its staff are routinely called "eco-terrorists" in the media and by conservative organizations, and the influential Center for Consumer Freedom has repeatedly labeled the Humane Society of the United States an "eco-terrorist" organization. This raises a challenge to a popular perspective among social movement scholars: the radical left flank. The idea is that radical groups make more moderate groups appear to be more reasonable and can therefore more effectively push the establishment to make greater concessions than they might otherwise offer. However, as we see both underground and aboveground environmental and animal-rights groups targeted as terrorist organizations by the state, this logic seems to break down. The radical left flank theory assumes that a moderately democratic government will respond with concessions to social movements, but that has not been borne out in this instance—an indicator of the political work the discourse of "eco-terrorism" does and authorizes.

Thinking about "eco-terrorism" in a larger historical and global context, in a very real sense, any revolutionary or nationalist movement seeking to (re)gain control over land and/or other forms of nonhuman nature might potentially be cast as "eco-terrorists" because their "crime" is seeking to challenge the dominant social order with respect to human/nonhuman relations. This would include freedom movements such as the anti-apartheid movement, the Basque National Liberation Movement (ETA), the Free Papua Movement, the Landless Workers' Movement of Brazil (MST), the Revolutionary Armed Forces of Colombia (FARC), the Irish Republican Army (IRA), and Puerto Rican, Palestinian, American Indian, and other indigenous, tribal, and peasant struggles for land, sovereignty, language, and identity. As the "war on terror" has gone global, so has the war on "eco-terror."

The emergence of Earth First! and the Earth and Animal Liberation Fronts (ELF and ALF) in the United States and United Kingdom in the 1980s and 1990s marked a new stage in the evolution of ecological politics. These were movements punctuated by a discourse of radical analysis and direct action that we had rarely seen in environmental or animal rights politics until that point. Through the use of tree spiking, equipment sabotage, vandalism, arson, and removal/theft/liberation of laboratory animals, these groups caused considerable economic damage to corporations, governments, and universities while questioning the violence of state policy and capitalism. By the late 1990s and early 2000s, segments of these movements were converging around new ideas and tactics, producing a broader discourse that linked ecology, social justice, anti-oppression, and animal liberation. Many scholars view the antecedents of "eco-terrorism" not only in the radical environmental and animal-rights movements but also specifically within the philosophies of deep ecology, ecofeminism, and green anarchy that took root in these movements. Accordingly, in the United Kingdom, the United States, Canada, Mexico, Australia, the European Union, and elsewhere, direct action by radical animal- and Earth-liberation activists has often been accompanied by messages directed at heterosexism, patriarchy, racism, class discrimination, state violence, capitalism, and hierarchy in all forms. As Plows, Wall, and Doherty write about actions by Earth First! in the UK,

> At the end of 2002 and in early 2003, EF! activists in these areas were heavily involved in protests against the threatened war in Iraq, some had been part of the International Solidarity blockades to protect Palestinians from the Israeli army, while others had covertly smashed the windscreens of Land Rovers in a Manchester car showroom in protest at Land Rover's sales to Israeli Special Forces. (2004, 211)

The response by states, corporations, and media has been swift, with surveillance, harassment, intimidation, imprisonment, and the passage of harsh new laws that criminalize pro-ecological activities that were previously protected or for which there were already legal prohibitions. Many observers have called these developments the "Green Scare."

Recent scholarship on the agency of nonhuman nature and "things" (Bennett 2010; Braun and Whatmore 2010)—inanimate objects—is instructive for how we might think about eco-terrorism's implications. In many ways these activists are "called" or interpellated by inanimate objects like bulldozers and backhoes and buildings to disable and destroy those objects in order not only to prevent short-term harm to ecosystems and nonhumans but also to send a message to other humans who routinely use those implements in destructive ways. In that sense, these nonhuman objects interact with both groups of humans (authorized and unauthorized users) in violent and destructive ways: in the first instance, they do so in order to harvest ecological and nonhuman wealth; in the second instance, they do so through their own destruction. Of course, in both instances, the humans interacting with these technologies/objects do so in the name of providing sustenance to some population (e.g., investors, shareholders, workers, future would-be beneficiaries of scientific experiments versus ecosystems and nonhuman animals). The agency of these nonhuman technologies, then, lies in their capacity—at least according to the respective user populations—to destroy or to prevent the destruction of life. Vanderheiden (2005, 427) describes the targets of radical environmentalists as "inanimate objects (machinery, buildings, fences) that contribute to ecological destruction"—suggesting an implicit view that these nonhuman technologies are complicit in ecological violence. But the strongest interpellation that occurs by nonhuman natures is the call that human activists feel they are answering in the name of protecting ecosystems and nonhuman animals.

27

Ecotourism

Robert Melchior Figueroa

Ecotourism often serves as an umbrella concept, and overlaps with other categories like *nature-based tourism, geotourism, adventure tourism,* or *responsible tourism* (Buckley 2009). For the United Nations the term "sustainable tourism" offers an even wider umbrella than "ecotourism," but still the two terms are often interchanged because ecotourism's popularity exploded after the advent of early sustainable development discourses (UNDSD 1992). By the 1992 Earth Summit in Rio de Janeiro, with its culminating policy Agenda 21, ecotourism became an operative concept for achieving sustainability (UNDSD 1992). A symbol of this relationship is the 2002 UN Year of International Ecotourism (UNDESA 1998), which sounded many lofty goals of economic, environmental, and cultural sustainability, and even "contributing to the strengthening of world peace" (UNDESA 1998). Hence, the terminological use depends heavily upon context and conveyor, and it should be expected that definitions differ according to locales. "Ecotourism," in other words, will have different meanings in places as distinct as India, Costa Rica, Australia, Chile, Kenya, and the Arctic Circle.

The coinage of the term is generally credited to Hector Ceballos-Lascurain, who, in 1983, deployed and defined the term in several presentations as the founding president of PRONATURA, a Mexican conservationist nongovernmental organization. In 1996, Ceballos-Lascurain refined his definition for the World Conservation Union (IUCN) as follows:

Ecotourism is environmentally responsible travel and visitation to relatively undisturbed natural areas, in order to enjoy and appreciate nature (and any accompanying cultural features—both past and present), that promotes conservation, has low negative visitor impact, and provides for beneficially active socio-economic involvement of local populations. (Ceballos-Lascurain 1996)

Despite many varying definitions, the most popular phrasing is the one offered by the International Ecotourism Society (TIES): ecotourism is "responsible travel to natural areas that conserves the environment and improves the well-being of local people" (TIES 2006; Honey 2008a). In affiliation with TIES, Martha Honey has broadened this simplified definition to identify seven crucial characteristics of ecotourism: it involves travel to natural destinations; it minimizes environmental impact; it builds environmental awareness through education initiatives; it benefits conservation financially; it benefits and empowers local people; it respects local culture; and it supports human rights and democratic movements (Honey 2008b).

Critical considerations of ecotourism come from three interrelated arenas: economic, environmental, and cultural-political critiques. In 2012, the number of international tourists exceeded one billion for the first time. Economically, the correlative receipts of international tourism rose to $US 1.075 trillion (UNWTO 2013). As the most steadily increasing form of tourism since the turn of the twenty-first century, ecotourism is extremely attractive for its significant market potential (TIES 2006). Environmental critique, which often takes the form of accusations of "greenwashing," is perhaps most pronounced and often aimed at major tourism corporations that attract conscientious tourists while failing to meet the goals of responsible ecotourism outlined by Honey. For instance, the local community must be able to claim direct economic benefits with capacity-building opportunities, substantive employment, and bargaining power. Yet tourism industry giants can redistribute resources from abroad rather than locally; offset good practices for bad ones elsewhere; import labor and management; and import goods from their industry chain rather than exchange with local producers. Local cultures, moreover, are more easily marketed for their stereotypes than for authentic local practices and values (Rozzi et al. 2010; Campbell, Gray, and Meletis 2008).

The tourism industry and agencies have developed official certification schemes to thwart greenwashing practices. However, greenwashing can evade even this certification remedy, as the industry has witnessed a flood of such schemes (Honey 2008a). To combat this proliferation, several agencies have succeeded in legitimizing "proper" certification with support from corporations that defy greenwashing practices (Center for Responsible Travel 2013a, 2013b; Honey 2002, 2008a, 2008b). Still, generating legitimate and overarching certification schemes is nearly impossible because of the enormity of global ecotourism, and problematic because regional nuances that determine the most appropriate practices in specific locales may be more appealing than those schemes claiming worldwide legitimacy.

Positive economic critiques point out that local initiatives, indigenous cooperatives, and significant instances of localizing economic benefits are evidence of success (Honey 2008a, 2008b; Maasai Wilderness Conservation Trust [MWCT] 2013). These successes model best practices but have still received scrutiny under the common market paradox of conscientious neoliberalism, in which the micro-economic aims of the conscientious ecotourist are potentially undone by the macro-economic requirement that ecotourism conform in some respects to the global marketplace. For even as the ecotourist may attempt to promote an alternative consumerism that supports local economies, local political empowerment, and respect for local livelihoods, while minimizing environmental impacts, he or she must often travel by means that bolster major airline and oil industries, as well as any number of global industrial market chains linked to the purchases of supplies, goods, and resources. This is further compounded as ecotourism moves into more remote interiors, such as the Amazonian rainforest, where road construction produces environmental destruction, whether initiated by oil exploration, as in Ecuador, or by ecotourism interests themselves, as in Belize (Gould 1999). Ecotourism may represent a positive market-shift away from the failures of earlier development models, but critics warn that global ecotourism and alternative consumption make capitalism "nicer" rather than challenging fundamental market assumptions (Campbell, Gray, and Meletis 2008). In response, sites that base ecotourism legitimacy upon environmental education to inform tourists about biocultural significance, resource use, and sustainable lifeways are defended for shifting the neoliberal global market agenda to improve local communities (Rozzi et al. 2010; MWCT 2013). For instance, organizations such as the Maasai Wilderness Conservation Trust model best practices, in which operators offset the carbon footprint of tourists with sustainable infrastructure and benefits to the indigenous community (Honey 2008b; MWCT 2013).

The ecotourism industry has also been applauded for utilizing ecologically sustainable materials, architecture, and landscape design that minimizes the ecological footprint of some forms of ecotourism. Furthermore, more natural areas have been designated as protected worldwide as a direct result of ecotourism (Honey 2008a; Campbell, Gray, and Meletis 2008). Critics argue, however, that some protected areas have been designated in direct response to market demands rather than ecological significance and that more deserving areas have in some cases been ignored. Additionally, despite the touted viable alternative that ecotourism offers against environmentally detrimental forms of mass tourism, including mass tourism to popular national parks, critics observe that ecotourists tend to expand their itineraries to include such sites (Campbell, Gray, and Meletis 2008). Thus, ecotourists may simultaneously choose *both* alternative and mass forms of tourism, which undermines the claim that the former necessarily avoids contributing to the latter.

Leaders in the industry have conceded that cultural-political gains have lagged behind the economic and environmental achievements (Honey 2008a). One conundrum is an underlying colonial double edge. The one edge neglects the colonial displacement of indigenous peoples from protected areas and parks, envisioning a human-free, pristine area. The other edge accepts human habitation on presumptions that dangerously reintroduce primitivism, rendering local peoples "exotic" for the sake of tourist experiences. This promotes a romantic wilderness fetish while simultaneously benefiting one local group's customs over another's according to which best conforms to the consumer's expectations and can simultaneously disavow the capability of local and/or indigenous people to combine traditional and current, even Western, lifestyles. In Montego Bay, for example, Jamaica's local traditional fishing industry, not necessarily deemed unsustainable, is pitted against the expectations of ecotourists who are permitted to fish in the park as fee payers while locals receive the management overwrite to fish elsewhere. Fishing locals thus fail to represent the expectations of "native" locals (Campbell, Gray, and Meletis 2008). Likewise in Kenya, Maasai have become the iconic imagery for the

cultural expectations of many ecotourism companies. However, the story of Maasai displacement as a result of parks, the modern reconfigurations of pastoral-nomadic traditions to relatively non-nomadic communities, and even the complicated blurring of staged traditional performances with traditional expressions that do indicate some cultural empowerment against the primitivism of more urbanized Kenyans goes underemphasized by touristic expectations of authenticity (MacCannell 1992; Akama 2002; Ritsma and Ongaro 2002). Figueroa and Waitt (2010) observe the double edge of colonial expectations at Uluru-Kata Tjuta National Park, where they find some tourists offended by the perceived absence of aboriginals. At the same time, the narrative delivery by aboriginal guides, as part of the overarching reconciliation effort in tourist education designed to disabuse ecotourists of the subtle double-edged colonial tendencies, risks the possibility that instead of the political realities of indigenous peoples being imparted upon the tourist's positive consciousness, tourists will be rendered defensive of their own lifestyles against cultural guilt.

Commercialization and exoticization of cultural practices for the benefit of tourist cultural experiences eclipse the modern political identities of indigenous and local communities. As Whyte (2010) points out, some forms of ecotourism carried out in the lands of some indigenous peoples might be more honestly defined as mutually advantageous exploitation, unless a robust form of direct participation and coalition is clear and present among communities, the industry, and ecotourists. Whyte argues that since communities would not engage in all these educational, time-consuming, and bureaucratic ventures unless they struggled under extenuating historical, cultural, economic, and environmental circumstances, it should be recognized as a mutual exploitation between communities and ecotourists. Gould (1999) has labeled this dynamic "tactical tourism," and it represents most historical narratives of indigenous communities and their acceptance of ecotourism for cultural, environmental, and economic survival, as with the Maasai, the Maya, the Maori, and the Inuits, to indicate only a few across the continents. It is possible for ecotourism to have transformative environmental justice results with indigenous communities if direct coalition building is at the center of the practice. Several model examples of successful coalition include the Little River Band of Ottawa Indians (LRBOI), which works with the U.S. Forest Service, the state of Michigan, and surrounding communities of the Big Manistee watershed in their ceremonial practices to restore nmé (sturgeon) through holistic management practices (LRBOI 2008). The Maasai Wilderness Conservation Trust is touted as a model of an indigenous coalition (MWCT 2013), as is the Community Baboon Sanctuary, a community-owned private reserve that has successfully protected a black howler monkey population and incorporated ecotourism in Belize (Gould 1999). These few examples indicate the critical need for direct coalition building to ensure that the principles of ecotourism are met. This transformative goal may not ultimately fit neoliberal models or the expectations of tourists, but it would be a direction for transforming ecotourism into the virtues its consumers, companies, and communities claim are industry's prized cultural benefit—sustainable self-determination for communities and their environmental heritage—while garnering mutual cultural respect between the stakeholders and their desired experiences.

28

Education

Mitchell Thomashow

The twenty-first-century planetary challenge orbits around three integrated Earth system trends—species extinction and threats to biodiversity, rapidly changing biospheric circulations, and altered biogeochemical cycles. These patterns are the template for the proliferation of natural resource extraction, accelerating consumer demand, and the challenges of global wealth distribution. Meanwhile, the extraordinary synergistic advances of global communications networks, computerization, miniaturization, and instrumentation provide humanity with the daunting prospect of simultaneously exacerbating these challenges while offering the means to solve them.

Since the late 1960s, and to some extent before then, environmental education has taken on this challenge, assuming that by expanding awareness of ecological relationships, natural history, and the human impact on natural systems, it would better equip people to perceive, understand, and manage these issues. Implicit in this assumption is the sense of grandeur and wonder that accompanies this awareness. With greater appreciation of the magnificence of the biosphere, people would be more inclined to protect and preserve what they have grown to love. Deeper awareness motivates ecological citizenship. The holy grail of traditional environmental education is formulating a robust ethic of care. Its great conceit (or naïveté) suggests that if they (those we educate) only knew what we know, they would act as we do. I have been privy to hundreds of academic and practitioner conversations that embody this attitude. Indeed, it is the foundation of hope that lies at the core of education and teaching.

Environmental education wanders through a wide assortment of ideas, approaches, and methods to implement this implicit mandate. Allow me to trace some of these connecting paths, which are all interested, in some measure, in how people think and learn about their relationship to the natural world. Hence they collectively comprise whatever we mean by "environmental education." Conservation psychology aspires to understand human behavior and motivations while applying its research to the practice of conservation biology. Place-based education and bioregionalism intend to build affiliation with local habitats and community. Natural history promotes familiarity, understanding, and identification of flora and fauna. The Wilderness Excursion promotes the virtues of outdoor expeditions. Urban ecology emphasizes understanding of the city as a complex ecosystem. The environmental arts and humanities engage diverse milieus to promote voice and imagination—the nature essay and poem, performance arts, acoustic ecology. Environmental justice links ecological challenges to issues of equity, gender, race, and diversity. Environmental ethics contemplates the virtues of ecological behaviors for individuals and communities. Spiritual ecology investigates how the world's wisdom traditions might incorporate ecological knowledge. Sustainability science studies how to minimize the human ecological footprint. There are literally dozens of fields of environmentally related inquiry—from green business to theories of the commons, from phenomenology to environmental security—that are relevant for environmental education. Most importantly, these approaches are implicitly concerned with how people apprehend the relationship between humanity and the biosphere.

This education agenda holds true for environmental organizations. Almost every regional and global environmental organization, from the Wildlife Conservation Society to Worldwatch to your local watershed association, considers the educational implications and applications of its work. Environmental education is a prominent concern for just about anyone who thinks about ecological issues—academics, practitioners, and concerned citizens alike.

There is a new trend that greatly impacts the theory and practice of environmental education—the proliferation of academic programs, organizations, businesses, and scholarship emphasizing sustainability. The practice of sustainability builds on the environmental education approaches described above. However, it brings many new emphases, stressing the built environment, especially energy, food, and water, and calling attention to how and whether organizations implement sustainability initiatives. Fields such as architecture, landscape design, planning, organizational behavior, finance and investment, and social entrepreneurship now have "green" orientations that also share educational missions. Sustainability studies contributes to an increasingly robust environmental education portfolio.

If environmental education is prevalent in all of these domains, what makes it distinctive as a field of study unto itself? What compelling synergy results from the combination of these words? Education connotes an emphasis on the teaching and learning process. The environmental prefix suggests that we are mainly concerned with teaching and learning processes that cultivate a deeper awareness about ecological processes and the biosphere. Ultimately, environmental educators are concerned with three curricular questions. What are the primary subjects of study? What are the most effective learning processes for teaching those subjects? How can those principles and processes be applied to real-world situations? In the remainder of this essay, I will present my view on these questions, but with an additional tweak. Given the planetary challenge described above, what is the foundation of a core environmental education curriculum? My suggestions are mainly organized for the college curriculum, although they can be readily adapted for a range of educational venues as developmentally appropriate.

I propose four broad categories as a curricular foundation—biosphere studies, social networking and change management, the creative imagination, and sustainability life skills. These are mutually reinforcing and reciprocal categories. They correspond with the classic formulation of natural and physical science, social science, the humanities, and professional practice. However, they are reconstructed to emphasize environmental education for the twenty-first-century learner. This formulation is derived from the tradition of approaches listed previously in the essay, and then synthesized and remixed for a contemporary perspective. For each foundation, I'll briefly present a core learning process, distinguish a personal and public dimension, and then suggest some areas for substantive inquiry and experimentation. These are intended as ways of stimulating discussion, controversy, and imagination for an environmental education curricular agenda. Consider them as a suite of emerging curricular design potentials, rather than a blueprint or manifesto.

Biosphere studies emphasizes an understanding of scale, learning how to interpret spatial and temporal variability, as informed by the dynamics of ecosystem processes and local ecological observations. The challenge is to develop a conceptual sequence that helps students perceive, recognize, classify, detect, and interpret biospheric patterns, what I'll describe as "pattern-based environmental learning." The purpose is for students to better understand and internalize global environmental change. The personal dimension involves the development of natural history observation skills so as to enhance appreciation of home and habitat. The public dimension involves how to contribute those observations and assessments to global networks of biospheric data collection. Examples for study may include biogeochemical cycles, atmospheric and oceanic circulations, evolutionary ecology, restoration ecology, watersheds and fluvial geomorphology, biogeographical change (species migrations, radiations, and convergences), plate tectonics, and climate change.

The field of social networking and change management describes how to enhance, cultivate, and utilize social capital. This includes a personal dimension—providing students with the ability to better understand how they learn and think, how they respond to stress, and how to maximize psychological and physical wellness. The public dimension promotes the ability to interpret collective behavior in organizational settings. The learning process involves how to integrate the personal and social dimension so as to maximize human flourishing in diverse institutional settings. Examples for study include cognitive theory, neuropsychology, organizational process, theories of social change, behavioral economics, ecological economics, social entrepreneurship, decision-making science, adaptive management, and networking theory.

The creative imagination entails the cultivation of an aesthetic voice, personal expression, and improvisational excellence to enhance the arts, music, dance, play, literary narrative, and philosophical inquiry. The personal dimension emphasizes how to use the creative process as a means to explore questions of ethics, meaning, and purpose, how to maximize aesthetic joy, and how to express emotional responses to challenging environmental issues. The public dimension develops the capacity for collective expression in social milieus—how to use public spaces such as buildings, parks, campuses, etc., to promote learning about sustainability and environmental issues. Examples for study include environmental art and music, acoustic ecology and sound design, biophilic design and architecture, environmental interpretation, environmental perception, environmental ethics, ecological identity, the aesthetics and epistemology of patterns, game design, information design, and biomimicry.

Sustainability life skills is the application of sustainability principles to the routines, behaviors, and practices of everyday life. The personal dimension involves the individual behaviors of sustenance, shelter, transportation, health, and domestic life. Further, it emphasizes how to incorporate sustainability principles into one's career and professional choices. The public dimension involves how to support organizational or regional sustainability efforts, including procurement, ecological cost accounting, recycling, health services, and/or other forms of community capacity building for sustainability. Examples for study include organic agriculture, nutrition, home building and engineering, construction, alternative energy, energy and water conservation, alternative transportation, sustainability metrics, habitat restoration, gardening, urban and regional planning, career development, reflective practice, and service learning.

For environmental education to be pertinent, enhancing, evocative, and ubiquitous, it has to continuously emphasize the importance of understanding the biosphere, but to do so in ways that correspond to how people learn, and finally to how they apply what they learn. The great new frontier in education is how brain research, evolutionary processes, and cognitive psychology inform perception and behavior. This has extraordinary implications for facilitating learning and teaching about the biosphere, and will be crucial for developing the skills and awareness necessary for the forthcoming decades—adaptation, resilience, versatility, anticipation, open-mindedness, self-reflectiveness, and clarity. The ultimate challenge for environmental education is posed with this question: how can humans come to recognize that they are the biosphere and their ecological behaviors and actions have significance for both human flourishing and the future of life on Earth?

29

Environment

Vermonja R. Alston

The first decade of the twenty-first century has been marked by several environmental disasters and debates about global climate change, which have compelled academics, policy makers, and grassroots organizations to reexamine the relationship between environment and development. But what exactly do these varied stakeholders mean when they use the term "environment"? For students and specialists alike, "environment" carries as much of a "complex and contradictory symbolic load" as "nature" (Soper 1995, 2). An overview of the etymology of "environment" brings some clarity and is a useful place to start if we are to understand the contrary ideologies and symbolic load the term carries.

The *Oxford English Dictionary* (*OED*) traces "environment" to the word "environs," from Middle French. Ironically, in contrast to twentieth- and twenty-first-century perceptions of environment as wide open natural spaces, early moderns understood "environment" as a noun—"the state of being encompassed or surrounded"—or a verb—"the action of circumnavigating, encompassing, or surrounding something." By the eighteenth century, "environment" designated "the area surrounding a place or thing." Not until the twentieth century did the term "environment" come to represent the natural world, but even then "environment" also designated physical surroundings "as affected by human activity." It is difficult to ignore the close proximity of the *OED* definitions of environment to the etymology of the terms "colony" and "colonial"—or the Latin *colōnia*—a public settlement of a newly conquered country, "the planting of settlements." Both "environment" and "colony" invoke a sense of enclosure, of space cordoned off by human activity. How might our understanding of "environment" and "the tragedy of the commons" be altered if we place the closing of the commons within the context of colonial settlement, as well as industrialization? Although the concept of common pasture lands and their enclosure dates to medieval land tenure systems, the term "the tragedy of the commons" is attributed to a 1968 *Science* article by Garrett Hardin, who argued that overgrazing will eventually deplete the commons unless its use is regulated by a private party or government. As a practical matter, the enclosure movement limiting open access to the commons resulted in the privatization of public space. What happens, for example, if we think of the tragedy of the commons in terms of Indian removal and the creation of the vast reservation system in the United States, reserves in Canada, or cotton plantations in the southern United States, sugar plantations in the Caribbean and Brazil, or tea plantations and poppy fields in India under the British raj? If we rethink environment as a history of enclosures, then we must also think about enclosures at the scale of the body.

In its broadest sense, the term "environment" connotes contested terrains located at the intersection of economic, political, social, cultural, and sexual ecologies. Scholars of postcolonial studies, indigenous studies, and globalism, as well as environmental activists from the Global South, point out that ideas about the relationship among culture, environment, and development are bound up with colonial ideologies of race, civilization, and progress. Until the last two decades of the twentieth century, the environmental perceptions of the poor, working-class, and marginalized minorities were not the subject of sociological or anthropological studies in the Global North or the Global South. Although environmental justice activism and research in North America, Latin America, and South Africa, "environmentalism of the poor" in South Asia, green movements in Africa, and "landless peasant movements" in Latin America are receiving greater scholarly attention, to a large extent, perceptions of the environment in the Global North have been shaped by a small group of writers with access to publishers of works of literature. As a consequence, nature writing in North America and Europe, influenced by the Romantics, has been associated with leisure activities, escapes from the world of work and interactions with other human beings. Moreover, American and English poets and novelists, when gazing east, frequently exoticized the "Orient" as a "natural" reservoir for their spiritual longings. Although mountains, rivers, and other waterways may hold religious significance for indigenous peoples, conservation politics (embedded in colonial and imperial expansionist regimes) are often inspired by interpretations of particular places and peoples as virgin territory, untouched by human hands, urbanization, and technological revolution. This sentimentalization of nature perpetuates the nature-culture dualism inherited from European and North American Romanticism, a response to industrialization in the Global North. The challenge of modern environmentalists, or "contemporary new world pastoralists," as Lawrence Buell puts it, is one of decolonization (1995, 55). How does one think through the "natural" environment without becoming entangled in the symbolic load of nature, or of nature-culture dualisms?

A focus on decolonization returns us to an understanding of "environment" as the enclosure of bodies of land, water, people, plants, and nonhuman animals in a colonial logic to exploit and appropriate biodiversity and indigenous knowledge. Vandana Shiva, physicist and India's foremost environmental writer and activist, identifies the privatization of water and the plunder of seeds and indigenous knowledge through the use of intellectual-property legal conventions as only the most recent form that enclosures have taken. "Patents on life," Shiva argues, "enclose the creativity inherent in living systems that reproduce and multiply in self-organized freedom. They enclose the interior spaces of the bodies of human beings, plants, and nonhuman animals. They also enclose the free spaces of intellectual creativity by transforming publicly generated knowledge into private property" (1997, 7). In articulating a concept of "living systems that reproduce and multiply in self-organized freedom," Shiva invokes an understanding of ecology that opposes itself to the imagined closed space of environment. "Biodiversity has been protected through the flourishing of cultural diversity," Shiva insists, citing the example of neem, *Azarichita indica*, a tree native to India and used for centuries as a biopesticide and a medicine, which has been patented by four companies owned by a U.S. multinational corporation (1997, 72, 69–71). This privatization (enclosure) is both a threat to biodiversity and an appropriation of indigenous environments and knowledge. Ecological economist Joan Martinez-Alier agrees, pointing out that chemical compounds, "such as morphine and quinine, [which] were originally discovered through their use by indigenous cultures," are just a few examples of the debt the North owes to the South (2000, 62). Martinez-Alier argues that the "main enemy of environmental justice in South Africa remains the globalized economy plus a local ideology which still sees the environment in terms of wilderness more than human livelihood despite the competent efforts of a very large group of activists and intellectuals who show in practice that environmental justice may become one main force for sustainability" (2003, 59). In her articulation of ecology as living systems inclusive of human knowledge, Shiva bridges the chasm between principles of deep ecology and environmental justice that have arisen in South Africa, which shares more ideologically with the settler colonies of North America than it does with India.

The *OED* offers a second definition of "environment," dating to the mid-nineteenth century, which comes closer to late-twentieth-century understandings of ecology. In the English language, environment as "the physical surroundings or conditions in which a person or other organism lives, develops, etc., or in which a thing exists; the external conditions in general affecting the life, existence, or properties of an organism or object," moves beyond anthropocentric or human-centered approaches; nevertheless, the retention of the term "surroundings" continues to connote enclosures or cordoned-off places rather than the more fluid concept of interdependent and interrelational ecological systems. Deep ecology or biocentric understandings of the environment arose out of a strand in the wilderness movement in American environmental thought and ecological movements in Norway. In the United States, deep ecology traces its roots to Henry David Thoreau (1817–1862) and John Muir (1838–1914). If for Thoreau preservation of wild nature is important for the preservation of civilization, even though his idea of the wild is pastoral, a middle ground between the encroaching cities and the savagery of the "wilderness" (Buell 1995, 54), for nineteenth-century ecologist John Muir, "nature has an intrinsic value and consequently possesses at least the right to exist" independent of the needs of civilization (Payne 1996, 5). Although Muir is credited with developing the idea of the interconnection of ecosystems for American ecologists, his ecosystems are purified through the occlusion of human inhabitants. And while the Norwegian Arné Naess's platform for deep ecology affirms that "the richness and diversity of life forms on earth, including forms of human cultures, have intrinsic worth," his support for "a substantially smaller human population" is controversial in marginalized communities, which have overall higher mortality rates and consume less of the world's resources than affluent northern nations (Rothenberg 1993, 127–28). Critics of Naess's biospheric egalitarianism, which seeks to place humans on an equal footing with other species, point out structural inequality and inequity within the greater human community. Environmental justice scholars and activists charge that deep ecologists—who reject anthropocentric or human-centered understandings of the environment—ignore social and economic inequality in both northern communities and the Global South (Guha 2000, 85).

Notwithstanding large and growing urban populations and the transformation of rural environments by agribusiness, mainstream environmentalism, by focusing exclusively on preserving "pristine wilderness" areas visited by those with leisure time and neglecting the environments and economic needs of workers, has long been vulnerable to charges of elitism. Although early-twentieth-century writers like Theodore Dreiser (1871–1945) chronicled the toxic environments faced by poor immigrant workers (Swedes, Poles, Hungarians, and Lithuanians) in an oil refinery in New Jersey, the urgency of industrial environmental problems did not transform environmental discourse in the United States. In short, green environments were subjects of environmental activism, while toxic brown environments became the purview of labor movements and occupational safety experts. It was not until the late twentieth century that the environmental justice movement linked issues of environmental equity with issues of social and racial justice. Environmental justice activists and scholars point out that the anti-urban bias of preservation politics has often resulted in the creation of toxic ghettoes in cities while scenic wonderlands have been cordoned off (Adamson, Evans, and Stein 2002). While the environmental justice movement has been influenced by Rachel Carson's (1907–1964) investigation of the relationship between the use of DDT and other pesticides and the health of human and bird populations in *Silent Spring*, few are aware of César Chávez's (1927–1993) struggles (as leader of the United Farm Workers) to end the spraying of banded pesticides over California fields where migrant farm workers toiled, grew sick, and died. The unequal enforcement of environmental laws and regulations, along with the creation of toxic wastelands in minority communities, raises issues of environmental equity. As sociologist Robert D. Bullard points out, "all communities are not created equal. Some are more equal than others" (2002, 90). "Environmental racism," Bullard argues, refers to "any policy, practice, or directive that differentially affects or disadvantages (whether intended or unintended) individuals, groups, or communities based on race or color" (2002, 90–91). In his research, Bullard investigates the ways in which seemingly innocuous municipal policy decisions such as where to locate municipal bus terminals have broad environmental justice implications for racialized communities often formed as a result of past restrictive housing and lending policies and practices. Environmental justice researchers and scholars point out that the exploitation of the environment and the exploitation of people are inextricably linked. Finally, a consideration of social, cultural, and sexual ecologies encourages investigations of the differing ways in which "culture may be said to 'work' upon nature" (Soper 1995, 135). Bodies do not simply exist as finished products, but are complex ecologies continuously in the making. Their overdetermination as natural ignores the ideological naturalization of what is socially and culturally construed (136).

If, as Bill McKibben asserts in his introduction to the Library of America anthology *American Earth: Environmental Writing since Thoreau*, "environmentalism can no longer confine itself to the narrow sphere it has long inhabited" (2009, xxx), then environmental scholarship and movements must return to the broader meaning of the term "environment." "If it isn't as much about economics, sociology, and pop culture as it is about trees, mountains, and animals," McKibben continues, "it won't in the end matter" (2009, xxx). The future of the environment depends on the dissolution of the border zones between the environmental idealism of naturalism and the passion for environmental justice. At the grassroots level, community organizers in New Orleans have begun to engage with organizers in the Maldives, and Canadian university students participate in exchanges and dialogues with their counterparts in South Asia. But they are not encouraged by decisions made by those at the highest echelon of power in both the North and the South, where governments continue to treat environments as if they are enclosed "surroundings" rather than complex interlinked ecological systems.

30

Environmentalism(s)

Joan Martinez-Alier

The three main currents of environmentalism could be named as the Cult of Wilderness, the Gospel of Eco-Efficiency, and the Mantra of Environmental Justice (together with the Environmentalism of the Poor). The religious overtones are justified by the fervor with which the faithful defend their own positions. Those currents are as three big branches of a single tree or three cross-cutting streams of the same river. There are many possibilities of collaboration among them, but there are also substantial cleavages.

If one goes to the World Conservation Congresses that take place every two years, organized by the International Union for the Conservation of Nature (IUCN), one hears the words "we the world environmental movement" or, synonymously, "we the world nature conservation movement." This is the core of the first current of environmentalism. In terms of the human and economic resources available, this movement is indeed large. Historically, since the nineteenth century, its main concern was to preserve pristine nature by setting aside natural areas from where humans would be excluded, and the active protection of wildlife for its ecological and aesthetic values and not for any economic or human livelihood value. Beautiful landscapes, threatened species, disappearing ecosystems such as coral reefs, mangrove forests, and tropical rainforests were and still are the main focus of this international movement.

In the United States, this movement has its origin in the work of Scottish-American naturalist John Muir and the creation of Yosemite and Yellowstone National Parks. There were similar movements in Europe and other continents. Even in India, where the doctrine of the "environmentalism of the poor" was put forward in the 1980s in opposition to the "cult of wilderness," there are great local traditions of bird watching and other forms of upper- and middle-class nature conservation.

This world conservation movement has been increasingly drawn to an economic language. Although many of its members believe in "deep ecology" (the intrinsic value of nature) and revere nature as sacred, the mainstream movement decided to join with the economists. The reports of the TEEB (Economics of Ecosystems and Biodiversity, a project supported by the World Wildlife Fund [WWF] and indeed the whole IUCN) in 2008–11, published under the United Nations Environment Programme (UNEP)'s auspices, follows this leitmotiv: to make the loss of biodiversity visible, we need to focus not on single species but on ecosystems, and then on ecosystem services to humans, and finally we must give economic valuations to such services because this is what will attract the attention of politicians and business leaders towards conservation. The world conservation movement is thus now closely allied to the Gospel of Eco-Efficiency. The TEEB, for example, enthusiastically praises mining corporation Rio Tinto's principle of "net positive impact." This principle suggests that nation-states or corporations can engage in open cast mining anywhere provided that the state or business support a natural park there or replant a mangrove forest yonder. John Muir would have been horrified.

This second current of environmentalism, the Gospel of Eco-Efficiency, is perhaps then the most powerful today. Its name recalls key words used in the title of Samuel Hays's 1959 book *Conservation and the Gospel of Efficiency: The Progressive Conservation Movement, 1890–1920*, explaining the early efforts in federal environmental policy in the United States to reduce waste, and also to conserve forests (or turn them into tree plantations) in contrast both to the reality of plunder and to the mystical proposals of nature activists like John Muir. One main public figure of Eco-Efficiency was Gifford Pinchot, trained in forestry in Europe.

The concept of "sustainability" (*Nachhaltigkeit*) had been introduced in nineteenth-century forest management in Germany, not to denote respect for pristine nature but, on the contrary, to indicate how monetary profits could be made from nature by obtaining optimum sustainable yields from tree plantations. This idea can be seen in today's panoply of recipes on sustainable technologies, environmental economic policies (taxes, tradable fishing quotas, markets in pollution permits), optimal rates of resource extraction, substitution of manufactured capital for lost "natural capital," valuation and payment for environmental services, dematerialization of the economy, habitat and carbon trading, and, in summary, sustainable development. The Gospel of Eco-Efficiency goes together with doctrines of "ecological modernization" and belief in so-called environmental Kuznets curves (EKC) showing that pollution first increases with income but later declines. This is indeed the case for sulfur dioxide, for instance, but not for carbon dioxide or for domestic waste or many other pollutants. Therefore, one cannot generalize the existence of EKC.

The words "sustainable development" became widely known in 1987 with the publication of the Brundtland report (the main author was a Keynesian social democrat politician who believed in economic growth). The first influential use of "sustainable development" came, however, from the 1980 IUCN's World Conservation Strategy trying to avoid, by the use of clever words, the clash between conservation and economic growth. Most governments and the United Nations (including UNEP more than any other institution) align themselves with the Gospel of Eco-Efficiency. Business leaders are divided between those who believe in the eco-efficient approach to the environment and those who despise environmental concerns and even subsidize think tanks denying climate change. The "green" faction suffered a jolt when in 2011, the former president of the Business Council for Sustainable Development, Stephan Schmidheiny, was convicted in a criminal case in Torino, Italy, because of his promotion over many years of the production and use of asbestos.

Meanwhile, the third current of environmentalism—the world Environmental Justice Movement (which is certainly not as well organized as the IUCN)—is an assortment of local resistance movements and of networks. The words "environmental justice" were first used in a sociological sense in the United States in the early 1980s, as an outgrowth of the civil rights movement. It was claimed that the burdens of pollution fell disproportionately on ethnic minorities, prompting the use of the term "environmental racism." From its origin in the United States, the words "environmental justice" spread to the world. There are now networks for climate justice and for water justice, and against the extractive industries. Poor and indigenous people (together, a majority of humankind) suffer disproportionately from pollution (including climate change) and from resource extraction. Many of these people may have never heard of the phrase "environmental justice" or "environmentalism," but they understand clearly that they are suffering from unjust environmental burdens. In the mounting resistance to these injustices, many people around the world are killed defending the environment. Very few come, however, from the ranks of the IUCN or from Eco-Efficiency groups.

There have been attempts to bring the conservation movement closer to the environmentalism of the poor and indigenous people who fight against deforestation, agrofuels, mining, tree plantations, and dams. For instance, mangrove forests can be defended against shrimp aquaculture because of the livelihood needs of women and men living there, but also because of their biodiversity and their beauty. Both sides could, for instance, be ready to respect an international convention on the Rights of Nature, following the lead established by the Constitution of Ecuador of 2008. Despite opportunities for bringing together the conservation movement with the environmental justice movement, this is often difficult not only because the first current of environmentalism, the conservation movement, consorts too closely with the second current, the engineers and economists, but also because, to judge from the financial sponsorship of the World Conservation Congresses, the conservation movement has sold its soul to companies like Shell and Rio Tinto. As a result, the alliance between the Cult of Wilderness and the Gospel of Eco-Efficiency is stronger than the possibility for alliance between the Cult of Wilderness and the movement for Environmental Justice and the Environmentalism of the Poor.

The Environmentalism of the Poor combines livelihood, social, economic, and environmental issues with involvement in extraction and pollution conflicts on the side of environmental justice. In many instances this movement draws on a sense of local identity (indigenous rights and values such as the sacredness of the land). It is on the left insofar as it positions itself in opposition to corporate power and to the coercive power of the state. However, the political Left (e.g., presidents Lula and Roussef in Brazil, the Communist Party in West Bengal in India, or presidents Evo Morales in Bolivia or Rafael Correa in Ecuador) does not like the environmentalism of the poor and the indigenous that explicitly opposes dispossession of land, forests, mineral resources, and water by governments or business corporations,

fighting against the inroads of the generalized market system and the growth of social metabolism. Such Latin American governments need the money from primary exports to sustain their policies for poverty alleviation. Sometimes they attack the environmentalists very openly, as when Correa calls them "infantile ecologists" or Garcia Linera (the Bolivian vice-president) discovers "environmentalist gringos" behind any complaint by Amazonian indigenous peoples. In West Bengal, after peasant opposition had prevented in 2007 the setting up of SEZs (special economic zones) at Nandigram and Singhur (for the chemical industry and for a Tata car factory), the Communist Party government was defeated in elections. The CP general secretary, Prakash Karat, complained in 2007 against "the modern-day Narodniks who claim to champion the cause of the peasantry" while neglecting the historic task of industrialization. He mentioned "the likes of Medha Patkar," a world-renowned popular environmentalist, among the Narodniks (Karat 2007). Poor people do not always think and behave as environmentalists would wish. To believe that poverty consistently results in responsible environmental behaviors would be blatant nonsense. But that said, the environmentalism of the poor arises from the fact that the world economy is based on fossil fuels and other exhaustible resources; it goes to the ends of the Earth to get them, disrupting and polluting both pristine nature and human livelihoods, and thus exacerbating poverty and leading to inevitable resistance by poor and indigenous peoples, who are often led by women.

Poor and indigenous peoples sometimes appeal for economic compensation (the internalization of externalities in the price system, one of the main platforms in the Gospel of Eco-Efficiency), but more often they appeal to other languages of valuation such as human rights, indigenous territorial rights, human livelihoods, and the sacredness of endangered mountains or rivers.

Despite the deep cleavages we have noticed between the three main currents of environmentalism, there is hope of a confluence among conservationists concerned with the loss of biodiversity, the many people concerned with the injustices of climate change who push for repayment of ecological debts and promote changes in technology towards solar energy, ecofeminists, and some socialists and trade unionists who are concerned about health at work and who moreover know that one cannot adjourn economic justice through promises of economic growth forever. There is hope of confluence between urban squatters who preach "autonomy" from the market, agro-ecologists, neorurals, the "degrowers," and the partisans of "prosperity without growth" in some rich countries, the large international peasant movements like Via Campesina claiming that "traditional peasant agriculture cools down the Earth," the pessimists (or realists, after Fukushima) on the risks and uncertainties of technical change, the indigenous populations who demand the preservation of the environment at the frontiers of commodity extraction, and the world environmental justice movement.

31

Environmental Justice

Giovanna Di Chiro

The quest for environmental justice is a social, political, and moral struggle for human rights, healthy environments, and thriving democracies led by residents of communities most negatively impacted by economic and ecological degradation. The term "environmental justice" emerged from the activism of communities of color in the United States in the latter half of the twentieth century and is now used by many to describe a global network of social movements fiercely critical of the disparities and depredations caused by the unchecked expansion and neocolonial logic of fossil fuel–driven modern industrial development. Activists and scholars of environmental justice challenge the disproportionate burden of toxic contamination, waste dumping, and ecological devastation borne by low-income communities, communities of color, and colonized territories. They advocate for social policies that uphold the right to meaningful, democratic participation of frontline communities in environmental decision making, and they have redefined the core meanings of the "environment" and the interrelationships between humans and nature, thereby challenging and transforming environmentalism more broadly. Tackling these bold social-change goals head on, environmental justice advocates work toward building diverse, dynamic, and powerful coalitions to address the world's most pressing social and environmental crises—global poverty and global climate change—by organizing across scales and "seeking a *global* vision" for healthy, resilient, and sustainable communities rooted in *translocal* "grassroots realities" (Lee 1992, v).

In contrast to the legacy of Anglo-American environmentalist concerns stemming from nineteenth- and twentieth-century aspirations to protect an external, nonhuman, and endangered "nature" from "humanity's" excesses, environmental justice (EJ) advocates focus on the everyday, embodied realities of *people* living in polluted "sacrifice zones." They argue that where you "live, work, play, worship, and learn" can predict your health outcomes, your quality of life, and the distribution of environmental benefits and burdens in your geopolitical location, be it your neighborhood, your village, your tribal reservation, or your part of the world (Bryant and Mohai 1992; Bullard 2005; Lerner 2010). Seeing the interdependence between environmental quality and people's everyday lives, the environmental justice perspective joins together familiar "environmental" concerns about the degradation of air, water, and land with what are typically considered more "social" issues: civil rights, public health, school and workplace hazards, land and resource rights, race, gender, and class politics, poverty and unemployment, abandoned lots and brownfields, incarceration rates, cultural rights, infrastructure disinvestment and deteriorating cities, residential segregation, and access to green spaces, safe neighborhoods, and affordable, healthy foods (Pellow 2007; Gottlieb and Joshi 2010; Sze 2007; Krauss 2009; Nelson 2008; Stein 2004). The expanding awareness and global diffusion of environmental justice as a social movement has been matched by its emergence as a topic of great interest to academic, policy, and community researchers, some of whom themselves identify as scholar-activists and have been engaged in the movement from the start. Decades of academic and community research from a wide range of disciplines have generated a substantial body of literature documenting historical patterns of environmental racism and disproportionate impact of industrial development and corporate globalization on the health and well-being of marginalized communities, as well as their resistances and struggles for change (e.g., Pellow and Brulle 2005; Sze and London 2008; Faber 1998; Adamson, Evans, and Stein 2002; Cole and Foster 2001; Pulido 2000; Schlosberg 2013; Agyeman 2013; McGurty 2009; Bullard 2008; Morello-Frosch and Shenassa 2006).

While environmental injustices based on racial, gender, and class inequalities have long been recognized and resisted by impoverished, colonized, and marginalized groups (uprisings of enslaved Africans, indigenous tribes' resource and land-rights struggles, and urban housing, public health, and labor and occupational reform movements) (Taylor 2009; Bullard 2000; LaDuke 1999; Gottlieb 1993), many U.S. scholars and activists point to several historical milestones as defining moments of the environmental justice movement in its contemporary form, one of the most important being the First National People of Color Environmental Leadership Summit held in Washington, DC, in 1991 (Bullard 2005; Cole and Foster 2001). The so-called EJ Summit was attended by over 650 community and national leaders from all fifty states, Puerto Rico, Mexico, Chile, Nigeria, and the Marshall Islands. Representations of this event appear in both activist and academic accounts, and it has been described and celebrated by many as a pivotal moment in the evolution of an intersectional eco-politics advancing environmental and social justice activism in the United States and beyond U.S. borders. Participants at the EJ Summit drafted a set of seventeen pathbreaking principles to help guide and coalesce the growing movement, and on October 27, 1991, these were adopted as the "Principles of Environmental Justice." The seventeen principles and the nearly

250-page proceedings published after the summit are essential texts that for over two decades have broadened coalition building and strengthened cross-racial, cross-class, and transnational alliances. Evoking the rhetorical style of the U.S. Constitution, the preamble to the EJ principles clearly lays out an agenda for a wide-ranging, just sustainability (Agyeman 2013), calling on leaders to adopt both historical ("looking back") and forward-looking perspectives, including (1) the recognition of the world-historical injustices of colonialism and genocide, (2) the need for healing and connection (physical, social, cultural, and spiritual), (3) the right to self-determination, (4) respect for both human rights and the rights of Mother Earth, and (5) the need to support the capabilities enabling healthy and sustainable livelihoods and economies (Lee 1992).

From the outset, the EJ principles embraced an analysis of interconnectedness and strove to dismantle the oppressive binary systems that construct divisions between "local and global," "economic and ecological," and "human and environmental." Using an intersectional approach to environmental coalition building, a group of EJ Summit participants, inspired by movement leader Dana Alston, distributed Spanish and Portuguese translations of the EJ principles at the United Nations Conference on Environment and Development (the "Earth Summit") held in June 1992 in Rio de Janeiro, Brazil. In Rio, participants from the U.S. EJ Summit joined together with activists from over fifteen hundred international NGOs to ensure that the intended outcome of these UN negotiations—an action agenda for global cooperation on "sustainable development"—would not reproduce the eco-neoliberal version of the idea of sustainability, that is, sustaining business-as-usual by setting aside the global commons to serve free-flowing capital circuits to keep intact northern standards of living (Di Chiro 2003). Meeting in tents located many miles away from the official UN sessions, the NGO "Global Forum" and the "Earth Parliament" (a global summit of indigenous peoples) mapped out the issues vital to building environmentally sound and socially equitable societies. Wangari Maathai, leader of the Kenyan Greenbelt Movement, summed up the core objectives of a global sustainability founded on environmental justice: "eliminating poverty, internalization of the environmental and social costs of natural resource flows, fair and environmentally sound trade, and democratization of local, national, and international political institutions and decision-making structures" (quoted in Athanasiou 1998, 10).

Activists deliberated on the intersectional vision embedded in the seventeen EJ principles and on other international NGO statements, including "Agenda Ya Wananchi: Citizens' Action Plan for the 1990s" ("*Ya Wananchi*" is Swahili for "children of the Earth"). This input from diverse grassroots representatives would add a critically informed NGO voice to the two major documents produced from the conference: the *Rio Declaration*, a set of general principles on environment and development, and *Agenda 21*, an eight-hundred-page, wide-ranging plan of action to address global environmental problems. Consequently, the vision of sustainability that would emerge from the Earth Summit—the familiar "three pillars" or "three Es" of sustainability (ecology, economy, and equity)—would at least rhetorically assume the mantle of environmental justice.

In 2002, at the second EJ Leadership Summit held in Washington, DC, attendees expanded on the original principles by detailing specific codes of practice for engaging in and building partnerships and coalitions. The product of extensive deliberation on the successes and challenges facing the EJ movement, the "Principles of Working Together" recommitted participants and allies to the goals of building organizing capacity and diverse leadership in the movement, including encouraging the empowerment of youth leaders, supporting the contribution of women organizers, and actively endorsing the traditional knowledges, ethics, and sustainable lifeways developed by indigenous communities from around the world ("Principles of Working Together" 2002). This document and the original seventeen EJ principles would also provide a conceptual and guiding framework for delegates attending the Rio+20 United Nations Conference on Sustainable Development in the summer of 2012.

The diverse practices of interconnectedness that have consistently been a part of the EJ movement and encoded in the EJ principles of "working together" are evidenced in the collaborative research and solution-based initiatives organized with engaged academic, government, legal, business, and movement partners. These proactive collaborations have strategically expanded the environmental justice frame to include both fighting *against* the power of the racial and colonial legacies of mal-development and generating power *for* community resilience and sustainability. In the United States, the dynamism and fruitfulness of these intersectional relationships comes to life, for example, in the work of ACE: Alternatives for Community and Environment, WEACT: West Harlem Environmental Action, Nuestras Raices, Inc., Communities for a Better Environment, Black Mesa Water Coalition, APEN: Asian Pacific Island Environmental Network, and in many other productive alliances building coalitions and actions for change (Loh et al. 2002; Quiroz-Martinez, Pei Wu, and Zimmerman 2005; Pastor et al. 2009; Corburn 2005). Both first and second versions of the EJ principles affirmed the rights to self-determination and provided guidelines for "participation as equal partners" in legal, political, economic, and academic research endeavors.

During the 1990s, the translocal and international networks of the environmental justice movement expanded, as did global public awareness of the dangers of accelerating climate change (Tokar 2010; Shiva 2008). While depictions of the science and economics of global warming are most often communicated in abstract, technical, and green-washed discourses referring to climate models, computer-generated graphs, ppms, CDM, and Cap & Trade, environmental justice leaders' explanations use idioms of "people's science" to make direct connections to people's lives and to document and make visible the *localized* effects of this planetary-scale environmental issue (Shepard and Corbin-Mark 2009). Activists argue that the effects of climate change are borne most severely by impoverished and marginalized communities across the globe, the people who are least responsible for creating the harmful pollution from fossil energy and who benefit least from the economic development and wealth it produces. Now organizing under the banner of *climate justice*, a growing international coalition of environmental and social justice groups has reframed global warming as a *grassroots* reality that puts at risk people's health, homes, and livelihoods, and further exacerbates the lethal patterns of public and private disinvestment in "social reproduction" (the social, economic, subsistence, and environmental factors necessary to maintain everyday life), which have only worsened in the last four decades of structural adjustments, austerity policies, and neoliberal reforms (Di Chiro 2008). Through a politics of articulation, the environmental justice–cum–climate justice framework changed the terms of the environmental and sustainability debates and demonstrated how global climate change is linked to the local bread-and-butter issues of concern to the world's most vulnerable communities.

By conceptually shifting scales—from household to planet—and making visible the often imperceptible, yet material (and "slow" or delayed) impacts of catastrophic climate change (Nixon 2011), EJ activists create an embodied understanding of global ecological decline; they connect high rates of asthma and type 2 diabetes in African American and Latino children who are living in contaminated inner-city New York neighborhoods breathing high levels of diesel particulate matter and whose families have limited access to decent grocery stores, with the plight of small-scale farmers in India, Brazil, and Indonesia unable to compete with unfair global trade practices and artificially suppressed agricultural commodity prices, with the extremes in weather from global warming resulting in record-breaking heat, flooding, and sea level rise in South Pacific islands, submerging peoples' homes and creating millions of "climate refugees," and with the extractive industries destroying people's health and livelihoods through oil drilling in the Niger Delta, mountaintop-removal coal mining in West Virginia, tar sands mining in the Beaver Lake Cree territory in Alberta, Canada, and hydro-fracking in rural Pennsylvania (Collectif Argos 2010; Klein 2013; Alkon and Agyeman 2011; Krakoff 2011; Steady 2009; La Vía Campesina 2013; Carmin and Agyeman 2011; Barry 2012).

The emergence of the rallying cry for climate justice in the 1990s as a "global vision rooted in grassroots realities" was further advanced in the 2000s with a Climate Justice Summit held in The Hague, Netherlands, in November 2000, which provided an alternative and parallel NGO forum to the official sessions at the COP-6 (6th Convention of the Parties) meeting of the UNFCCC (United Nations Framework Convention on Climate Change), and in 2002 with a series of global grassroots organization meetings at UN conferences focusing on climate and sustainable development in Bali, Indonesia, and New Delhi, India. From these gatherings, an international coalition of groups, including National Alliance of People's Movements (India), Corp-Watch (United States), Third World Network (Malaysia),

Indigenous Environmental Network (North America), and groundWork (South Africa), drew up the "Bali Principles of Climate Justice," redefining climate change from the perspective of environmental justice and human rights (Bali Principles 2002). Inspired by the seventeen EJ principles, these expanded principles provided a guiding vision for the growing climate justice movement in the 2000s, which continues to flourish through the international and translocal organizing efforts of groups such as the Durban Group for Climate Justice, Climate Justice Action Network, Environmental Justice and Climate Change Initiative, the World People's Conference on Climate Change and the Rights of Mother Earth, and Idle No More, among many others. These diverse declarations, collaborative agreements, and organizing efforts draw upon the original EJ principles of (1) affirming the rights of all peoples, in particular those most adversely affected by climate change, to represent and speak for themselves, (2) affirming the sacredness of Mother Earth and the interdependence and unity of all species, (3) affirming the rights of all people, including women and low-income, rural, and indigenous people, to affordable and sustainable energy, (4) affirming the rights of youth to participate as equal partners in the movement, and (5) recognizing the ecological debt that rich countries and corporations owe to the rest of the world, including compensation, restoration, and reparation for loss of land, livelihood, and ecological damages.

This flourishing of a new environmental imagination committed to concrete action to curb global warming is grounded in the mobilization of what La Vía Campesina calls the "people's solutions." An international network representing millions of peasant and small-scale farmers and indigenous communities, La Vía Campesina calls for people's solutions to "cool down the earth" (Desmarais 2007; Walsh 2011), recognizing that traditional knowledges and environmental ethics "counsel that the wind and sun are beneficial forces in the lives of human beings, supporting renewable energy projects" (Tsosie 2009, 254) and other practices of sustainability in agriculture, forestry, and energy production. Rejecting the racist and anti-immigrant underpinnings of the siege mentality of "homeland security" (Hartmann 2004), environmental justice activists argue that people's solutions tackle the *real* risks facing many communities to the everyday needs of social reproduction wrought by climate change; they instead organize for *genuine* "hometown security" rooted in living systems: renewable energy, sustainable agriculture/permaculture, and community economies (Carter 2010; Shiva 2008; Gibson-Graham, Cameron, and Healy 2013). The global environmental justice movement's valuing and proliferation of people's sustainability solutions that bring together struggles for everyday life in one's *hometown* with global ecological change on *Earth* embodies an emergent "grassroots ecological cosmopolitanism" (Di Chiro 2013). As the global poverty gap continues to widen (while the public discourse on poverty appears passé or old-fashioned) and the devastating impacts of climate change *hit home* every day, our world sorely needs new "worlding" imaginaries that might ignite what long-time activist Grace Lee Boggs calls "militant time"—everyday revolutionary-solutionary action at every scale (Boggs 2012).

Environmental justice advocates remain true to the original purpose of making visible the disproportionate effects on poor and low-income communities and communities of color of the negative externalities of modern extractive fossil fuel–based economies—now most terrifyingly represented by extreme climate change. Guided by the "original instructions" inscribed in the principles of environmental justice (and their ever-adapting translations, revisions, and enrichments), activists promote community building, action research, and sustainable development strategies grounded in grassroots organizing while simultaneously imagining and enacting transcommunity solidarities, alliances, and partnerships. David Suzuki, elder statesman in the global environmental justice movement, maintains that *active* hope is required to imagine and make manifest a just and sustainable future for all. Echoing one of the fundamental principles of environmental justice and honoring the Haudenausaunee's Great Law of Peace and original instruction to care about future generations (Lyons 2008), Suzuki argues that an imaginary of hope must be driven by the primary commitment to *intergenerational justice*: the eco-politics of *caring* about the futures of all children, all their children's children, and all the great thriving diversity of earthly life. Proposing another conceptual triangle (in addition to the three "Es") of best practices for promoting sustainability and environmental justice, he asserts, "We can't just sink into despair and denial and then say to our kids and grandkids, 'I'm sorry.' To make a better world we need to *care, think, and act*" (Suzuki 2013). This tripartite agenda has long been at the center of the history of activism, scholarship, and cultural innovation, joining together environmental sustainability and social justice in the United States and worldwide.

32

Ethics

Hava Tirosh-Samuelson

Ethics answers the question "how should one live?" It involves both exhortations to right action (i.e., *normative ethics*) and reasoning about the status and nature of ethical claims (i.e., *meta-ethics*). "Environmental ethics" refers both to how humans should interact with the environment and to the theoretical justifications of these directives. As a branch of philosophy (Zimmerman 1993; Jamieson 2001), environmental ethics consists of the "greening" of ethics by extending the scope of moral considerability to nonhuman nature (Sylvan and Bennett 1994). For example, Tom Regan (1982) argued that intrinsic worth inheres in nonhuman individuals that possess self-awareness as "subjects-of-a-life" toward which human moral agents have duties. Alternatively, Peter Singer (1990) located intrinsic worth in the capacity of animals to suffer: humans have an obligation to serve the interests of, or at least to protect, the lives of all animals who suffer or are killed, whether on the farm or in the wild. While these philosophers held that standard moral philosophy is sufficient to ground environmental ethics, most environmental ethicists sought a new, nonindividualistic, holistic, biocentric, and egalitarian paradigm. In general, environmental ethics is a critique of anthropocentrism, the belief that only humans are valuable in and of themselves and that nonhumans possess only instrumental value. As the field became theoretically more sophisticated, environmental ethicists were disengaged from pressing environmental problems that gave rise to this academic discourse. In critique, Bryan Norton (1991) argued that these theoretical debates make little difference in practice; in public policy the anthropocentric and nonanthropocentric perspectives converge. Norton's Convergence Hypothesis was hotly debated (Callicott 1995 [2002]), but it gave rise to environmental pragmatism as a major strand of environmental ethics (Minteer 2009).

Much of environmental ethics is axiological, namely, it seeks to identify objective and universal ground, or grounds, for environmental value (Hay 2002; Keller 2010). This results in distinct strands of environmental thought, each with its own analysis of the causes for the ecological crisis, the salient ethical dimension, and the solution to the crisis. Deep ecology (Naess 1989; Sessions 1995; Devall and Sessions 1985; Fox 1990) articulated egalitarian biocentrism and metaphysical holism as it advocated deep rethinking of the place of humanity in the ecological whole. By contrast, social ecology (Bookchin 1990, 2007) holds that humans are indeed part of nature but also a "quantum leap" within the natural process, because only humans actualize the "potentiality for nature to become self-conscious and free." Humans have a responsibility toward nature that could be carried out only if human first eliminate practices of exploitation, domination, and hierarchicalization by developing communitarianism. Dismantling the logic of domination is also the core of ecofeminism, but for ecofeminists the paradigm of domination is not the control of one class of people over another, but the control of men over women, namely, patriarchy or androcentrism. Ecofeminists, however, are deeply divided about how to accomplish the egalitarian goal or how to justify it philosophically (Warren 1994). Some ecofeminists align themselves with social ecology (Biehl 1991; Salleh 1997), while others are more attuned with deep ecology (Spretnak 1986, 1997; Christ 1997, 2003;

Starhawk 1979), and still others (Plumwood 1993) identify ecofeminism with an ethics of care that calls for the cultivation of character traits that lead a person to care for the environment. Despite the diversity, "a common theme in ecological feminist scholarship is the desire to combine ecofeminist theory with strong ecofeminist activism" (Davion 2001, 247).

Feminist ethics of care is an example of virtue ethics, according to which the ethical question "how ought I to live my life?" cannot be separated from the question "what kind of person ought I to be?" Environmental virtue ethics (Sandler and Cafaro 2005) holds that standard interpersonal virtues (e.g., honesty, temperance, and compassion) should be considered normative in the environmental context. Since the environmentally virtuous person is disposed to recognize the right thing to do and does it for the right reason, the cultivation of character leads to proper action. Environmental virtue ethics thus shifts the focus from meta-ethical issues to pragmatic concerns about environmental action and is similar to environmental pragmatism. Environmental pragmatism (Light and Katz 1996; Minteer 2012) calls for civic engagement with practical issues, prefers a problem-oriented approach to environmental issues, looks for theories that are closely related to social experience, and endorses environmental activism that yields factual results. Pragmatism offers strategies of practical rationality to organize environmental issues within moral experience (De-Shalit 2000; Light and De-Shalit 2003).

Like environmental pragmatism, postmodernist ethics (Oelschlaeger 1995) calls for the reconceptualization of environmental ethics, but pragmatism is less political, more communitarian, and primarily attentive to language. Human beings are linguistic creatures: they perceive the world and make sense of it through the mediation of language. Since language constitutes the meaningful world that humans inhabit, concepts such as "nature" or "wilderness" are no more than linguistic constructs; they represent not independent reality but rather the stories that humans themselves construct (Cronon 1995). Postmodernist environmental ethics emphasizes the performative aspects of language and its ability to transform human social experience. Effective discourse means a discourse that moves people to action, which means acting sustainably. Postmodernist ethics thus focuses on how humans have constructed their locales through shared stories in a particular place (Light and Smith 1998; M. Smith 2001). The emphasis on "storied living" means attention to communities and their collaborative discussion. Humans judge what is good and bad within specific, local, discursive communities and in the context of certain narrative traditions that make the judgment possible. By analyzing these stories, postmodern environmental ethicists tend to move society in the direction of sustainability.

Postmodernist environmental ethicists have acknowledged that, for millennia, humans have framed the meaning of nature in terms of religious narratives (Oelschlaeger 1995). Whether animistic or theistic, sacred myths have defined the moral status of nature and human obligations toward nature without providing philosophical justification. As scholars began to explore world religions in light of the environmental crisis, a new academic discourse emerged within the discipline of religious studies. Whether defined as "religion and ecology," "religion and nature," or "religious environmentalism" (Taylor 2005; Gottlieb 2006a, 2006b; Bauman 2011), the new religiously informed environmental ethics maintains that (a) the environmental crisis has an ethical dimension because it is human induced; (b) humans have a moral obligation to address environmental degradation they have brought about by overuse of natural resources, greed, and shortsightedness; and (c) religion (however defined) frames the moral situation and the appropriate social action. Religious environmental ethics is both theoretical and action oriented, it is carried within a religious narrative defined by sacred texts, and its norms and values relate not just to humans but to an ultimate source of value: "God" by whatever name. In the West, three strategies—stewardship, eco-justice, and eco-kosher—illustrate how religious communities address the environmental crisis. These concepts were based on the biblical creation narrative (Genesis 1–2) and the biblical notion that God instructs His creatures how to conduct themselves in all aspects of life, including treatment of nonhuman nature.

"Stewardship" emerged in the 1980s to support Christian environmentalism, especially associated with evangelical Protestantism (Jenkins 2008). In contrast to secular, environmental ethics, Christian environmental ethics framed environmental problems in terms of Christian obedience or as a unique Christian response to the environmental crisis. The environmental crisis was seen as a challenge to the faith and as a sign of God's ongoing call to turn toward repentance. Stewardship was theologically justified by appeal to God's command to the first human to "till and protect" the Garden of Eden (Genesis 2:15), interpreted to mean that humans are mandated to care for, watch over, cultivate, govern, and/or improve the Earth "on behalf of God," that is, as God's stewards of the created world. In Christian (as well as Jewish and Islamic) environmental discourse, the natural world is referred to as "creation," a term that pertains both to the divine act of bringing the world into existence and to the outcome of that act (i.e., the created natural world). Both the act and its outcome are viewed as gifts of a benevolent creator who commands His creatures to take care of His work; creation care thus entails conforming to God's will rather than to nature's orders (Winright 2011). From this religious perspective,

the current environmental crisis is interpreted as a sign of human sinfulness, and stewardship is the call to renounce environmentally exploitative practices. While the strategy of stewardship does not erase the boundary between God and creation or between humans and other creatures, it calls on humans to develop loving attentiveness to ecological others, understanding their own integrity and needs.

Eco-justice is a second strategy for religiously based environmental ethics. If the secular environmental justice movement exposed the inequality underlying toxic waste disposal (Westra and Wenz 1995), religious eco-justice insisted on a global commitment on behalf of all creatures. Seeking to build ecological ethics applicable to the entire world and instructive for all persons, eco-justice rallies the global community around an ideal of harmonious coexistence throughout the community of life. Eco-justice is rooted in the doctrine of creation: God created the world, made the world good, and made human beings to live in harmony with the world. Eco-justice illuminates the significance of nature and its distress within Christian experience by the way creation's integrity summons and shapes forms of response (be they love, preservation, protection, or liberation). What eco-justice means in practice remains vague and open to conflicting interpretations, but the key insight is that when creation suffers diminishment (for example, in loss of biological diversity), the relationship with God suffers distortion and diminishment as well.

The ideal of eco-justice was given a more specific content in the Jewish tradition in the concept of eco-kosher. Narrowly defined, the term "kosher" means "ritually clean," and it pertains to the animals that Israel, the Chosen People of God, is allowed to consume and the specific way in which these animals should be slaughtered and prepared for human consumption. In the early 1970s the concept of eco-kosher was popularized by the Jewish environmentalist Arthur Waskow (1996), combining left-leaning social ecology with the teaching of traditional Judaism. The term "kosher" now meant not only the animals fit for consumption but an entire range of practices related to food production: foods can become ritually unclean if produced by unjust, exploitative practices toward humans (e.g., child labor or adult exploitation) or to animals (e.g., industrial animal farming). Eco-kosher (and its Muslim equivalent, eco-*halal*) thus specifies in concrete, legal terms what it means to treat animals justly and what is the connection between justice toward animals and justice toward humans. This notion coheres with the Jewish religious prohibition on causing distress to animals (*tza'ar ba'aley hayyim*). Hence Jewish religious sources forbade hunting for sport and devised slaughtering methods that ensure quick and painless death. Nonetheless, it is mistaken to conclude that Jews as a social group have been more concerned with the welfare of animals than other groups. Although the term "eco-kosher" has made no impact on the secular discourse of environmental philosophy, it offers an insightful way to bridge the gap between theory and praxis, duty and virtue, law and ethics.

"Stewardship" (or "creation care"), "eco-justice," and "eco-kosher" illustrate how world religions can, do, and should play a crucial role in environmental ethics precisely because religious traditions directly address the key questions of environmental ethics: "how ought I live my life?" and "what kind of person ought I to be?" The contribution of religion to the history of environmental ethics is now recognized along with the realization that most of the world's population lives by religious convictions, values, and norms. Whether religious environmentalists *recover* environmental resources of their tradition, *critique and reinterpret* their tradition in light of environmental sensibility, or *replace* existing tradition with environmental spirituality, they all mobilize the resources of their tradition to address the environmental crisis. In so doing, they may be much more influential than professional, secular environmental ethicists. Despite their differences, both types of ethics recognize the moral standing of nonhuman nature, advocate respect for nature on its own terms, and recognize the interdependence of humans and nonhumans. Whether secular or religious, environmental ethics cannot be separated from a broader set of metaphysical, cosmological, and epistemological assumptions and must lead to social and political action. In environmentalism, theory and praxis converge.

33

Ethnography

Deborah Bird Rose

"Ethnography" literally means "writing" (*graphy*) about the culture of a group of people in their particularity (*ethno*). The term became attached to the discipline of anthropology in the early 1900s, and refers both to a method *for* research and an outcome *of* research. The ethnographic method is characteristically empirical, interpersonal, predominantly qualitative, and holistic (e.g., Peña 1997; Burawoy 1982). Once closely aligned with anthropology, the term is now used to describe qualitative research in many disciplines. In general, it aims for in-depth description and analysis, "thick description," in Clifford Geertz's memorable phrase (1973). An ethnography (as outcome) offers an account of a way of life. The most inspiring ethnographies bring readers into the experience of life within a world of meaning, generating empathy as well as understanding, and raising questions as well as answering them. Questions of who we are as humans are given breadth and depth through ethnography, causing us not only to reflect upon humanity but increasingly to consider our entanglements within the nonhuman world. Once the domain strictly of inquiry into humanity, "a new genre of research and writing" is, in Eben Kirksey's bold words, currently taking form as "multispecies ethnography" (2010, 545).

Shaped in the early twentieth century by German American anthropologist Franz Boas, ethnography was marked from the start by a concern for social justice: in its core it opposed the prevailing racism of the time, and its overall approach was therefore attentive to the fluid, dynamic, adaptive, and hybrid qualities of human culture. Over the course of subsequent decades, "ethnography" has encountered and engendered vigorous critique, and in the twenty-first century it has become diverse, eclectic, responsive to the changing world, and in dynamic dialogue with its own traditions. At its best, ethnography is provocative; it unsettles self-satisfied certainties, expands the range and substance of the questions we ask ourselves and others, and opens space for challenging encounters across multiple human and nonhuman cultures.

Boasian fieldwork typically requires the researcher to live with the research subjects, learning languages, local customs, and cultures. It thus requires intersubjective methods. Anthropology's major methodological innovation was participant observation. To become a participant in the lives of others is to set aside the notion of subject-object research. If the research goal is to understand the cultural and social lives of others, one must necessarily work with the subjectivity both of others and of one's self. At the same time, the term also indicates "observation." The researcher is always attempting to make sense of life not only within others' terms but within his or her own terms as well, to ensure that the results are empirically valid, and to conduct research that is even-handed in ways that research subjects themselves are not necessarily required to be. An ethnographer inhabits a marginal zone, moving between worlds, seeking dialogue and translation, and aiming in due course to write a rich account in which one shares one's own learning with others.

Ethnographic practice has generated excellent critique that has revealed that violence has also marked the endeavor in spite of its underlying commitment to knowledge and justice. Critique has focused on unexamined assumptions that ethnographers bring into their analysis, including assumptions about gender, sexuality, and colonialism, among others. One outstanding, ongoing trouble is the slip toward an "us" and "them" divide. For example, Johannes Fabian's well-known study *Time and the Other* (1983) examines the tension between the shared time and place of ethnographic research, and writing that has the potential to represent others as if they were fossilized at the margins of the contemporary world. The postcolonial critique of ethnography has made it clear that the condition of moving between worlds and engaging in translation is an endemic part of our increasingly globalized world, and critique arising from indigenous scholars delivers incisive perspectives on ethics and practice (Smith 2012). In this era of decolonization, land rights, demands for sovereignty, and other political matters that involve identity, the ethnographic gaze that focused so effectively on colonization now includes decolonizing practices as well (Povinelli 1993, for example).

At this time ethnography is a field of immense activity. A continuing concern is ethnographic writing, including the poetics and politics of representation (Clifford and Marcus 1986; Starn 2012). The genres of representation are expanding, and so is the subject matter. Ethnographies now include studies of life in contemporary globalization, tourism, television, internet, fine art, and a myriad other arenas of invention and connection. Another important development is "multisited ethnography." Moving away from classic community-based studies, this innovation is driven by the knowledge of global systems, and by the desire to investigate the movement and circulation of people, things, and ideas (Marcus 1995).

One fascinating issue in ethnography involves taking seriously that which the research subjects take seriously. What if their understandings of the world unsettle those of the ethnographer? What if the ethnographer

is transformed (Goulet and Miller 2007)? The newly formalized project "multispecies ethnography" connects with most of the issues mentioned above, and is strongly influenced by people who have done research with indigenous people for whom the world is always already made up of multispecies communities. Eduardo Kohn is a good example; on the basis of his research in the Amazon, he writes about an "anthropology of life," by which he means "an anthropology that is not just confined to the human but is concerned with the effects of our entanglements with other kinds of living selves" (Kohn 2007, 4).

Ethnographers who have spent years of their lives with people whose kin groups include nonhumans—animals, plants, landforms, winds, and more—have sought to consider the sociality involved in a world made up nonhuman persons as well as human persons (see for example Rose 2009). The recent revitalization of the term "animism" indicates that intersubjective research outside the human domain is being pushed along by people who inhabit multispecies social worlds (see G. Harvey 2006). At the same time, Western science now recognizes the fact that many nonhuman creatures lead cultural and social lives (see "Animal," this volume). Multispecies ethnography engages with this complex field in which entanglements within webs of life are shared (Haraway 2007), and "'human societies' are always made up of many natures" (Lestel, Brunois, and Gaunet 2006, 164).

Inevitably, the focus on the lives of nonhumans must engage with the fact that many of them are not faring well in this era of waves of extinction cascades (see "Extinction," this volume). In Australia, for example, which has the highest rate of mammalian extinctions in the world, multispecies ethnography bears the inflection of the ecological humanities, holding a focus on life, death, and the generations of living beings in this time of enormous loss. There is an emphasis on temporal trajectories as well as multisited zones of community formation. The approach to multispecies communities is therefore broadly inclusive: it encompasses not only humans and nonhumans but also the dead and the yet to become. Methods for research within such communities are interdisciplinary. Some of the most exciting points of connection are those that bring ethnography into conversation with ethology, philosophy, and ecology. These fields include the humans (who are by no means a homogeneous group); the animals and other others (also not homogeneous); philosophical questions of time, death, and ethics; and ecological questions of entangled histories and futures within the cascading effects of loss (see Rose and van Dooren 2011).

Multispecies ethnography is not, however, confined to animals or animists. In Kirksey's words, "multispecies ethnographers are studying the host of organisms whose lives and deaths are linked to human social worlds" (2010, 545). Scholars are addressing issues of extreme significance, including genetic engineering, bionic humans and nonhumans, and the viral entanglements of zoonotic diseases (for example, Lowe 2010). Multispecies ethnography is thus deeply engaged with contemporary challenges. Accordingly, it is situated within the Anthropocene, this new era that shatters anthropocentric disciplines (Dibley 2012; Chakrabarty 2009) (see "Anthropocene," this volume). At the same time as the disastrous effects of our impacts are being forced into visibility, the vulnerability of all life on Earth is also becoming ever more evident. Boundaries once taken for granted are shown to be porous or nonexistent.

Ethnography has always worked productively with the tension between even-handedness and commitment. In this time of increasing awareness of peril for so much of life on Earth, many ethnographers are explicit in their commitment to passionate engagement with the big questions of life and death. Anna Tsing writes that there is "a new science studies afoot . . . and its key characteristic is multispecies love. Unlike earlier forms of science studies, its *raison d'être* is not, mainly, the critique of science. . . . Instead, it allows something new: passionate immersion in the lives of the nonhumans being studied" (2011, 19).

This is a time for working toward ethnographies that will have the power to awaken and change people, to call humans into heteronymous proximity with "Earth others" (Plumwood 2000, 294). As ethnographies become more inclusive, questions of justice become more complicated and contentious. The provocation of ethnography will continue as its practitioners take up questions that cut through boundaries in modes that are relational, ethical, inclusive, open, and responsive to the vulnerability of the entangled loops of earthly life.

34

Evolution

Dorion Sagan

In its broadest formulation, evolution is change over time. Massive evidence—from fossils, biology, morphology, organismal behavior, biogeochemistry, ecology, and cosmology—reveals life evolving since its inception over 3.5 billion years ago. Evolution may be seen as ecology operating over vast periods of geological time (Hutchinson 1965), little of which—less than .00028 of which—has been graced or marred by the presence of humanlike beings. Recent efforts in anthropology to grapple with interspecies communities may in part reflect the doubling of the human population in the last half-century (Sagan 2013; Kirksey 2014). Such growth, tentatively connected to global climate change and more definitively to a human-caused mass extinction (Kaufman and Mallory 1993), has been historically supported by both religious anthropocentrism and secular politics advocating economic growth. However, over the evolutionary long run, very few life forms have grown exponentially and maintained their population numbers. One exception is the cyanobacteria, whose rapid growth some two billion years ago led to an increase of free oxygen from under 1 to 20 percent of Earth's atmosphere (Margulis and Sagan 1997). Descendants of these organisms are now symbiotic as the green parts, or plastids, of algae and plants. An eco-evolutionary perspective suggests that for humans to survive in the long run we must become more attentive of, and connected to, the planetmates that support us within the long-lived global ecosystem.

Nonetheless, as a scientific worldview, the introduction of evolutionary thinking in Western culture is relatively recent. The Newtonian scientific revolution connected falling bodies on Earth to the movement of planets in the heavens, giving the entire cosmos a character of eternal cycling. But Galileo's scientific observation of the changing moons of Jupiter emboldened him to proclaim that the entire universe is characterized by change:

> [I]f the Earth were not subject to any change, I would consider the Earth a big but useless body in the universe, paralyzed . . . superfluous and unnatural. Those who so exalt incorruptibility, unchangeability and the like, are, I think, reduced to saying such things both because of the inordinate desire they have to live for a long time and because of the terror they have of death . . . they do not realize that if men were immortal, they would never have come into the world. (Galilei 1632)

Evolution repositions humans as mortals within a temporal continuum; it connects us integrally, not only with all animals but also with our microbial ancestors, and with the chemical composition of an evolutionary hydrogen-rich cosmos. Rather than anthropomorphizing the cosmos, evolution naturalizes humanity, unhinging humans in a manner similar to Copernicus's decentering Earth from the center of the solar system.

"Evolution" is not one subject or idea among others but a privileged realm of interdisciplinary scientific explication and story making with multiple flavors and investigative protocols the cumulative effect of which is to give, in the popular scientific consciousness of the West, a description of reality that authoritatively challenges monotheistic religion, and produces a cross-cultural, "mononaturalist" description of humanity and its origins.

Evolutionary theories were hotly debated well before Darwin's time (Desmond 1992). Although Darwin did not use the term "evolution" until later editions of *Origin of Species*, his term "descent with modification," understood as "change over time," has been ascendant since its November 24, 1859, enunciation in *On the Origin of Species by Means of Natural Selection; or, The Preservation of Favoured Races in the Struggle for Life.*

Religiously raised Charles Darwin, the ship's naturalist on the *Beagle*, returned from the Galapagos convinced (he said it was like "confessing to a murder") that species were not created once and for all but rather change via a process he called "natural selection" (which, although he said it "occurred" to him upon his return, he was likely to have been exposed to, for example, in his grandfather Erasmus's poem "The Temple of Nature," and his two-volume medical work, *Zoonomia*; Darwin did not discover natural selection, but he presented it methodically, and with multifarious evidence in a way that brokered its acceptance). Darwin was emboldened by James Hutton's "uniformitarian" school of geology picturing change occurring slowly over vast periods of time. Darwin and his crew were meticulous, but over the duration of the voyage they had eaten thirty-eight giant turtles because they assumed that collection from different isles would yield members of the same, eternally created, species; after consulting with the respected ornithologist John Gould upon his return, however, Darwin realized that the turtles he collected from different islands belonged to different species. Darwin's scientific epiphany as to how species could change was triggered by reading Malthus's essay on population, which asserted that populations tend to grow exponentially except where they are checked by disease, war, and famine. Applied to evolution, this suggested that more organisms are born, hatched, sporulated, or otherwise reproduced than can possibly survive. This differential

reproduction *is* natural selection: slight changes can be "adaptive" (provide survival advantages) and eventually create new species.

While Darwin's "bulldog" (because he defended him so adamantly), the popularizer Thomas Huxley, exclaimed how "stupid" it was not to "have already thought of" the great idea of evolution by natural selection, critics appeared early on. Darwin's childhood neighbor, the Victorian novelist Samuel Butler (whom cyber-ecologist Gregory Bateson called "Darwin's most able critic"), argued that while natural selection was necessary to explain the change of organisms over time (i.e., evolution), it was insufficient, and moreover, it had been put forth by others such as Lamarck and Buffon, about whom Darwin remained silent in early editions of *Origin of Species* and mentioned only meagerly in later editions. (Somewhat cattily Butler, who felt himself to be victimized by Huxley, remarked that it was very likely that Darwin came across natural selection perusing the library of his father, Robert, which must have included the aforementioned natural selection–containing *Zoonomia* by Charles's grandfather Erasmus Darwin.)

Butler argued, as others do to this day (Hustak and Myers 2012), that real-time volition and affective involvements among masses of organisms with distributed agency cannot be ignored in any thoroughgoing, phenomenologically accurate account of evolutionary change. The great model for scientific revolutions had been that of Newton, whose gravity mathematically connected the heretofore changing Earth with the eternal heavens, thus in a way scientizing and mathematizing the universe. Such an approach explained changing phenomena, and human lives, in terms of eternal laws, tending to run roughshod over the primordial data of living sensation and willful action. Although Darwin (1871) in his theory of sexual selection argued that female choice of mates could shape evolution, he was criticized for presenting natural selection as intrinsically automatic, and for ignoring the inner worlds and complexity of sentient beings (Uexküll 1934 [2010]).

Nonetheless, the 1859 publication of *Origin of Species* was a landmark event, comparable to Galileo's *Dialogue concerning the Two Chief World Systems*, in that it was not published in specialized scientific jargon, or in Latin, but was made directly available to the reading public. Like heliocentrism, Darwinian evolution presented humankind not as theologically privileged, but as a material system; if Earth was no longer central, now humankind was no longer the endpoint in a vast natural process.

By the late 1930s, Darwin's account of speciation by the gradual accumulation of random mutations was combined with Mendelian genetics and statistical tools to study population dynamics, forming the "modern" or "new synthesis." This synthesis explained evolution on the basis of small successive mutations that proved advantageous under different environmental conditions. This was the mode or "law" (giving it a Newtonianlike mechanicity) that accounted for the vast diversity of present life on Earth, a diversity of which the lush biota of the Galapagos served as a fortuitous microcosm.

Darwin's mathematical followers, whom Gould (1997) tagged "Darwinian fundamentalists" and "ultradarwinists," have been criticized for a totalizing view that explained virtually all organismic traits in terms of the advantages they conferred. But as Gould, for example, argued, male nipples are not "adaptive," but the result of embryological symmetry of males to females who do need to breastfeed. Similarly, the bright colors of our innards, while striking, were not selected for by females preferring to mate with colorful males.

Today the relationship between Darwin's nineteenth-century descriptions and evolution as understood by molecular biology is uncoupling, with some naturalists and scientists underscoring that the modes of evolutionary change stressed by Darwin are largely mistaken, and thus sometimes even disavowing the terms "Darwinism" and "Darwinian," while in no way endorsing creationist accounts. For example, the evolutionary biologists Lynn Margulis and James Shapiro have emphasized that natural selection has, respectively, more of an editorial or stabilizing function—that it does not and cannot in principle "create" anything and is not the source of novelty in the creation of new species, which was the main problem Darwin set out for himself to solve. For "splitters" like Shapiro, and others such as historian of biology Jan Sapp, the cumulative evidence of "non-Darwinian" evolutionary modes is so compelling that the notion of identity between "Darwinism" and "evolution" should be dropped. For Margulis and many other "lumpers," however, Darwinism, which highlights descent with modification from common ancestors, remains apt as shorthand to describe evolution and its mechanisms of action, whereas "neo-Darwinism" is intrinsically narrower. Neo-Darwinism (sometimes called "the modern synthesis") refers specifically to the union of Mendelian genetics and natural selection theory, and emphasizes natural selection over Lamarckian inheritance of acquired characteristics (which, ironically, Darwin entertained in his theory of pangenesis). Here neo-Darwinism refers to evolution narrowly focused on mathematical models of changes in gene frequency in populations of animals and tends to exclude broader studies that attempt to integrate cell biology, planetary biology, geochemistry, epigenetics, natural history, microbial ecology, and biochemistry.

The biggest commonsense problem with neo-Darwinism from an evolutionary viewpoint is the obvious evidence, over roughly 3.5 billion years of evolutionary time, for the appearance of individuality at ever greater levels of inclusion; thus from archaea and eubacteria, the first cells to appear on Earth as adduced from their distinct RNA sequences, came the eukarya, cells such as those of amoebae and our own, which are distinguished by nuclei and chromosomes. The evolution of eukaryotes from prokaryotes is arguably the single greatest morphological transition in life's history, although it occurred well over two billion years ago; today it is accepted to have been the result of symbiotic mergers of metabolically very different archaea and bacteria (Margulis 1992; Sapp 1994).

Multicellular clones of eukaryotic cells then evolved to become the first protoctists (multicellular "protozoans"), modern relatives of which include kelp, slime molds, and seaweeds. It is from this ancient lineage, in which reproductive sex, aging as we know it in animals, animals themselves, fungi, and plants evolved. Thus prokaryotes of different kinds merged to become eukaryotic cells, which evolved in groups to become the first plants, animals, and fungi. Eusocial animals, such as termite colonies, social ants, and bees, and even mammals such as the naked mole rats, then evolved societies in which only some members of tightly knit populations were biochemically permitted to reproduce.

Human beings, too, with our technologies, evolve not just individually but as societies (Wilson 2012). The reappearance of individuality at ever more inclusive levels argues plainly against the neo-Darwinist assumption that evolution is merely a competitive battle among individuals, be they animals or genes: genes work interdependently and not at all removed from the context of cells, which are themselves open thermodynamic systems, capable of trading genes and evolving to whole new levels of organizations, up to and including and no doubt surpassing human beings.

The average human body contains more bacteria than eukaryotic "animal" cells by number and dry weight; most of these prokaryotes are not pathological but neutral or helpful. Vitamins K and B12, for example, cannot be produced without them; acidophilus and other lactobacteria help regulate *Candida albicans,* a normally occurring skin and gut fungus that can, however, become systemic (and is a common cause of death in AIDS patients). The human organism, in other words, is always more than human, and this situation is by no means unique to our species: cows would not be able to digest grass or provide us with milk or meat were it not for the methanogens and microbial communities in their specialized rumens that help them break down tough cellulose of grass; and termites and other animals would be similarly disabled without their normal swarms of hindgut "germs."

As a scientific worldview with a "storylike" character, evolutionary theories offered a narrative distinct not only from monotheism, with its great but all-too-human stories of life's creation by an anthropomorphic deity. Evolution also differs from once-popular narratives of classical thermodynamics, which pictured the entire universe eventually "dying" from a lack of usable energy (the so-called heat death of the universe, popular in the 1850s).

More recent accounts emphasize that evolutionary complexity, far from violating the second law of thermodynamics, is driven by energetic dispersal; sugar-seeking bacteria and light-seeking plants, for example, are phenomenologically connected to the breakdown of energy gradients, and produce more entropy (basically, a measure of energy's spread) than less organized or long-lived natural complex systems (Jantsch 1980; Prigogine and Stengers 1984).

The increase in complexity over 3.8 billion years of evolutionary time is increasingly viewed not as exceptional but as a particular instance of a natural thermodynamic phenomenon, the breakdown of ambient gradients, aided and maintained by naturally occurring, materially cycling complex systems of which life is one example (Schneider and Sagan 2006; Rossini 2012).

The status of life's essential units, metabolizing cells, as open thermodynamic systems capable of merging and trading genes, recasts the tree of life as something more like a web. Especially at its base, evidence abounds for rampant gene trading and cellular mergers among ancient organisms. Orthodox biology has admitted this startling fact (Sapp 2012) but has been slow to see that such gene-trading and symbiogenetic evolution continues. Theodosius Dobzhansky, a neo-Darwinist who famously said, "Nothing in biology makes sense except in the light of evolution," bred new species of fruit flies by separating populations and exposing them to different temperatures. Eventually the hot-bred and cold-bred fruit flies no longer interbred. A new species had thus evolved according to the "biological species" concept (Dobzhansky 1935; Mayr 1942). But the only difference was that one "species" was infected with *Wolbachia* bacteria, and the other was not (Eldredge 2012; Sagan and Margulis 2013). As Sapp (2012, 66–67) summarizes, "Bacteria of the genus *Wolbachia* are inherited through the eggs of at least 25–75 percent of all insect species and those of nematodes, too. And far from being slaves, they manipulate the development of their hosts, causing parthenogenesis and cytoplasmic incompatibility, and can turn functional males into functional females." That *Wolbachia* sp. or "mycoplasmas" are associated with multiple insect species, leading in some cases to all-female populations, led Margulis and Cohen (1994) to suggest that nature's "inordinate fondness" for beetle species may be the result of *Wolbachia* infections.

Modern biology remains zoocentric in the sense that the modular unit of life on Earth is the cell, not the animal, and most organisms on Earth do not reproduce sexually, so an animal-based definition of species does not apply to them. Sonea and Panisset (1983) chronicle plasmids, viruses, conjugation, and other processes that allow bacteria to trade genes, directly adapting by promiscuously trading favorable genes. Retroviral genes are required for implantation of mammal placentas, and a large proportion of "our" human genes may be not just bacterial but viral (Carter 2010; Sagan 2013). The "intelligent" conferring of antibiotic resistance across bacterial strains, and other evidence of promiscuous genetic interchange, has led some to adopt the notion of the pan-genome; similar to a bank account, whose money you can access even though it is not on your person, pan-genomes include genes that organisms access outside their bodies (Mira et al. 2010).

James Shapiro has compiled copious evidence for massive, real-time rearrangements of genomes in response to environmental pressures such as mutagenic radiation that causes DNA damage. Apart from evolution by symbiosis (Margulis and Sagan 2003), there have been transfers of genes for making cytoplasmic proteins to the nucleus, coordination of cell cycles in the wake of symbiotic mergers, DNA proofreading itself during and after replication, rapid duplication of entire genomes at the base of our own vertebrate lineage, and many other modes of quick, responsive change by mobile genetic elements. Stressed cells are not just passive subjects of natural selection but can respond actively, drawing on a bag of genetic tricks to participate relatively immediately in their own genetic and epigenetic evolution (Shapiro 2011).

From a social sciences perspective, critique of the monolithic or mononaturalist notion of "origin" with its putative explanatory power outweighing a structural or poststructural account that refuses to consider precedence causal, a notion or rubric or evolutionary logic that has been used to justify oppressive regimes against indigenous peoples (historically considered evolutionarily "lower" or closer to nature) with their own distinct culturally valid stories is now being joined from within molecular biology, which shows that evolution can occur quite quickly, and that the sources of our biological being are multifarious and sometimes, even from a solely genetic standpoint, strikingly contemporary (Tallbear 2012). From an evolutionary view, individuality is not exclusively animal but arises at ever more inclusive scales as symbiotic bacteria become ameba-like cells, and then these ameba-like (eukaryotic) cells become protist communities (protoctists) that are essentially individuals at a higher order of integration, societies becoming organisms, including the ancestors of the three more familiar kingdoms of visible, macroscopic organisms—animals, fungi, and plants. On this extended timeline, animals are not models but examples of the recrudescence of individuality, which comes to resemble more a taxonomic abstraction than a biological reality (Gilbert, Sapp, and Tauber 2012). As Shapiro argues, biology as we know it today is evolving so fast it may be unrecognizable in fifty years.

35

Extinction
Ursula K. Heise

Most life forms that have ever existed—over 99 percent, according to some scientists—are extinct. Extinction is, therefore, one of the most basic characteristics of the planet's ecology. Species disappear because they change through gradual adaptation to such a degree that they can no longer be considered the same species, or because all individuals die off before they can reproduce. Adaptation, consisting of the combined processes of mutation and natural selection as theorized by Darwin and his twentieth-century successors, results in both extinction and speciation, the emergence of novel species. Die-offs tend to affect small populations and/or those with a very limited geographical range—especially island populations—that are vulnerable to unusual climatic events or outbreaks of disease. Extinctions in the normal course of evolution occur at the so-called background rate, which is computed either as the number of species that go extinct in a particular number of years, in Million Species Years (MSY), or as the time intervals during which species survive (Lawton and May 1995).

Mass extinction events, during which a majority of existing species dies out, are much rarer than individual species extinctions. They have occurred five times in the 3.5 billion years of life on Earth, the most famous example being the demise of the dinosaurs and up to 80 percent of contemporary species due to a meteor impact sixty-five million years ago. Since the 1980s, theorists of evolutionary change have tended to attribute greater importance to such contingent but massive extinctions than earlier accounts with their emphasis on the logic of adaptation (Raup 1991; Sepkoski 2012). Some biologists consider the anthropogenic biodiversity loss currently underway as the sixth mass extinction, as it might lead to the disappearance of up to 50 percent of currently existing species by 2100 (Wilson 2002). Biodiversity usually takes millions of years after a mass extinction to return to precataclysmic levels, with a considerably different set of species. For this reason, mass extinctions shape evolution in crucial ways.

Extinction came to be established as an ecological fact in the early nineteenth century, but was not accepted without resistance. So long as species were believed to have been divinely created, it seemed implausible that God would let them die out. Fossil discoveries, the postulation of geological time spans, and the advent of evolutionary theory helped extinction gain cultural acceptance (Barrow 2009). The identification of dinosaurs, named in 1842 by Richard Owens, generated a wave of public interest in 1850s England and in the United States and even led to the "Bone Wars," feuds between competing paleontologists, in the United States during the 1880s and '90s (Mitchell 1998). Concurrently, growing awareness of the endangerment of contemporary species led to the foundation of wildlife conservation associations in Europe and America in the early twentieth century, often with a colonially inflected focus on African fauna (Barrow 2009). The endangerment of the American bison, the disappearance of the Carolina parakeet, and the death of the last passenger pigeon in the Cincinnati Zoo in 1914 also galvanized concerns over species loss in the United States. Species extinction and conservation is therefore one of the longest-running concerns in (proto)environmentalist thought. Overshadowed by pollution and demographic growth in the 1960s, it came to the fore again

in the 1980s with the coinage of the term "biodiversity" (Myers 1979; Ehrlich and Ehrlich 1981; Wilson and Peter 1988). The idea of an ongoing mass extinction caused by humans competes with climate change in contemporary environmental discourse as one of the unfolding ecological mega-crises of the twenty-first century (Wilson 2002).

Anthropogenic species extinction is not, for all that, a recent phenomenon. As paleoanthropologists and paleobiologists have shown, large-bodied animal species have tended to die out after the arrival of *Homo sapiens* in many locations around the globe, especially islands (Boulter 2002; Godfrey and Rasoazanabary 2012). But if direct human predation caused most of these historic extinctions, current anthropogenic extinction triggers are far more diverse and tend to be summarized with the acronym HIPPO: habitat destruction (by far the most significant cause of contemporary extinctions), invasive species, (human) population growth, pollution, and overharvesting (Wilson 2002). As a consequence, current extinction rates may exceed the normal background rate by fifty to five hundred times, and if species whose fate is not known with certainty are included, one hundred to one thousand times (Baillie, Hilton-Taylor, and Stuart 2004). This rapid pace of extinction has led to fears of far-reaching consequences, from ecosystem collapse and the loss of energy, food, and medical resources to the vanishing of cultural anchoring points and assets (Wilson 2002). Some scientists and environmentalists have even claimed that biodiversity loss might ultimately lead to the extinction of *Homo sapiens* (Leakey and Lewin 1995), though this outcome is not widely considered probable.

Scientifically, extinctions of individual species are often hard to ascertain beyond doubt, especially since current knowledge of species endangerment is heavily biased toward vertebrates; birds, mammals, and amphibians have been extensively studied, but far less is known about the state of invertebrates, plants, and micro-organisms (Hilton-Taylor et al. 2008). The diagnosis and relevance of the current mass extinction of species are further complicated by the fact that the total number of species on Earth is unknown (May 1989), that species represent only one of the levels of biological diversity, which ranges from genes and populations to ecosystems, and that biodiversity as a concept is hard to define, difficult to measure, and blends scientific observation with value judgments (Takacs 1996; Maclaurin and Sterelny 2008; Maier 2012). Given these conceptual difficulties, it is all the more significant that rapid biodiversity loss has become an integral part of the cultural awareness of ecological crisis in many parts of the globe. In large part, public consciousness of species loss has been galvanized by the high visibility of "charismatic mega-fauna" in popular science and the media: accounts of endangerment and extinction tend to focus on relatively large animal species that are appealing to humans and often culturally significant, such as gorillas, bears, wolves, pandas, tigers, rhinos, whales, raptors, or parrots, whereas smaller, lesser-known, more unsightly, or culturally marginal species—including most plants—tend to receive less attention and conservation funding. Growing public concern over vanishing species led to the implementation of laws and treaties for the protection of endangered species in numerous countries around the globe, as well as to a vast spectrum of conservation efforts in the second half of the twentieth century. In the cultural sphere, the vanishing of species tends to be portrayed through the literary genres of elegy and tragedy in works of popular science, literature, and art, with an accompanying mobilization of well-known tropes of loss, mourning, and melancholy (Mortimer-Sandilands 2010; Heise 2011). Indeed, the ecological facts of extinction often become

culturally meaningful in the first place because an endangered or extinct animal is perceived as a symbol for broader upheavals in a particular culture's relationship to the natural world during processes of modernization and colonization (Heise 2011).

Because biological extinction is often understood through the lens of its cultural significance, it has sometimes been associated with the extinction of languages, which is also progressing at a rapid pace (Grenoble and Whaley 1998; Crystal 2000; Nettle and Romaine 2000). In some cases, the connection between the two types of extinction is material: as habitats and species are destroyed that a particular society, often an indigenous one, relied on for its subsistence, its culture and language are imperiled. In other cases, species loss and language extinction are portrayed as parallel rather than causally related reductions of knowledge about the natural world. The languages that paradigmatically feature in such accounts are indigenous languages associated with deep histories, distinct sets of cultural practices, and an aura of "authenticity" (cf. Harrison 2007). By contrast, threats to pidgins and creoles, and more generally to dialects and languages that have emerged from shallower histories, patchwork cultures, hybrid identities, and colonial encounters, tend not to be invested with similar tropes of loss and biological comparisons (Garrett 2012). Such comparisons, furthermore, tend to deemphasize the speakers' agency in shifting to other languages as their principal medium, and to overlook revivals of threatened languages that have taken place in the twentieth century, from Hebrew and Cornish to Native American languages. Even though certain shared factors play a role in both biological and linguistic reductions of diversity, therefore—such as increased global connectedness and colonial domination—many linguists are wary of drawing close analogies between the two processes.

For environmentalists, pointing to the reduction of linguistic diversity as a process that parallels biodiversity loss is often a convenient tool for highlighting the human costs of environmental destruction. Both language and species loss, in this context, come to form part of a more general narrative of decline that has characterized environmentalist discourse in the West for over two centuries. According to this story template, natural systems were harmonious, homeostatic, and self-balancing until the advent of modern human society, whose interventions into nature caused a long and slow destruction of nature punctuated by occasional cataclysms. While many scientists and environmentalists have criticized the dominance of this storyline and called for alternative visions of humans' history with their natural environments, it remains a powerful counterweight to mainstream narratives of progress and has catalyzed a wide spectrum of political, legal, and economic efforts for the conservation of species.

36

Genome

David E. Salt

To many scientists and nonscientists alike the term "genome" sounds quite modern and perhaps even a little like science fiction. The contemporary glow surrounding "genome" may stem from its connection in people's minds with the recent unraveling of the sequence of the human (*Homo sapiens*) genome—a "man on the moon" moment for most biologists. The controversies surrounding genetically modified foods (GMO) and human "gene therapy" also add a slightly sinister nature to this glow. However, as with many things that we perceive as modern, the term "genome" is in fact much older than we imagine, having arisen almost a century ago in the writings of Hans Winkler, professor of botany at the University of Hamburg, Germany (Winkler 1920). In this primary work, "genome" is defined as "the haploid chromosome set," and, along with factors from the cytoplasm, it is described as encoding the complete biological framework of a species. A current definition of the term "genome" would read more like "the complete DNA sequence of one complete set of chromosomes of an individual organism."

Understanding the original meaning of "genome" and what transpired between 1920 and 2012 to alter its meaning requires that we go back in time through the pages of a book written in Hamburg almost one hundred years ago to reveal one of the twentieth century's greatest discoveries, which now seems so obvious it feels like we always knew it: *characteristics of organisms are inherited from generation to generation by the passage of genes from parent to offspring.* You may look and think like your parents (or perhaps grandparents) because the entities (genes) that define these characteristics are passed from your parents to you during the reproductive process. It is amazing to think that there is an unbroken chain from you to the earliest human ancestors (and beyond) in which genes from one generation have been copied and passed to the next. What this statement does, beyond providing a stunning explanation for biological inheritance, is to bridge two different human worlds. It takes us from the seventeenth century, in which heritability of biological features had no clear meaning, and leads us to the twentieth century and the discovery that the genes (in the form of the sequence of nucleotide bases in deoxyribonucleic acid [DNA]) that define the biological framework of almost all living organisms on Earth are passed unidirectionally across generations and allow characteristics to be stably passed from one generation to the next. Such a profound leap has had many implications, including placing a solid mechanistic foundation under Darwin's and Wallace's hypotheses in the 1850s of evolution by natural selection. Evolution by natural selection now provides a profound understanding of the origin of species (including our own *Homo sapiens*), and is the foundation for the explosion in our mechanistic understanding of Earth-bound biology and our tentative steps to explore for extraterrestrial biology.

This journey from the seventeenth to the twentieth century was taken across the stepping stones of *heredity*, *chromosomes*, *genes*, and *DNA*, and on into our current era of genome-enabled biology. In this current era, biological sciences are again being revolutionized—this time by our rapidly growing ability to unravel the sequence of the genomes of all extant species, and even some that are extinct (Miller et al. 2008; Reich et al. 2010). This enablement of biological science is reflected in the increasing use of the term "genome" in the scientific literature. The use of "genome" was sporadic at first, with only a couple of dozen published examples by 1960. In the decades following 1960, its use leapt from close to 1000 between 1960 and 1970 to over 175,000 between 2000 and 2010. Interestingly, the use of the term "genome" has paralleled the increasing speed with which we can read the sequence of base pairs in DNA. But before we get caught up in the excitement surrounding genome-enabled biology, we need to first take a closer look at those four key stepping stones of *heredity*, *chromosomes*, *genes,* and *DNA* that got us first to the origins of genome and led us to its current meaning.

As you might expect, as far back as the Greeks (and probably back into prehistory), people had tried to explain the similarities and differences between parents and offspring—though, as we all know, the evidence is often contradictory and inconsistent. Children sometimes look like one parent, sometime another, sometimes a mixture, and even like their grandparents. This makes it hard to uncover any underlying patterns of heredity or provide a testable hypothesis that might explain these patterns. As the scientific revolution ushered in a more systematic approach of experiment, observation, and theory, our understanding of heredity grew, but it was not until Gregor Mendel's work between 1856 and 1863 on the inheritance of characteristics (traits) in pea plants (e.g., plant height, seed coat color, and texture, pod color) that we had a firm scientific understanding of heredity. In his monastery's experimental garden at the Augustinian Abbey of St. Thomas (Brünn, Austria), Mendel experimented with crossing (hybridizing) peas with different traits and then observing the same traits in the offspring. From such experiments Mendel developed the concept that there are certain "factors" present in the plants and responsible for their external forms. A copy of each of these "factors" is provided by each parent during fertilization, leading to the new offspring having two copies of each factor. The type of factor inherited from each parent determines the external appearance of the offspring. Importantly, Mendel also proposed that the factors determining each external trait (plant height, pod color, etc.) are not physically linked, and so are inherited separately—though we now know that genes that are close to each other physically on chromosomes have a higher probability of being inherited together.

It wasn't until the independent work of Theodor Boveri and Walter Sutton that Mendel's "factors" were determined to be on chromosomes and it was discovered that chromosomes distribute between parent and offspring according to Mendel's laws of inheritance. What had been a virtual "factor" had now become associated with a physical object, as chromosome—something that could be observed and manipulated. What became known as the Boveri-Sutton Chromosome Theory was born at the beginning of the twentieth century and was strongly supported by the elegant experiments of Thomas Hunt Morgan on fruit flies in the following decade. By crossing fruit flies with different visible traits (e.g., eye color, wing shape, etc.), Morgan was able to test Boveri and Sutton's hypothesis that the chromosomes were the site of Mendel's "factors." Morgan went on to establish that chromosomes actually contain independent elements that follow Mendel's laws, and these elements he called genetic "factors" or "genes." Morgan's statement that "units [genes] that segregate are themselves only parts of a whole which is the sum total of all the units" (Morgan 1917) clearly set the stage for the emergence of "genome" as a term to describe this "sum total of all the units." But by discussing genes instead of chromosomes, Morgan was in fact already moving toward our modern view of the genome even before Winkler in 1920 had proposed the term with a more Boveri-Sutton type of definition with a chromosome focus.

Within forty years of Mendel's hypotheses on inheritance being published (Mendel 1866), evidence-based reasoning had delivered genes within chromosomes as the units of inheritance. The mystery of the molecular nature of these genes now became the burning question, though by 1940 elegant experiments by Oswald Avery, Colin MacLeod, and Maclyn McCarty had delivered the answer (Avery, MacLeod, and McCarty 1944)—and it was DNA! Using Rosalind Franklin's x-ray diffraction data to help them, Crick and Watson proposed the double helical structure of DNA in 1953 (Crick and Watson 1953), providing a solid mechanistic basis to explain how DNA could both carry information and replicate. From the folklore statement that "like breeds like," human ingenuity has uncovered the molecular mechanism of biological heredity and placed that unit of heredity squarely in DNA, divided into genes, arranged in chromosomes that are inherited as described by Mendel, which together form the genome of an organism. The sequence of chemical units in DNA provides the source of the information needed to build the biological framework of an organism, and it is variation in this sequence, manifest as changes in the physical nature of the organism, that natural selection acts upon to drive the evolution of new forms and species. DNA forms the heart of the genome, this is clear. However, realizing the centrality of DNA and its structure to our understanding of the genome and its function is one thing. To truly understand how the genome works, we need to know the full sequence of the DNA that makes up the genome; knowing this will lay bare the inner working of heredity.

This is the situation we now find ourselves in. Advances in the technologies used to determine the sequence of DNA delivered the first full genome sequence of a virus in 1978, a microbe (brewer's yeast) in 1996, the first multicellular organism (a free living nematode) in 1998, the first full genome of a plant in 2000, and by 2003 the full human genome had been sequenced, which is 260 times the size of the first microbial genome sequenced with 3,200,000,000 base-pairs of sequence. There are now thousands of species that have had their complete genomes sequenced, and the number is growing daily. What do we see when we look into the heart of the genome now? Amazing complexity, and unraveling this complexity has given rise to new research fields with new vocabulary to describe them. The study of genomes has been born—*genomics*.

Genomics is discovering many striking aspects of the genome. Comparing genomes across microbes, plants, and animals is revealing an amazing amount of conservation, with similar genes appearing across all groups of organisms. Life is far more unified than we think. The size of a genome does not correlate with the perceived complexity of the species. For example, the plant *Fritillaria assyriaca* and the lungfish (*Protopterus aethiopicus*) have genomes that are forty times larger than the human genome—does this mean that these organisms are forty times more complex than humans? Scanning sequenced genomes for genes also reveals that the number of genes found appears to also not relate to the perceived complexity of the organism or its genome size. The black cottonwood tree has twice as many genes as humans (approximately 45,500 compared to 20,000) contained in a genome six times smaller. Clearly, there is a lot to be learned about the relationship between genes and their biological function. Perhaps even more surprising than these assaults on our preconceived notions of genome function is the fact that up to 99 percent of an organism's genome appears to not be coding for genes—or at least not in the way they are currently defined as a unit of DNA that encodes for a protein. This 99 percent of the genome has been mischievously

termed "genomic dark matter," after the mysterious material physicists have proposed to account for the mass that appears to be missing from the universe. Genomic dark matter is perhaps not that dark. Recent experiments are starting to suggest that the DNA in these regions is likely to be controlling the function of the genes that we can identify. Whether we can currently "see" genes in the genome or they still lie in the shadows does not really affect Thomas Hunt Morgan's description of genes, summed up simply in his statement, "The germ plasm must, therefore, be made up of independent elements of some kind. It is these elements that we call genetic factors or more briefly genes" (Morgan 1917). What it does reflect is our current poor understanding of the way heritable DNA sequences control the nature and functioning of an organism.

It took thirteen years to sequence the first human genome, but dramatic improvements in sequencing technologies have now reduced this time to around one day and made it over one hundred thousand times cheaper. You can now get your genome sequenced for a thousand dollars. So, now, rather than having a human genome, or a maize or rice genome, we have hundreds or soon to be thousands of genomes from individual people, maize, rice plants, etc. Now we are dealing in genomes from individuals, not simply species. What this has revealed is diversity within the genomes between individuals of the same species. The human genome contains roughly three billion base-pairs. The identity of 99.9 percent of these base-pairs is conserved across the human population; they are the same in all people. This means that we are all perhaps more similar than we like to think, yet we also differ in about 0.1 percent of our genome. These differences represent the complex migration patterns and ancestries of the human population, and their study is leading to new insights into the diverse nature of human physiology and a deeper understanding of human

GENOME DAVID E. SALT

history and evolution. In fact, such genome-wide variation exists across all species and is being used to help us further understand how organisms adapt to their environment and evolve over time.

The original construction of the genome captured the great biological framework that heredity rests on and that has allowed life to evolve into a myriad of complex forms occupying almost every corner of the Earth. The modern construction of the genome has revealed the deep unity of life across kingdoms and back through time to the origins of life on Earth three billion years ago. Our current understanding of the genome has also revealed the diversity among individuals, a diversity necessary to create resilience in the face of an ever-changing environment.

Heredity to *DNA* is the story of the genome—so far.

37
Globalization
Arthur P. J. Mol

It is nowadays hard to imagine that less than three decades ago the notion of globalization did not figure in the vocabularies of academics, policy makers, protesters, and business leaders. Globalization as a notion is rather new. It has replaced concepts such as internationalization and transnationalization (common until the late 1980s) because the notion/idea of globalization better reflects a new condition in worldwide economic, political, and cultural relations. Two clear differences mark this shift in conceptualization.

First, the traditional concepts of internationalization and transnationalization were considered as being too narrow for the current times, as they focused too much on the increasingly interwoven nature of *national* economies and *nation-states* through inter*national* trade and political relations. In the last decade of the twentieth century it became increasingly accepted that nation-states are not the only—and according to some, not even the most important—actors in global processes. Multinational companies of various kinds, financial organizations, global (networks of) NGOs, and subnational regions and cities play an increasingly important role in shaping worldwide developments.

Second, globalization is better suited to frame and conceptualize the social developments of ever-intensifying and ever-extending networks of cross-border human interactions. These interactions cross borders and bridge increasingly large distances in ever-decreasing time, distinguishing our present era of globalization from the era of internationalization. Or as David Harvey (1990) and Anthony Giddens (1981) put it: globalization is time-space compression. According to Ulrich Beck (2005), this makes "methodological nationalism," i.e., taking the nation-state as the unit or container of analysis, no longer an option.

With globalization, the idea of the nation-state as the rule, the organizing principle and unit of analysis, and everything outside it as the exception that proves and fortifies this "rule" has to be discarded. Globalization is not an additional layer that can be studied for social processes above and beyond nation-states and societies. Rather, it should be understood as a fundamentally new social constellation (i.e., global modernity) that completely alters the research agendas of the social sciences and humanities. Building on Manuel Castells's (1996, 1997a, 1997b) influential trilogy on the Information Age, prominent social theorists such as Beck (2005), Saskia Sassen (2006), and John Urry (2000) have argued and illustrated that this means that states and societies—as core units of social science analysis for the past century—need to be replaced by networks and flows. This is particularly valid where the social sciences and humanities study the environment. If we are to understand how globalization structures environmental deterioration and environmental reform, Gert Spaargaren and colleagues (2006) argue, we have to move beyond studying states and societies and their environmental problems and prospects. Instead, we should focus on transboundary pollution and waste, global mobility, global green product flows, labeling and environmental information flows, environmentally sensitive trade and investment, carbon credits, global biodiversity and invasive species, and the global networks that structure and govern these "environmental" flows: networks of global cities, media networks, networks of state authorities, networks of environmental NGOs, networks of academics (or epistemic communities) such as those that cohere around climate change, businesses in value chains and networks, and all kinds of hybrid networks. Access to and exclusion from these networks, power in these networks, the rules governing these networks, and the resources used by actors in these networks are all vital to understanding environmental processes and outcomes of globalization.

But what are the substantive environmental "outcomes" or consequences of globalization? Although the notion of globalization has been widely accepted in the environmental social sciences and humanities, there exists continuing debate on the environmental assessment and consequences of globalization. Globalization is often contested by environmentalists and environmental advocates because it is associated with the dynamics of global capitalism. The World Trade Organization and global financial institutions, the profit-making priorities of carbon markets, the operations of multinational extractive companies, and the expansion of China's search for natural resources into Africa are all interpreted as signs and materializations of an environmentally destructive globalization. In such assessments globalization is equated with—and limited to—neoliberal global capitalism. As the emergence, shape, and dynamics of globalization are explained by the internal dynamics of the capitalist mode of production, globalization will lead to the same kind of social and environmental disasters that befell capitalism, as has been argued so strongly by (neo)Marxist scholars in the 1970s and 1980s (e.g., Schnaiberg 1980). Globalization—as global capitalism—is then interpreted as the root cause of a new round of environmental destruction, with climate change as flagship.

At the other side of the spectrum, there are scholars whom Held and colleagues (1999) would call "hyperglobalists." Hyperglobalists see an evolutionary process of ever increasing globalization and judge this to be both inevitable and beneficial for conquering sustainability problems. "McDonaldization" and Castells's (1996) space of flows are no longer analytical categories but have become substantial preferences for addressing the global environmental challenges of tomorrow. Local place-based green alternatives (such as local food markets and Local Exchange Trading Systems, or LETS) are the last remnants of an old age, to be replaced by global harmonized and universal standards, norms and values of sustainability that become fully institutionalized in globally operating public and private networks and redirect global production and consumption processes towards sustainability.

Both sides of the globalization-cum-environment spectrum are far too simple and monolithic. As many scholars have correctly pointed out, globalization should not be equated with global capitalism, as it reduces the idea of globalization to a simple economic neoliberalism operating beyond national borders. As Jan Nederveen Pieterse (1997, 372) puts it, "If the target is neo-liberalism and the unfettered market-economy, why attack globalization?" Globalization has a much wider meaning and should refer to a diversity of social processes, including increased environmental information exchange among environmental NGOs around the world, enhanced global environmental politics, global diffusion of environmental norms and values, global value chains and networks in organic and fair trade products, to name but a few. At the same time, one cannot reduce globalization to an unproblematic unfolding of a historical necessity that will incorporate and "solve" the sustainability challenges of today and tomorrow. Far too much empirical evidence points at the troublesome functioning of all kinds of global networks and institutions, such as carbon markets, multilateral environmental agreements, global transparency and disclosure

regimes (Gupta 2010), and the failures of these institutions to address climate change.

Hence, as I have argued elsewhere (Mol 2001, 2008), the relationship between globalization and environment is multidimensional and works through markets, politics, culture, and civil society. Sometimes global processes work against environmental preservation (as the global capitalism scholars have forcefully shown); sometimes global processes enable and enhance environmental protection and reform (in cases of global environmental NGO networks or Forest Stewardship Council and Marine Stewardship Council certification). But in most cases globalization consists of a complicated and mixed picture, where final conclusions are not easily drawn. Even global markets (condemned by many environmental scholars for advancing unsustainability) work sometimes in favor of the environment, as ecological modernization scholars have forcefully argued (Mol, Sonnenfeld, and Spaargaren 2009). And global civil society actors and institutions, as advocates of sustainability, sometimes undermine the sustainability agenda, as with Greenpeace in its 1995 Brent Spar campaign.

Should we then fall back to a postmodern position on globalization-cum-environment, where "all that is solid melts into air"? I don't think so. Although it is not possible to draw general conclusions on celebrating or condemning globalization for the way it relates to environmental sustainability, we have been able to gain specific insights on how globalization transforms the power structures, actor constellations, and institutions of environmental deterioration and reform, in comparison to the preglobalization era. And if scholars want to advance sustainability under current conditions of global modernity, they have to take these—analytical and substantive—globalization insights into account. In that sense, there is no sustainability without globalization.

38

Green

Stephanie LeMenager and Teresa Shewry

Green emerges through arrangements that involve human and nonhuman agencies, including the movement of light through an object, a body that measures and interprets the light, and the social worlds in relation to which such interpretations take form. Green shares affective space with ideas about nature, the countryside, fertility, and life. But profound disjunctures mark the imaginative and political legacies of green. These disjunctures exist even at the level of language. The Germanic languages, including English, associate "green" with vitality (*Oxford English Dictionary* 2012). There is no end to green life, if we conceive of the color partly as an English word, thriving within Anglophone print culture. Even the sea, which first lent its name to "the Blue Planet" in the era of the Apollo 8 photographs of Earth, is green, at the bottom of English. "Designating the water of the sea," "green" served English writers as an epithet of Neptune (ca. 1450). The development of "green" as a signifier within a web of lively, embodied signifieds propels it toward Anglophone usage as a hip synonym for "ecological" and "environmental" (ca. early 1970s). The darker aspects of green as a multivalent sign enter the word "green" largely through the ancient Greek "χλωρός," meaning "green" as in "pale." Ancient and medieval medical traditions, and the classical Latin "*viridis,*" connote a greenish complexion taken as a sign of illness or an excess of bile (*OED* 2012). In its associations with figures of alterity, green points to the in-between

psychological space of existential nausea. Green is slime, the monster, the sterile light of the surgery. The color of lively matter, it invites both the aesthetic and the abject.

Ideological usages of green, which have striven to narrow the color's broad band of resonance, pose problems for Anglophone environmentalism. We might begin with the social complexity of the green place, sunk deep in the English language through the chiefly British usage of "green" for "a piece of public or common grassy land situated in or near a town or village; a village green" (ca. 1190–1200) (*OED* 2012). Should green places be conceived in terms of commons or in the more precious sense of the *locus amoenus* and place apart from time? Although a complex concept, the pastoral has been associated with the latter drift toward the ahistorical, relatively unpeopled place. The subversive potentiality of Shakespearean pastoral, imagined within a context of performance, narrows within later visual representations of a balanced stand-off between Nature and History. Consider *Dream of Arcadia* (1838) by Thomas Cole, leader of the famous Hudson River School of landscape painters. Environmentalists' preoccupation with green as an ideological marker reflects historical implementations of ideology in space. Some green places elaborate the idea of commons, as in the common forests described in the second charter of the Magna Carta. Other spatial instantiations of "green" reflect aspirational desires for a world free from social conflict and unruly life, for example, the classical palace gardens of the eighteenth century, Caribbean estates, and the suburban lawns of Levittown, New York. Green has oriented imperial projects of possession and transformation. In Australia settlers sought to install green in the arid landscapes of the desert (Lynch 2007, 74). Settlers systematically eliminated the forests and wetlands of New Zealand's low-lying plains to make way for sheep and cattle farms, creating the "geometric checkerboards of greens and browns" that stretch across much of the archipelago today (Park 1995, 16).

Institutions and individuals sought to blast green landscapes out to the scale of the planet during the twentieth century. The green revolution (ca. 1968) in agriculture involved an assemblage of projects and discourses directed at making agriculture everywhere "advanced" through technological and economic transformation, including through heavy use of fertilizers, pesticides, and crop monoculture. Associated particularly with the period from after the Second World War through the 1970s, the green revolution was supposed to increase food production and make food cheap, save on land and labor, and ensure a poor industrial labor force (Escobar 1995, 128–29). The name "green revolution" was coined by the director of USAID, William Gaud. Green is imaginable as a global aspiration because it connotes vitality and nature, yet many people critically engage visions of green transformation. Arturo Escobar writes that the ambitious interventions of the green revolution separated peasants from the land and undermined their livelihood (1995, 129–30).

As an emblem of political will, green is a work in progress. To draw from Anna Lowenhaupt Tsing's discussion of universals, green is not a completed imaginary or object but must move across differences and be activated in relation to "a heterogeneous world" (2005, 8). In 1970 in British Columbia, a group was forming in protest of a planned United States high-yield nuclear test on Amchitka Island. The protesters debated what to name the campaign and the boat on which they would sail towards the bomb site, and the words "green" and "peace" emerged. "Obviously," someone said, "it has to be a '*green peace*'" (Bohlen 2001, 30)—thus launching the global NGO Greenpeace. Environmentalists have complicated the Euro-American histories that wove together green with ideas of nature as something different

from human life. They have associated green explicitly with social justice. The Green Belt Movement (ca. 1977), an indigenous NGO founded by Wangari Maathai in association with the National Council of Women of Kenya, engages varied injustices related to access to land and other resources, including education and political participation (Maathai 2004).

With a platform that includes grassroots democracy, nonviolence, and social justice in addition to ecological wisdom, the Green Party (ca. 1977) often presents a stark ideological contrast to neoliberal capitalism. The term "greens," when used to indicate political actors, developed in West Germany in the early seventies, with the first use of "green party" in English coming about in 1977, three years before the formation of West Germany's Green Party in 1980. The name "Green" spread like wildfire from the German "*die Grünen*," such that the British Ecology Party and the New Zealand Values Party changed their names to Green. In an era when planetary scale has been powerfully represented by free market logics, the world's Green Parties offer strategies for attunement to the smaller scales of people and the other life that sustains human habitat.

Its multitude of vital referents have lifted the word "green" toward ecology and global alternatives to supercapitalism. Yet "green" also has been associated with the universal commodity, money. "Greens" is a slang term for money in the United States (ca. 1901). In the 1980s, when the Reagan and Thatcher regimes in the United States and the U.K. led much of the world toward policies of deregulation and privatization, some environmentalists chased the conciliatory idea of a green economy (ca. 1986) devoted to products and services intended to minimize and even remediate harm to the environment. "Free Market Environmentalism," a generous moniker for Thatcher's program, complemented movements in the popular culture of North America,

Britain, and parts of the West toward green technology (ca. 1983), green consumerism (ca. 1988), and green marketing (ca. 1988). The twentieth anniversary of Earth Day in 1990 saw the supposed "mainstreaming" of green interests with those of corporate culture, as companies like Dow Chemical turned out for Earth Day. Such corporate enthusiasms were discredited by environmentalists as greenwash (ca. 1987).

A discussion of "green" within the context of environmentalism must consider that environmental thought has not simply been dominated by green. Environmentalists have conceived their projects in different colors, to engage ecosystems that do not meet hegemonic aesthetic criteria for protection. In the Pacific, activists have drawn heavily on black to engage the testing of nuclear weapons. The fallout of the atomic bombings of Hiroshima and Nagasaki appeared in the form of precipitation, as black rain (*kuroi ame*), and this motif was taken up by Masuji Ibuse in his novel of Hiroshima, *Black Rain* (1966 [1969]), as well as by Shōhei Imamura in the film adaptation (1989). Drawing on Pacific indigenous cosmologies that associate darkness with creativity and with everyday life, New Zealand Māori artist Ralph Hotere created a series of lithographs named "Black Rainbow/Mururoa" to protest French nuclear testing in the Pacific. The insensible beauty of oil on water has preoccupied many artists. The decay of modern industrial infrastructure like Nigeria's oil pipelines may produce new environmental colors, like rust or the black-red of tailings in the Canadian photographer Edward Burtnysky's images of nickel mining. Perhaps green will recede from the color palette of the Anthropocene, our current era (ca. the eighteenth-century industrial revolutions).

39

Health

Alissa Cordner, Phil Brown, and
Rachel Morello-Frosch

Environmental health is increasingly a topic of international sociological research. Although the environment and human health are inextricably connected, the social and environmental contributors to population health, disease, and wellness are too often ignored in the social and medical sciences in favor of a more individualized focus on behaviors and illnesses. This represents a shift from previous modes of inquiry that emphasized environmental links to public health. As early as the nineteenth century, Karl Marx and Friedrich Engels connected population health with harsh labor conditions, unfettered industrialization, and capitalist oppression. In the United States in the early twentieth century, urban public health practitioners and activists highlighted the health problems associated with urban environmental conditions and chemical exposure (Gottlieb 1993).

During the 1960s, environmental and antitoxics activists brought health more directly into the consciousness of environmental movements and researchers. The mainstream environmental movement was inspired by increased scientific knowledge about environmental problems linked to industrial activity, including industrialized agriculture and the widespread use of petrochemicals, attention to ecological limits, and economic growth and increased affluence (Dunlap and Mertig 1992). Environmental justice activism changed the face of environmentalism in the 1980s, growing out of the civil rights and labor movements, black religious organizations, indigenous movements, and the farm worker rights movement (Bullard 2000; Shepard et al. 2002). The antitoxics movement was originally composed of place-based campaigns that mobilized against industrial contamination and the location of hazardous facilities, and has since grown into a more inclusive movement focused on regulatory and corporate reform (Edelstein 2004; Freudenberg and Steinsapir 1992; Levine 1982). Toxic waste activists have brought attention to the health hazards of chemicals and radiation, helped secure the federal Superfund Program to clean up contaminated sites, and secured regulations and bans on many toxic substances (Brown and Mikkelsen 1990).

Social science research on health and the environment grew up alongside these environmental-health social movements. Researchers have investigated how communities respond to the experience of living in contaminated environments (Edelstein 2004; Lerner 2005) and, more recently, how individuals develop an embodied exposure experience connected to knowledge about their personal exposure to chemicals (Altman et al. 2008). Sociologists have studied examples of hazardous workplace exposures for many types of workers, including factory workers (Fox 1993) and farmworkers (Harrison 2011; Pulido 1996), and have shown that framing environmental issues as health issues facilitates coalition building between environmentalists and unions, known as "blue-green alliances" (Mayer 2009). Empirically, environmental sociologists and geographers have quantitatively documented that racial-ethnic minorities and the poor are significantly more likely than whites or the affluent to live near environmental hazards such as hazardous waste plants, nuclear reactors, and other facilities (Downey 2006; Mohai, Pellow, and Roberts 2009).

Social movements focused on embodied health issues have led to major shifts in illness experience, disease definition, challenges to professional standards of care, and health-care-system reform (Brown 2007). These embodied health movements challenge existing medical and scientific knowledge and practices, and often collaborate with scientists and health professionals in pursuing treatment, prevention, research, and expanded funding. These movements generally work on contested illnesses, diseases whose etiology, form, or symptoms are disputed, with public debates centered on environmental causes. For example, asthma is a growing epidemic, especially among children in urban areas, and increasing evidence links the illness with environmental exposure to pollution (Brown et al. 2003). Community health activists have focused on asthma as an example of environmental and social injustice, and have organized campaigns to reduce local pollution from city and school buses, conduct health studies to determine rates of asthma, and improve health education. Individual illness problems are transformed into a "collective politicized illness experience" when organizers make direct links between their experiences with asthma and the social determinants of their health (Brown, Morello-Frosch, and Zavestoski 2011).

Many environmental health social movements deal with emerging contaminants, chemicals of concern because of their widespread use, their environmental ubiquity, toxicological and epidemiological evidence of adverse health effects, and advances in analytical chemistry that make detection in human tissues and the environment possible. The science around emerging contaminants is often contested because they have numerous sources of uncertainty and risk, such as inadequate or absent safety information or regulatory standards, despite their widespread presence in consumer products. Social movements seeking precautionary regulation of emerging contaminants frequently juxta-pose the interests of the public and consumers against the interests and policies of industry and the state, and are sometimes successful in reforming corporate prac-tices or tightening regulations. For example, flame-retardant chemicals are widely used in consumer and household products, and have been the subject of controversy and activism in the last ten years due to widespread exposures and emerging scientific evidence about their associated adverse health effects (Brown and Cordner 2011).

From environmental disasters like the deadly gas leak at a Union Carbide plant in Bhopal (Zavestoski 2009) to political economy studies of the international elec-tronics industry (Smith, Sonnenfeld, and Pellow 2006), research increasingly recognizes that environmental health is a global issue. The health and social-equity impacts of global climate change are a significant area of emerging social science research (Confalonieri et al. 2007; Shonkoff et al. 2012).

40

History

Stefania Barca

The concept of nature and the terms that are associated with it are, in many senses, historical. Their meanings—and the words themselves—are historically constructed and change over time, at once reflecting and constituting social change. As Raymond Williams noted, not only does "the idea of nature [contain], though often unnoticed, an extraordinary amount of human history," but it also contains the very "idea of man in society, indeed the ideas of kinds of societies" (Williams 1980, 67–71). Take the English word "wilderness," for example: from signifying the barren and uncultivated "waste" of the eighteenth-century English landscape, awaiting enclosure and improvement, by the end of the nineteenth century it had become one of the most celebrated concepts in U.S. culture, associated with the aesthetic of the sublime and with national identity (Cronon 1995). A consideration of how people engage the idea of wilderness around the globe reveals that many cultures understand it in significantly different ways while others do not appear to use an equivalent concept at all (see the entry "Translation" in this volume).

Ideas of what nature is, how it works, and how it re-lates to human societies are typically associated with political discourses (Radkau 2008). The words "nature" and "natural," for example, were central in the birth of modern Europe's economic thought, from cameralism to political economy, with "nature" mainly referring to the living world of plants and animals and "natural" to the physical properties of matter and the laws of mechanics (Worster 1985; Schabas 2005). By the early nineteenth century, "nature" and "history" had become two major pillars upon which modern European states were being built: national identity came to be constructed upon celebrations of a nation's natural beauties and resources, while the ability to dominate and improve the natural features of both continental and colonial territories became essential prerequisites of the sovereign (Pádua 2002; Cederlöf and Sivaramakrishnan 2006; Armiero 2011). In *Enclosing Water: Nature and Political Economy in a Mediterranean Valley, 1796–1916*, I show how the philosophers of the Neapolitan Enlightenment, preoccupied with assessing the "backwardness" of the country in respect to northern European nations, looked at environmental degradation as a manifestation of the inefficiency typical of ancient regimes, and used it as a powerful argument against the feudal order and in favor of private property. These philosophers supported that argument with the belief that "history"—namely, twenty centuries of political "barbarism"—had ruined the originally harmonious and beneficial relationship between the country and its natural resources (Barca 2010). Some decades later, in his renowned *Man and Nature* (still viewed as one of the founding texts of conservationisms), American ambassador to Italy George Perkins Marsh popularized this "declensionist" narrative of the Mediterranean as a fragile environment ruined by a long history of human mismanagement (Hall 2005).

That not only the idea of nature but nature itself, or the "environment," contains a good amount of human history is the foundational belief of Environmental History—a field of studies that has reached worldwide attention in the past two decades. EH is based on three theoretical premises, built upon an articulation of ecological and historical facts. The first premise is that nature is not a fixed, immutable entity, as change

is a constitutive element of terrestrial ecosystems and living organisms—not to mention the universe itself (Merchant 1989 [2010]). What interests environmental historians, however, is that portion of environmental change occurring in human history, both affecting and being affected by human agency, and the human ability to accelerate or substantially alter biophysical, chemical, and genetic processes occurring in the rest of the natural world. It is widely accepted that no Earth ecosystem can be found today that has been left untouched by human intervention. Even though the human ability to alter and disturb nature is limited and should not be overestimated, especially when performed by low-density populations through nonmechanized technologies, there is no question that the last 150 years—what is now called the Anthropocene—have constituted an epochal fracture in the history of humankind and of planet Earth, due to an exponential increase in anthropomorphic alterations of the biosphere and even of the atmosphere (Crutzen and Stoermer 2000; McNeill 2000). Two concepts are key to a historical view of human-induced environmental change: "social metabolism," i.e., the energy and materials flow that societies exchange with their environment (from resource extraction and processing to waste disposal), and its corollary, "entropy," i.e., the measure of how human use of natural resources accelerates the rate of energy dispersal and degradation that characterizes the universe (Martinez-Alier and Schlupmann 1990).

In looking at environmental change in human history, however, environmental historians have a structural interest in understanding the social implications of human/environment relationships at different scales. A second foundational idea of EH holds that different ways of interacting with nature correspond to different types of societies, and that, in the act of modifying the natural world, humans also modify their own nature and social relationships (Worster 1985). In her environmental history of New England, Carolyn Merchant theorized environmental changes through the concept of "ecological revolution," namely, contradictions arising historically among ecology, production, reproduction, and consciousness. She identified two such revolutions: the colonial, by which Native American matriarchal society and agriculture were broken down and replaced by northern European patriarchy, while nature became objectified into a bundle of "natural resources"; and the capitalist, by which the early colonial dual economy of subsistence and commerce was replaced by a fully industrialized and commodified social relationship to nature. She concludes by noting that New England is today in the middle of a third, global ecological revolution, as it is now one of the core economies in the world system, deeply interconnected with natural resources and labor coming from the new peripheries (Merchant 1989 [2010]).

A third, fundamental premise for EH is that scientific understandings of nature, including the science of ecology, also evolve historically, in a symbiotic relationship with social evolution, and influence dominant ideas of nature in society. Not only are scientific ideas of nature profoundly historical—as Donald Worster's seminal *History of Ecological Ideas* made clear (Worster 1985); they also have important repercussions upon environmental policies and natural resource management. One example is the concept of environmental restoration, which stems from what experts and politicians perceive as a "natural" order of past ecosystems disturbed by human agency. What point in the past one should choose as a manifestation of the "natural," or the "pristine," to restore is of course a complicated matter, and restorative projects may even turn into something totally new (Lowenthal in Hall 2010). The point is, however, that restoring a certain "natural" order is a halfway cultural

practice, because one can hardly restore the social order that was attached to it.

Finally, environmental historians are profoundly influenced by contemporary scientific ideas. Viewing the Earth as a self-regulating system, for example—probably the most powerful ecological idea of our time—does not imply humanizing nature but does challenge our understanding of agency as an essentially human ability. Overall, environmental histories show that nature does have historical agency, even though few environmental historians would call themselves "determinist." EH is simply much more likely akin than other kinds of historical writing to recognize and speculate upon the acting of environmental forces—viruses, germs and bacteria, climate, geology, etc.—within human history (Hughes 2000). Moreover, we (humans) are not alone: other living beings and living communities have co-inhabited the Earth since the beginning of life. Many have been domesticated, a great number have become extinct, some are now protected. Many have been and continue to be used as knowledge instruments, tortured and experimented upon for the sake of human well-being or profit (Ponting 1992). All have co-evolved with us and made us what we are today, and would deserve a larger consideration within historical narratives.

Even nonliving things such as toxins or POPs (Persistent Organic Pollutants) are to be considered historical agents, acting upon and reacting to human agency: from Rachel Carson's *Silent Spring*, environmental historians have learned about the importance of those micro-particles that constitute and circulate through matter, crossing the invisible border between living and nonliving, natural and artefact, while constantly altering what we call the "environment" and our own body (Langston 2010). Nonliving matter is thus an active member of what Carson described as "that complex web of life—and death—that scientists call *ecology*" (Carson

1962). Never entirely conquered by the "linguistic turn" in the social sciences, environmental historians maintain that matter does *matter* in the understanding of environmental change, which they see as a product of active interactions between the discursive and the material, and between human and more-than-human agencies—a position they share with material feminists and recent trends in political ecology (Alaimo and Heckman 2008; Bennett 2010).

By taking into account nonhuman agency—from micro-organisms and micro-particles to plants and animals—EH ultimately aspires to enlarge the dominant twentieth-century idea of social history, by truly accomplishing the mission that *Annales* scholar Lucien Febvre called for, that of writing a *histoire à part entière*.

41
Humanities
Joni Adamson

Rumors of "fading interest" in the humanities at institutions of higher education appear frequently in the media. However, statistics show that the number of students taking degrees in the disciplines traditionally focused on the study of human culture has remained constant since the 1980s (Bérubé 2013; Lewan 2013; Paul and Graff 2012). Moreover, in the first decade of the twenty-first century, business and education leaders are declaring the "environmental humanities" (history, philosophy, aesthetics, religious studies, literature, theater, film and media studies informed by the most recent research in the sciences of nature and sustainability) crucial to addressing the anthropogenic factors contributing to increasingly extreme weather-related events (drought, fire, hurricanes, melting glaciers, and warming and rising oceans) (Nye et al. 2013; Braidotti et al. 2013).

While stereotypes associated with humanities scholarship (dry discourse analysis, esoteric debates) may have once made these disciplines seem ill suited to addressing crises outside the walls of academe, the alteration of every biophysical system on the planet has led humanists, with their colleagues in anthropology, cultural geography, political ecology, and science, to research and write about changing environmental processes that are "inescapably entangled with human ways of being in the world" (Rose et al. 2012, 1). Humanists are confronting the perception (common in the second half of the twentieth century) that the humanities and sciences constitute two widely separate "cultures" (Snow 1959). Since the early decades of the twentieth century, they have been pointing out that this culturally constructed "bifurcation" between "Culture" and "Nature" began emerging at the end of the seventeenth century with the creation of the highly specialized and disciplined methods of knowledge production categorized as "science," which, in turn, was often used to authorize the terms of colonial expansionism and, later, neocolonial development schemes (dam construction, deforestation, resource extraction) that displace human populations and exacerbate species collapse (Whitehead 1920; Latour 2010, 479–80; Rocheleau and Nirmal, this volume).

"The Future We Want" (a document produced at the 2012 Rio+20 United Nations Conference on Sustainable Development and signed by all UN member states) confirms a broad general agreement that global society should strive for a high quality of life that is equitably shared and sustainable for all species (Costanza et al. 2014, 284). However, social and environmental justice and sustainability are not new desires among humans. As cultural theorist Raymond Williams has noted, any full history of the uses of the keyword "nature," would "be a history of a large part of human thought" (1976 [1983], 219, 221), and many of the oldest recorded oral story cycles and written texts created by people living in Egypt, Rome, China, Europe, or the Americas illustrate that humans have long been raising questions about "quality of life" in stories and collected archives of information that explicate sophisticated understandings of human relations to other human groups and other species. Some of these mixed-genre texts, referred to as "almanacs," intermingle practical advice on the agricultural arts, mathematics, and astronomy with imaginative social commentary that takes the form of creation or migration narratives, songs, or prayers (Merchant 1989 [2010]; Adamson 2001, 133–34).

For example, in *Works and Days* (ca. 700 BCE), Greek poet Hesiod offers advice to farmers dealing with extreme weather and poor soils. Imaginative social commentary is interwoven into the myths of Prometheus, who created humans from clay, and Pandora, whose curiosity impelled her to open a jar from which malicious spirits escape. These "evil" spirits, which are linked to Greek "seizure" of wealth from their "neighbors" (1914, ll. 320–41), raise questions about how humans might live well during times of scarcity. By paying attention to the skies and the activities of animals, humans might work ethically for comfort rather than engage in life-extinguishing plunder (1914, ll. 342–51; Adamson 2014). Similarly, social commentary is woven into the first printed versions of the Popul Vuh (ca. 1500 CE), described by Quiché Mayan scribes in the Americas as a "seeing instrument" or "complex navigational system for those who wish to see and move beyond the present" (Tedlock 1985, 32, 71). In this corpus of mytho-historical-astronomical narratives, the gods of the sea and sky make four attempts to create human beings. In one failed attempt, they create a being of mud that is washed away by the rain. Mayan elders suggest that this story is not a paraphrase of the Greek story of Prometheus or the biblical story of Adam's creation from mud, but a "negation of Adam," and probably a direct critique of the beliefs and practices forced on Quichéan peoples by Spanish colonizers who did not believe the Maya to be human (Tedlock 1983, 263–64, 270–71). In previous work, I have suggested the phrase "seeing instrument," long associated with the Popol Vuh, to describe cultural productions and humanities scholarship—from ancient indigenous story cycles and farmers' almanacs to contemporary novels, blockbuster film, street theater, ecocriticism, and ethnography—that makes the complex connections between biogeochemical processes and the "well-being" of all life on the planet more visible to a broad general public (Adamson 2001, 145; Adamson 2012a; Rose, this volume). These ancient narratives, out of which most humanities fields (philosophy, ethics, history, and literature) emerged, have long worked as an imaginative force for thinking about "the origins and [ongoing evolutionary] transformations of the world and its inhabitants" (Cruickshank 2005, 99; Adamson 2014).

Other texts that have profoundly shaped the humanities are also being reevaluated for the ways they muse on anthropocentric values and beliefs that unleash "malicious spirits" (or what sustainability scientists today refer to as "unintended consequences"). In *Song of the Earth*, literary critic Jonathan Bate rereads some of the most iconic nineteenth-century British literary works for what they can tell us about human understanding of the weather as the "primary sign of the inextricability of culture and nature" (2000, 102). He argues that, in the Anthropocene, there is much we can learn from literary works that attend "to the weather" (Bate 2000, 102). For example, Mary Shelley's novel, *Frankenstein; or, The Modern Prometheus* (1818) portrays a young scholar turning away from the understandings of "natural philosophy" found in ancient almanacs and embracing "modern science" as he creates an ungovernable monster from the "mud" of human flesh. Shelley began writing the novel, she explains in the preface, during a "cold and rainy season" (1998 [1818], 14). As Bate establishes, the stormy weather of 1816, often referred to as the "Year with No Summer," was the result of the catastrophic 1815 explosion of Tambora, an Indonesian volcano, which affected the atmosphere of the entire northern hemisphere for the next three years, leading to worldwide crop failure, riots and starvation, and a global cholera epidemic (Bate 2000, 97; Wood 2014). Today it is known that a tropical eruption paradoxically cools the planet with a blanket of volcanic dust but drastically warms the Arctic owing to changes in wind circulation and north Atlantic ocean currents (Wood 2014). As a result, the British Admiralty began receiving reports of a remarkable loss of sea ice around Greenland and began planning a new age of polar exploration. This helps to explain why Shelley sets the opening scenes of *Frankenstein* in a frozen North, where a young ship's captain first encounters the monster and its creator. Famous for its laboratory scenes, *Frankenstein* has often been dismissed as a "romantic reaction" at odds with science, a categorization that veils the brilliance of Shelley's critique of science as a discourse that authorized colonial activities on a scale that today can be recognized as having altered planetary systems. As well, the novel itself is a literal record, or "seeing instrument," into a two-hundred-year-old event, proving that a "changing climate changes everything" (Wood 2014).

Mary Shelley's contemporary, Alexander von Humboldt, a Prussian geographer, naturalist, and explorer who traveled throughout Amazonia at the end of the 1700s and wrote the best-selling five-volume work *Kosmos* (1845–62), might also be seen as having anticipated and modeled the research approaches that, today, are being referred to as the "environmental humanities." Humboldt defined "nature" as a "planetary interactive causal network operating across multiple scale levels, temporal and spatial," and was the first to warn about the links among "deforestation, environmental change and depopulation" (Walls 2009, 11, ix). His scientific observations profoundly imprinted Charles Darwin's theories of evolution, and his views on the immorality of slavery and the intelligence and agency of indigenous peoples influenced Henry David Thoreau, whose writings are considered a touchstone for what Lawrence Buell has called the emergence of the modern "environmental imagination" (Walls 2009; Buell 1995). Later, at Columbia University in the 1920s, Humboldt's devoted disciple Franz Boaz, known as the "father" of anthropology, and Ruth Benedict, an ethnographer who wrote one of the first scientific cases against racism (Benedict and Weltfish 1943), would teach another writer whose work would open the environmental imagination to considerations of race, gender, and economic status. Zora Neale Hurston's ethnographic research among Caribbean and southern indigenous and African people would become the basis for her most famous novel, *Their Eyes Were Watching God* (1937). Hurston intermingles human imagination with the processes of weather and Earth as she depicts a historically accurate 1928 hurricane that killed thousands of predominantly black migrant workers living in the lowlands of Florida. The novel is significant for anticipating twenty-first-century understandings of "natural disaster" that link structural racism, poverty, and increased vulnerability to "climate events" (such as Hurricane Katrina) to growing global social inequality between rich and poor (Parrish 2013; Agyeman, Di Chiro, Martinez-Alier, and Priscilla Wald, this volume).

Like Hurston, more recent writers, such as Nigerian (Ogoni) Ken Saro-Wiwa (1941–1995), who was "judicially murdered" for his writing and activism, have woven complex indigenous scientific literacies concerning weather and Earth found in indigenous oral story cycles into literary and filmic works that contribute importantly to contemporary understandings of anthropogenic causes of catastrophic social breakdown linked to environmental change. Saro-Wiwa's novels, screenplays, and memoir draw attention to the "slow violence" of decades of oil spills in Ogoniland and dramatically illustrate how the humanities can make events unfolding at multiple temporal and biogeophysical scales— from the "spectacular" (hurricanes) to the "slow" (toxic drift)—imaginatively perceptible and tangible to a broader public (Nixon 2011, 103–27). Saro-Wiwa's work

acts as a "seeing instrument" into the reason why delegates from one hundred Global South countries met in 2010 in Bolivia for a World Conference on the Rights of Mother Earth and Climate Change after the failure of the 2009 United Nations climate talks in Copenhagen. In their declaration, they stated they would not wait for the world's richest countries to enter into binding agreements on climate. Rather, they would begin working immediately to shift the world's attention from "living better" to "living well," which they defined as societies that are based on "environmental justice" and that see "life as [their] purpose" (Universal 2010).

The failure of the Copenhagen talks (COP 15) illustrates that there are significant barriers to achieving this goal, including bureaucratic inertia and the tendency of governments, academia, and other groups to work in isolation. To break through this inertia, humanists are developing university courses with titles such as "The Cultures of Climate Change," "The Political Ecology of the Imagination," or "Science Fiction to Science Fabrication" that focus on the ways in which disciplines across the campus, not just the humanities, might employ ancient almanacs, classic or contemporary literature, film and media, or recent documents such as "The Future We Want" as "seeing instruments" that encourage students to imagine, as Bruno Latour phrases it, "the shape of things to come" and the radical transformation of notions of "progress" into "prospects" (Latour 2010, 486). In these courses, students might read and discuss recent speculative fiction, which has been described as "cli-fi" or an "emerging climate canon," which includes works such as Nathaniel Rich's *Odds against Tomorrow* and Ian McEwan's *Solar* (Gunn 2014). Then, teaming up with engineers, economists, and scientists, they might be given assignments that encourage them to imagine our prospects, or to plan "plausible futures" built on social, technological, and ecological systems agile enough to adapt to coming weather extremes (Pérez-Peña 2014).

Set into a history of a "climate canon" that includes recent speculative fiction but goes back in time to include much older texts as well, the narrative and networking possibilities for an environmental humanities that seeks creatively to shift public imagination towards a "future we want" begins to emerge. Without abandoning the subject-matter strengths and specific tools that are the hallmark of their discipline, such as critical analysis, humanists are actively reimagining "the proper questions and approaches" of their fields (Rose et al. 2012, 3). Extensive global networks to do this work are emerging in the United States, the UK, Australia, China, India, Korea, Scandinavia, and Taiwan. Nurtured first by the Association for the Study of Literature and Environment (ASLE) and the Australian Academy for the Humanities, and more recently by the International Social Science Council and the Consortium for Humanities Centers and Institutes (CHCI), these organizations are becoming the "spokes" of global research "hubs" (Nye et al. 2013, 22–28; Braidotti et al. 2013). As they have for thousands of years, humans are reading "the signs of the times in the signs of the skies" (Bate 2000, 102) and employing the humanities as an imaginative force for thinking about the ongoing evolutionary transformations of the world and its inhabitants.

42

Imperialism
Ashley Dawson

The great waves of European expansion that began in 1492 were characterized by fundamental ideological and material transformations in people's relations to land. In his *Second Treatise of Government* (1689) the English philosopher John Locke, who owned plantations in both Ireland and the American colonies, wrote, "He who appropriates land to himself by his labor does not lessen but increase the common stock of mankind" (Locke 1988, 293). Later in the *Treatise,* Locke explicitly references the Native Americans as an example of people who, appearing to him to do no labor, consequently fail to develop the Earth according to god's plan and therefore can stake no legitimate claim of ownership to the land on which they live. Locke's argument was central to the creation of private property and, more broadly, to the process of enclosure of communal lands that unfolded in the British Isles and in England's far-flung colonies in the centuries before and after Locke wrote. Marx called this "primitive accumulation," a process through which people were separated from their land and their means of production. Fundamental to the birth of capitalism, this history was "written in the annals of mankind in letters of blood and fire" (Marx 1976 [1990], 875).

Enclosure, which had been taking place in England since the late medieval period, laid the foundations for the modern capitalist system and for the notions of "development" that emerged as part of this system, fundamentally transforming people's relations to land and to one another. For example, in his *Utopia* (1516), Thomas Moore offers a powerful indictment of the depopulation of the countryside catalyzed by enclosure. According to Moore's character Raphael Hythloday, the feudal nobility, driven by greed that left them unsatisfied with the rents they obtained from serfs living on their land, enclosed the open fields and common land where peasants had grazed animals and farmed. They used this land as pasture for sheep whose wool could be used for lucrative textile production. The upshot, as Hythloday explains, was that sheep were turned into "devourers of men," the countryside was depopulated, and farmers, deprived of their means of subsistence, were turned into brigands and beggars. As the commons were enclosed, women lost much of the autonomy they had struggled to maintain in the medieval period; the pan-European witch hunts of this period destroyed the vestiges of this autonomy, relegating women to the nascent domestic sphere and subordinating their reproductive capacity to patriarchal control (Federici 2004). Capitalist enclosure of the commons and commodification of nature thus went hand in hand with the consolidation of gender hierarchy and sexism.

The enclosures in England took place in tandem with corollary forms of dispossession around the globe, including the growth of the transatlantic slave economy and the establishment of private property rules by British colonial ventures such as the East India Company in Bengal. Notions of the land as virgin terrain waiting to be exploited were used to mask the bloody violence through which this appropriation of the commons was accomplished. Writing in the same century as Locke, the metaphysical poet John Donne offered a revealing glimpse into the pervasively gendered ideologies that underpinned early European imperial exploits. In "Elegy XX: To His Mistress Going to Bed" (1633), Donne penned a bawdy erotic poem in which the speaker describes the disrobing of his mistress and, when she is stripped naked, compares her to one of the recently founded English colonies: "O, my America, my Newfoundland, / My kingdom, safest when with one man mann'd, / My mine of precious stones, my empery; / How am I blest in thus discovering thee!" (Donne 1896, 149). Donne's poem is important not simply for its forthright acknowledgment of the imperial enterprise but also for its clear admission that this enterprise involves resource extraction, in this case in the form of the mining of precious stones. Furthermore, the metaphorical link between the speaker's mistress and a New World colony underlines the gendered nature of imperial ideology: from the outset, empire shored up the sovereign male European ego. Donne's poem also lays bare the insecurity on which that imperial ego was founded by entertaining jealous thoughts about other men who might also "mann" this colonial terrain. This concern with the control of women's sexuality works to allegorize fears of competition between imperial powers for colonial territories and resources in the early modern period.

Such fears about inter-imperial competition loomed even larger by the late nineteenth century, when, according to Raymond Williams, the term "imperialism" entered into common usage in English (Williams 1976, 159). Imperialism came to signify three interwoven forms of power: political, to the extent that an imperial center ruled over distant colonial territories; economic, insofar as this political rule guaranteed access to the natural resources and markets in the colonies; and cultural, since the previous two forms of domination were legitimated through racist notions of Europe's tutelary responsibility for its various putatively uncivilized colonial subjects. In the period after 1870, imperialism became an increasingly central facet in each of these spheres as Britain faced increased competition not simply from traditional rivals such as the French but also from newly consolidated powers such as Germany. In this high imperial era, despite attempts to mediate inter-imperial competition such as the Berlin Conference (1884–85), enmity between the leading European powers led to a shift from informal spheres of political and economic influence to the ruthless annexation of entire regions of the planet, as in the infamous "scramble for Africa."

The rush to carve up Africa in order to control the continent's natural resources and markets led not just to despoliation of the environment and genocide in colonies such as the Belgian Congo and German Southwest Africa but also to increasing tensions between the imperial powers that led ultimately to warfare in Europe. Seeking to make sense of the mutual destruction of the European imperial powers, theorists such as J. A. Hobson, Rudolph Hilferding, and V. I. Lenin posited that imperialism had to be understood as a specific stage of capitalism. Lenin, for example, argued that modern imperialism resulted from a crisis of surplus accumulation in the core imperial powers (Lenin 1996). Faced with declining domestic profits, industrial cartels merged with banks in order to export capital to the colonies, where underdevelopment and monopolies of labor and natural resource exploitation guaranteed super-profits for this novel configuration of capital, which Lenin, following previous critics such as Hobson, termed "finance capitalism." Under finance capitalism/imperialism, the national bourgeoisies of the European powers and the United States mobilized their respective states to carve up the planet into exclusive, protectionist economic spheres, with each power jostling for greater possessions in order to guarantee augmented profits for its capitalist class.

The upshot of this system for those in the colonial periphery was a continual state of underdevelopment in which they furnished agricultural products to the industrial center for consumption (e.g., sugar, coffee, tea, vegetable oils) and industrial expansion (e.g., cotton, timber, minerals) but never themselves developed into competitive, industrialized powers. In many cases, this segmentation of the planet into core and periphery intensified processes of what David Harvey, rethinking Marx's model of primitive accumulation, terms "accumulation by dispossession" (Harvey 2005). In countries such as Liberia, for instance, the peasantry and their subsistence agricultural practices were displaced to make way for vast rubber plantations furnishing the raw materials out of which industrial machines, telegraph technology, medical equipment, and the automobile industry were built (Tully 2011). If the early phases of European expansion saw widespread examples of ecocide such as the destruction of the bison in the Great Plains of the United States and the wanton slaughter of big game in Africa, this imperial phase saw the devastation of local economies, with formerly autonomous peasants reduced to landless laborers or poverty-stricken sharecroppers working on massive mono-crop plantations. Since these plantations exported the vast majority of their produce to the imperial center, the denizens of colonized nations experienced periodic famines while surrounded by food destined for the tables of Europe. In Bengal, for example, three million people starved to death in 1943 while British troops protected well-stocked granaries.

The anticolonial revolutions that followed the Second World War were an attempt to overturn this imperial system. Significant confusion has been generated by the fact that while the political sway of the European imperial powers over much of the rest of the planet was dissolved as a result of these successful national liberation struggles, the economic and cultural domination of the European powers and of the new global hegemon, the United States, was not in the least eclipsed. Terms such as "neocolonialism" came into use to describe the system of economic domination and indirect political and military control that unfolded during the era of the Cold War. While many liberation movements sought to redistribute land to the peasantry in this period, most postcolonial regimes followed the model of industrialized agriculture that characterized both the Soviet Union and the United States (Moyo and Yeros 2005). The aim was to spawn domestic markets for fledgling industries through the wealth generated by increased agricultural production, but the bulk of this production continued to be exported to global markets, where a glut of commodities drove prices down year after year, trapping nations in their subordinate, peripheral role. The trend towards large land holdings begun under European colonial rule continued, as did the dependency of postcolonial nations. By the 1980s, fifty countries that had been self-sufficient before independence had become net importers of food. According to the United Nations Food and Agriculture Organization, 925 million people, one in every seven human beings alive today, is undernourished, a figure that has increased significantly over the last forty years.

The Green Revolution, exported by the United States in the 1960s and '70s to head off radicalization in postcolonial nations, intensified these trends by substituting resource- and energy-intensive capitalist agriculture for the sustainable subsistence farming of the traditional peasantry (Shiva 1992). While the Green Revolution did increase food production significantly, it also generated massive waves of urbanization as the global peasantry was displaced by industrial agriculture. In addition, the Green Revolution increased the amount of land under cultivation and the intensification of that cultivation, leading to deterioration of natural ecosystems, including widespread soil erosion; the contamination of watersheds, communities, and human bodies

with tons of pesticides and persistent organic pollutants; and massive deforestation in tropical regions such as South America, West Africa, and Southeast Asia. In addition to establishing a destructive, inefficient, and unsustainable model of "development" based on the use of finite and polluting fossil fuels, the Green Revolution also pushed many postcolonial nations into the debt trap that led to the structural adjustment policies of the neoliberal era.

The offshoring of industrial production after the 1960s brought a reconfiguration of the international division of labor. National borders that had long constrained capital seemed to melt away, generating the sense of malaise and disorientation articulated by the deacons of postmodernism. Talk of imperialism receded to a certain degree during the initial decades of neoliberalism as many in the educated classes, including those of postcolonial nations, bought into notions of global competition and flow (Lazarus 2011). By the first decade of the twenty-first century, however, the spread of export-based capitalist industrialism to East and South Asia and portions of Latin America not only catalyzed a deep debt crisis in the former core nations but also sparked fears of renewed inter-imperial competition that resemble those of Lenin's day all too closely. In addition, neoliberal policies such as the eradication of tariff barriers in the name of global competition dramatically increased the processes of land alienation and accumulation by dispossession evident in previous decades. Attempts by imperial powers such as the United States to maintain global hegemony through the imposition of intellectual property agreements over things such as the genetic makeup of seeds led to the sweeping commodification of nature and, indeed, of life itself (Rosset 2006). The resulting crisis of the countryside throughout the Global South has catalyzed the rise of militant rural movements around the globe, including national organizations such as Brazil's Landless Worker Movement and transnational groups like Via Campesina, as well as the more spontaneous protests against food-price rises in Tunisia, Jordan, and Egypt that helped catalyze the Arab Spring (Bush 2010).

The enclosure of the global commons has proceeded to such a pitch that survival has become the principle horizon of contemporary politics (Abélès 2010). Where extraction once affected those in the colonial periphery or in the rural hinterlands of the imperial powers, today extreme extractive industries such as hydraulic fracturing threaten the drinking water of key global cities such as New York. Where commodification of the environment once was constituted principally by plantation agriculture, today all of nature has been turned into an object of speculative investment, further concentrating the control of nature. Under the guise of saving the planet, institutions of global governance such as the United Nations Environmental Program have launched programs to quantify and commodify nature's "environmental services" (Bond 2012). Yet while the green economy grows apace, so do the spiraling carbon emissions that threaten to pitch the planet into devastating climate chaos. Green capitalism may be the highest and final stage of imperialism.

43

Indigeneity

Kyle Powys Whyte

Ecological scientists often define "indigeneity" as a species' ecological nativeness to a place. A species is indigenous or native when its presence in a region stems from natural processes and not human ones. Indigenous species are not necessarily unique, or endemic, to a particular region. Points of human influence distinguish indigenous (prior) from nonindigenous (newcomer) species (like nonnative invasive species). Wild rice in the western Great Lakes region of North America, for example, has long been considered a native species whose ecological significance concerns the way it contributes to supporting diverse biological communities. It is a food source for waterfowl, muskrats, and various invertebrates, and a provider of roosting areas, loafing areas, and brood cover for waterfowl. Human communities can reduce wild rice populations through damming waterways, mining, or importation of nonnative invasive species for fishing, ornamental, and other purposes. Nonnative invasive species like common carp, rusty crayfish, or purple loosestrife can outcompete wild rice. Changes in indigenous wild rice species, then, have consequences for the other species to which it is related (e.g., waterfowl) in the region's ecology ("Manoomin [Wild Rice]" 2013; David 2008; Minnesota Department of Natural Resources 2008; University of Wisconsin Extension 2007).

The concept of indigeneity, or indigenousness, does not always exclude humans. Humans who identify themselves as indigenous often seek to express a prior or more original claim to a place in contrast to individuals they consider to be settlers or newcomers. Such claims are often expressed through place-based descriptions of relationships. Anishinaabe people in the Great Lakes region of North America, for example, have been in a relationship with wild rice, or "*manoomin,*" as it is called in the Anishinaabe language, across many generations. Wild rice is a spiritual food (gifted by the Creator) that figures crucially in Anishinaabe origin and migration stories. Anishinaabe people consider themselves in a relationship with wild rice that has evolved across many generations. The relationship is moral because Anishinaabe persons consider themselves to have responsibilities for taking care of wild rice habitats and honoring the plant through ceremonies; wild rice, in turn, has responsibilities to nourish and bring together human communities, among other responsibilities it may have to other species. Both humans and wild rice are ascribed forms of agency that engender mutual but differentiated responsibilities across the species (Foushee and Gurneau 2010; "Manoomin [Wild Rice]" 2013; LaDuke 2003; LaDuke and Carlson 2003; David 2008; Adamson 2011; Vennum 1988; Johnston 1993; Benton-Benai 1988; Andow et al. 2009).

When Anishinaabe people refer to their own nativeness or indigeneity or that of wild rice, they are often referring to a more complex intergenerational system of their place-based relationships connecting humans and nonhuman beings (e.g., plants and animals), entities (e.g., spirits and sacred shrines), and systems (e.g., seasonal cycles and forest landscapes) in the region. Human communities, then, are an integral part of the ecological system. Moreover, human communities via their cosmologies ascribe agencies and responsibilities to the different beings and collectives in the region. These moral relationships between, for example, humans and wild rice are both intrinsically and instrumentally valuable (Whyte 2013). They are intrinsically valuable as part of, for example, Anishinaabe identity, and they are instrumentally valuable as sources of goods such as (1) nutrition, (2) motivation for protecting against environmental degradation, and (3) knowledge of the region's ecology. According to this understanding, indigeneity refers to systems among humans and nonhuman operatives in particular places over many generations.

Other conceptions of indigeneity that are ecologically relevant are, at first glance, primarily political. Indigenous peoples often define themselves as the pre-invasion and/or precolonial inhabitants of territories currently dominated by nation-states like the United States (Anaya 2004; Sanders 1977; Weaver 2000). Many such communities exercised their own forms of governance prior to invasion, colonization, or settlement and have yet to consent to the sovereignty of nation-states in their territories (Turner 2006). Many indigenous peoples seek to reestablish their own forms of governance in their territories. Indigenous peoples are also characterized politically insofar as they are typically communities who share sufficiently similar experiences of colonial oppression, including territorial dispossession, economic marginalization, racial discrimination, and cultural imperialism (Niezen 2003; Byrd and Rothberg 2011). Global political actions identified with indigenous peoples' movements have been growing stronger since the latter half of the twentieth century and on through the present (Adamson 2012c; Niezen 2009; Cadena and Starn 2007; Allen and Xanthaki 2011; Joffe, Hartley, and Preston 2010). Ostensibly, the movements seek to redress colonial and settler oppressions and reestablish acceptable forms of political self-determination. Examples of these movements include global networks like the Asian Indigenous Women's Network, which works to protect the subsistence traditions of indigenous communities living in forests (Tebtebba 2011), as well as organizations occurring within the United Nations like the Permanent Forum on Indigenous Issues (UNPFII). These movements have produced multiple declarations, such as the UN Declaration on the Rights of Indigenous Peoples (UNDRIP 2007) and the Mandaluyong Declaration (2011).

While this conception seems overtly political, the indigenous movement focuses considerably on ecological concerns. Indigenous peoples' political self-determination is often centered on protecting intergenerational systems of placed-based relationships from being obstructed by globalization and other political, social, and economic forces. For example, Anishinaabe people have engaged in multiple political actions to protect their community ricing systems from a range of actions by U.S. settlers in Minnesota, such as risky university research programs and impacts from mining and other industries (Adamson 2011; Andow et al. 2009; LaDuke and Carlson 2003). Māori organizing, including the Waitangi Tribunal, has helped establish the legal personhood of the Whanganui River in the the New Zealand settler nation-state, thus establishing respect for the river's rights and interests as a living system (Postel 2012; Te Aho 2010). Indigenous peoples in Ecuador played an important role in the nation-state's new constitution, which includes legal rights to tropical forests, islands, rivers, and air (de la Cadena 2010). The United League of Indigenous Nations has a climate-change working group that seeks to protect culturally significant species from alterations such as sea level rise and glacier retreat (Grossman 2008). Indigenous peoples in Alaska have engaged in political actions to hold industries accountable for contributing to climate change that will force their communities to permanently relocate (Kronk 2012; Shearer 2011; Osofsky 2006). UNDRIP's preamble recognizes "that respect for indigenous knowledge, cultures and traditional practices contributes to sustainable and equitable development and proper management of the environment" (United Nations 2007). Indigenous groups have used other parts of UNDRIP, such as the idea of Free, Prior and Informed Consent (FPIC), as a way to protect communities from being potentially displaced or discriminated against with respect to the UN program Reducing Emissions from Deforestation and Forest Degradation (REDD), which aims to create commercial markets for forest conservation. UNDRIP affirms the value of indigenous consent and conservation practices when there is concern that state-based policies like REDD+ will commodify forests in exploitative ways or fail to respect the conservation practices indigenous peoples have already been doing for centuries (Alexander et al. 2011; Griffiths 2007; Corbera 2012; Van Dam 2011). Finally, indigenous people who are long established in urban areas or who have been permanently relocated also use concepts of intergenerational systems of placed-based relationships and responsibilities as the foundation for indigenous education and research (Bang et. al 2010; Smith 2012).

The concept of indigeneity also modifies knowledge. Indigenous knowledges often refer to observations of species and the environment over a long time scale in a particular place. Indigenous peoples use this knowledge to support their susustenance and self-determination. Examples range from taxonomies of local plants to knowledge of burning practices to the creation of forest islands for the production of fruit and attraction of game (Turner et al. 2011; Kimmerer and Lake 2001; Berkes, Colding, and Folke 2000). Though historically scientists recognized the importance of indigenous knowledges (though often without crediting indigenous peoples in ethical ways or at all), more scientists are beginning to recognize the importance of many examples of indigenous knowledges for improving research on topics ranging from the nutritional properties of plants to understanding climate-induced environmental changes such as retreating sea ice (Williams and Hardison 2013; Reidlinger and Berkes 2001; Anderson et al. 2012). A host of UN documents as well as guides to scientific research include sections on indigenous knowledges or their synonyms, such as "traditional ecological knowledge" (Nakashima et al. 2012; Berkes 1993). Unfortunately, the guiding assumption is often that what makes knowledge indigenous is its being a collection of related observations over a long scale of time in a particular place. Yet many indigenous community members, scientists, and scholars contend that indigenous knowledges must include "knowledges of" one's responsibilities to the human, nonhuman beings and entities, and systems that make up the places where one works, lives, and plays (Reo and Whyte 2012; McGregor 2008; Pierotti and Wildcat 2000; Kimmerer 2002; LaDuke 1994). It is the knowledge of how one is situated as an agent in relation to other beings, entities, and systems that exercise different and similar forms of agency. Indigenous knowledge refers to a person's actual participation in that system. Collective observations over time are just one component.

The concept of indigeneity, then, can refer to the aboriginality of many possible individuals or groupings, from particular species to governments to knowledges. It is important to recognize that indigeneity is seldom used to express "coming before" in a basic sense; rather, it is more often used to express intergenerational systems of responsibilities that connect humans, nonhuman animals and plants, sacred entities, and systems.

44
Landscape
Dorceta E. Taylor

In *The Environment and the People in American Cities* (2009), I argue that landscape is a socially constructed entity that is an important element of nation building. It is not simply an object to be viewed, depicted, cultivated, or manipulated, but as W. J. Thomas Mitchell (2002) contends, it is an instrument of cultural force central to the creation of national and social identities. The *American Heritage Dictionary* (2009) defines "landscape" as an expanse of scenery that can be seen in a single view. "Landscape" is also commonly defined as a depiction of scenery in a picture or painting, or as a branch of art. The term, first recorded in 1598, arose from the Dutch word *"landschap,"* meaning "region, tract of land." It took on artistic overtones in English usage, where it came to mean "a picture depicting scenery on land."

However, civilizations of the Near East began designing and managing their landscapes before the birth of Christ. By about 700 BC, Assyrian noblemen practiced riding, hunting, and combat skills in reserves earmarked for training purposes. The Persians were influenced by the design of these reserves and used them to help in the development of royal hunting enclosures. Royal hunting grounds flourished in Asia Minor between 550 and 350 BC. The Greeks were the first to landscape public places (the plazas, or *agoras*). The tree-shaded plazas, adorned with fountains, were used for public gatherings, rest, and relaxation. The *agora* could be seen as a precursor to the modern city park (Runte 1987).

Positionality in the landscape can be a reflection of the political and social hierarchies and inequalities in a society. For instance, in medieval Europe, landscapes characterized by open spaces were maintained exclusively for the use of the ruling classes to be used for private estates, hunting, and other recreational purposes. Landscape gardening emerged in England in the early sixteenth century. These early parks were characterized by broad stretches of greensward (lawn) framed by sparsely distributed trees to convey a pastoral image of the landscape. Many of the parks that sprang up all over Europe were created by evicting the poor from the land to create what Raymond Williams calls a "rural landscape emptied of labour and labourers . . . from which the facts of production had been banished" (Williams 1973). Hence, Michael Conzen (1994) sees landscapes as comprehensive and cultural. They represent a visual observation of long-term relationships (of alterations and creations) between humans and the environment. John Wylie (2007) agrees with this interpretation of landscapes. According to him, landscapes reflect the relations among the natural, cultural, spatial, and social realms. Wylie also makes the important point that a landscape is not just an image of a scene; it is also way of seeing things.

Landscapes also served as a mechanism to stimulate environmental awareness and challenge power relations. Nowhere is this more evident than in the United States. During the 1840s, New England transcendentalists such as Ralph Waldo Emerson and Henry David Thoreau implored Americans to cultivate a greater concern for the land and become better stewards of it. The transcendentalists provided a counterframe to the widespread commitment to conquering and exploiting resources and racial minorities in the name of progress. Notwithstanding, environmental activists weren't very sensitive to racial inequalities when it came to landscapes of interest. Hence, activists such as John Muir helped to frame the landscape in Romantic terms that promoted wilderness as national treasures and culturally revered spaces that were "untouched by human hands." Hence, differences in the European American view of the wilderness and the Native American view of the land came to a head when tribal lands were taken and designated as national parks to preserve iconic and "endangered" American landscapes. Social inequities were apparent as elites supported and helped to establish vast national parks and forests from which Native Americans were expelled and prohibited from engaging in traditional practices (Taylor 2009).

While the preservation of scenic and unique wilderness landscapes was a feature of rural America, the urban landscape was manipulated too in order to improve living conditions and re-create wilderness in the city. Hence, landscape architects such as Frederick Law Olmsted and Calvert Vaux articulated the cultural and civic benefits of clearing more than seven hundred acres of Manhattan inhabited by poor Black, German, and Irish families to build the nation's first landscaped park. They were supported by other elites who expressed cultural nationalistic sentiments promoting landscape improvements in urban and rural areas of America. Olmsted and Vaux—who propagated their landscape designs around the country—helped to introduce and spread pastoral and picturesque landscapes in America. To develop these designs, Olmsted drew on influences from European, Asian, and Latin American landscapes gleaned from his travels in these regions (Taylor 2009).

Landscapes are still important today; they are still hotly contested too. Rachel Carson (1962) called into question the mass poisoning and degradation of the American landscape with agricultural pesticides in her groundbreaking book, *Silent Spring*, in 1962. Vandana Shiva accomplished a similar feat in 1988 with her book *Staying Alive*, detailing the destruction of India's landscapes for large-scale agricultural production and women's resistance to it. The emergence of the global environmental justice (EJ) movement in the 1980s has expanded the range of landscapes that activists and policy makers are concerned with. The EJ movement put the spotlight on industrial landscapes. The movement presumes that landscapes are racialized spaces; hence, activists focus on identifying social inequalities that arise in the siting and operation of industrial facilities in poor and nonwhite countries and communities and the environmental degradation that arises from industrial operations. However, environmental justice activists are not concerned solely with derelict landscapes; they are also focused on transforming the landscapes in their communities through the establishment of forests, parks, urban farms, and community gardens. Hence the work of activists such as Wangari Maathai (2006) of Kenya's Green Belt Movement encapsulates such efforts.

45

Natural Disaster
Priscilla Wald

There is a redundancy in the expression "natural disaster," a double disavowal. "Disaster": "bad star," circuitously, through the Italians and the French, from the Greeks and Romans, for whom fate was written in the stars. Nature spoke in the voice of the gods; expertise in reading omens allowed anticipation, maybe even placation, but human influence was secondary. Human agency was never fully absent from these events. They might be punishments for the commission or omission of a deed—for a violation, intentional or otherwise. But the event itself—the overly strong winds or the lack thereof, the flood or fire, the earthquake or drought, a destruction or a withholding of necessities or bounties—bespoke a higher agent and the expression of displeasure.

As the definition of "disaster" moved away from its etymology, "natural" kept the emphasis on causes of the calamitous event that superseded human agency and responsibility: the devastation of a tsunami rather than insufficiently constructed and unwisely located housing, the tragedy of a drought resulting from unaccountable weather fluctuations rather than an accumulation of greenhouse gases. "Natural disaster" signals the nontraversable distance between human beings and "acts of God." It maintains that humankind's most primal struggle is with environmental forces beyond human control. In the (unwittingly) animated universe of the modern world, "natural disaster" continues to manifest Nature's displeasure with humanity—or worse, her indifference.

Having witnessed such self-evidently unnatural disasters as death camps and atomic warfare, the philosopher Max Horkheimer traced "the disease of reason" to "man's urge to dominate nature" (1947, 176). His contemporary, the political philosopher Hannah Arendt, similarly posited the source of human alienation from nature in the human mastery that made "the destruction of all organic life on the earth with man-made instruments . . . conceivable and technically possible" (1948 [1973], 298). But she observed as well the pairing of the adversarial urge to dominate nature with its tacit deification. Having exiled God and king alike in that Age of Reason, humanity turned to nature as the source of the Rights of Man: when "rights spring immediately from the 'nature' of man . . . it makes relatively little difference whether this nature is visualized in terms of the natural law or in terms of a being created in the image of God, whether it concerns 'natural' rights or divine commands" (1948 [1973], 297). Disaster ensued, she observes, when the need for those rights was most urgent: when, between the two world wars, populations deprived of their rights found nature as absent as any divinity, and humanity confronted the mirror of its own creation.

If nature could not forestall human-generated disasters, it could still author disasters of its own. Unlike their human counterparts, natural disasters were often said to be great equalizers. After all, floods and tornados, influenza and cancer could surely not distinguish among bank accounts or racial backgrounds. If a distinction could be made, it was the difference in frequency of natural disasters among populations characterized by their proximity to nature and those that measured their technological and civilizational sophistication precisely by their successful control over, and harnessing of, their environment. The distinction, however, presumed a tacit agreement on the meaning of the term "natural disaster," and words are notoriously unstable.

Widespread attention, particularly in the United States, turned to the idea of a natural disaster and the consequences of the continuing antagonism humankind expressed for its environment when, in 1962 in the *New Yorker* magazine, a biologist penned a fable about a proliferation of dying birds and a world without their song. Rachel Carson's *Silent Spring* dramatized the logical consequences of the compulsion to dominate the environment in her portentous account of a growing planetary toxicity and the eventuality of an uninhabitable planet. Populations worldwide, especially in the Global South, had long known of the increased hazards they endured in the name of "development" benefiting primarily developers in the Global North. But for the *New Yorker* demographic (white, middle-class, American) exposure to the consequences of "civilization," the dangers lurking in their air, water, food—and genes—were news. Seven years later the United Nations Economic and Social Council issued a report that complemented Carson's analysis and fueled the growing movement to change the terms of humankind's relation to the planet. The report proclaimed a "crisis of the human environment" and warned that the continuation of "current trends" could endanger "the future of life on earth" (United Nations Economic and Social Council 1969) Politics was slowly catching up to science fiction, which had long depicted the disastrous consequences of humanity's planetary violence.

The movement was part of a general shift in thinking not only about violence against nature but also about the nature of violence in terms that further challenged the idea of a natural disaster. Writing in 1969, the political scientist Harold Isaacs called attention to a radical shift in the geopolitics of the decolonizing world in which "some 70 new states carved out of the old empires since 1945 [was] made up of nonwhite peoples newly out from under the political, economic and psychological domination of white rulers" (1969, 235). The change had left the former colonizers "stumbling blindly around trying to discern the new images, the new shapes and perspectives these changes [had] brought, to adjust to the painful rearrangement of identities and relationships which the new circumstances compel[led]" (1969, 235). Theorists of decolonization advocated a conceptual reorientation. They offered new insight into the strategies that perpetuated social inequities: demographic analyses of the disproportionate distribution of hardships—including vulnerability to natural disasters—among populations.

The Norwegian sociologist and peace activist Johan Galtung used the phrase "structural violence" to describe these distributional inequities. Structural violence "shows up as unequal power and consequently as unequal life chances. . . . [I]f people are starving when this is objectively avoidable, then violence is committed, regardless of whether there is a clear subject-action-object relation, as during a siege yesterday, or no such clear relation, as in the way world economic relations are organized today" (1969, 171). Unequal life chances appear in boldface in the demographics of deaths and injuries caused by "natural disasters." Conditions of poverty significantly increase susceptibility to their ravages by influencing factors such as the ease with which people can leave a hazardous area temporarily or permanently; their access to information and communication; their access to education, shelter, nutritious food, and health care; their political clout; and the resources that facilitate recovery following a calamitous event.

Galtung's contemporary, the Black Power activist Stokely Carmichael, underscored the racialization of structural inequities when he distinguished between individual and institutional racism. "When unidentified white terrorists bomb a black church and kill five black children," he explained,

that is an act of individual racism, widely deplored by most segments of the world. But when in that same city, Birmingham, Alabama, not five but 500 black babies die each year because of lack of proper food, shelter and medical facilities; and thousands more are destroyed and maimed physically, emotionally and intellectually because of conditions of poverty and discrimination in the black community, that is a function of institutionalized racism. (1968, 151–52)

Marking the perpetuation of structural inequities as a form of violence, Galtung and Carmichael underscore the disavowed human agency and responsibility that enable that perpetuation. The differential impact of ostensibly natural disasters across populations exemplifies structural violence and displays what the term itself masks: the human influences on the environment that contribute to tornados, flash floods, or earthquakes, to pandemics and rising cancer rates, and the socioeconomic conditions that turn such hazards disastrous, measured by the incalculable loss of lives and livelihoods that follow individual events or the slower erosion measured in mortality statistics over time. The adjective "natural" obscures the human contributions to environmental hazards and the conditions that transform them into disasters.

The anthropologist Anthony Oliver-Smith roots "the nature of disasters . . . in the co-evolutionary relationship of human societies and natural systems." Calling natural disasters the "sentinel events of processes that are intensifying on a planetary scale," he observes, "the interpretation of the messages brought by these sentinels remains a crucial issue" (2004, 24). Nature speaks in Oliver-Smith's formulation, but not in the language of God. Rather, the disaster speaks uncannily, the return of the repressed human agency that has turned the hazard into a disaster.

NATURAL DISASTER PRISCILLA WALD

The term "natural disaster" has nonetheless proved resilient in the specialist literature, mainstream media, and popular culture, which suggests a resistance to analyses of "how," as Dennis S. Mileti puts it, "human systems create and redistribute hazards" (1999, 105). The recognition that "demographic differences play a large role in determining the risks people encounter," that race and class predict susceptibility and, in turn, "disasters exacerbate poverty" (1999, 6–7), comes with an ethical responsibility to make profound structural changes that is obviated by a "natural disaster." But the term also has staying power because it names a longstanding relationship between human beings and the environment. The long history of displaying human achievement through the domination of nature makes it difficult to recast the terms of that naturalized relationship.

The expression "natural disaster" functions alchemically, converting a hazard into a naturalized disaster, the social and environmental intricately intertwined. The metamorphosis illustrates the power of language to obscure human agency and structural inequities and to place a hazard into a context in which the conditions of its possibility and its impact are (temporally) collapsed, folded into a single event rather than a process. The alchemical process draws attention to the rhetorical and narrative strategies through which "nature" operates to turn social inequities that can—and ought to—be addressed into immutable conditions of existence. It is in that sense the cornerstone of environmental and social justice. The crucible of the expression manifests the bonds by which the ideas of social and environmental justice are inextricably bound.

46

Nature

Noel Castree

There are close connections between scientific claims about the contemporary world and wider shifts in the terms of societal discourse. As history demonstrates time and again, scientists change our *actualité* not only through their technological inventions but also through the vocabularies and methods they employ to persuade those outside science to pay attention. Consequently, when *Time* magazine recently informed its many readers that "[n]ature is over"—one of "[t]en ideas that are changing your life"—it came as no surprise to discover science as its inspiration. In his article, *Time* journalist Bryan Walsh pointed to the idea—first advanced by Nobel Prize–winning atmospheric chemist Paul Crutzen and freshwater biologist Eugene Stoermer (2000)—that we now live in "the Anthropocene." But he could just as easily have referenced the research of laboratory scientists, such as the world-famous geneticist Craig Venter (2013). Where Crutzen and Stoermer suggested that the supposed ontological divide between humans and nonhumans has been unintentionally breached on a planetary scale, the likes of Venter have for many years sought to dissolve the divide in more controlled, localized circumstances. In different ways, these spokespeople for the material world tell us that nature, in its various forms, has lost its former naturalness—either by accident or by design. Whether "the Anthropocene" concept enters public discourse worldwide remains to be seen, but the current familiarity of "genetic modification" and "synthetic biology" reminds us of how esoteric scientific neologisms can, given time, become keywords in the everyday arenas of politics, commerce, and civil society.

If Walsh is right, then the things that certain biophysical scientists are saying (and doing) are pitching one of the Western world's most foundational concepts into a state of crisis. For better or worse, "nature" appears to have more of a past than a future (hence the scare quotes). We have been here before, of course. In years gone by, some environmentalists proclaimed "the end of nature" (McKibben 1989) because of anthropogenic climate change, while others—more focused on "human nature"—worried that invasive biotechnologies (like gene splicing) might lead to a "post-human future" (Fukuyama 2002). Meanwhile, less pessimistic commentators insisted that nature's disappearance was relative, not absolute. For instance, journalist Michael Pollan's 2001 book *The Botany of Desire* detailed humans' co-evolution with four important plant species. In the period since these and other attempts to shape public thinking at the *fin-de-millennium*, it appears that few serious efforts have been made to slow down the evident worldwide denaturalization of nature. Equally, though, there's no compelling evidence that "nature" has lost its semantic importance as a key signifier in both expert and lay discourse. It still performs very important work in various cognitive, moral, and aesthetic registers; a great deal is still said and done in its name. To announce that "[n]ature is over" is hardly to secure the case. Instead, it is to suggest that a changing "reality" demands alterations in those key terms designed to hold a conceptual mirror up to it. Profound questions consequently arise. Should humanity act rapidly to restore Earth surface systems to their Holocene state before we cross "planetary boundaries" and enter biophysical *terra incognita*? Like Venter, should we—instead—celebrate the liberatory potentials of a postnatural world in which

we can, through ingenuity, edit out those formerly uncontrollable and negative aspects of nature (like "genetic diseases")? Or should we internalize Pollan's argument that even a thoroughly "anthropogenic nature" retains plenty of wildness and agency, such that the ecological anxieties expressed by latter-day Fukuyamas and a die-hard McKibben (recent author of *Eaarth* [2010]) are overstated and geneticists' dreams of mastery hubristic? Is the death of nature real or exaggerated, a matter of degree or kind? Which scientists, if any, might we turn to for robust evidential answers to this question?

In addressing these and other critical concerns, most people outside universities are not—to hazard a large generalization—accustomed to hearing the voices of social scientists or humanities scholars. Yet at least thirty years before Crutzen, Stoermer, Venter, and other scientists were suggesting nature's incipient end, a few such voices (later growing to a chorus) were insistent that nature is not quite as natural as it seems. But their arguments were rather different from those advanced by McKibben, Fukuyama, Pollan, and other public commentators fifteen to twenty-five years later. In the mid-1970s a number of anthropologists, human geographers, science studies scholars, media analysts, sociologists, and others critically examined claims being made about supposed "facts" of nature by members of the scientific community. For instance, there were high-profile neo-Malthusian assertions of imminent "limits to growth" posed by finite natural resources in the context of global human population growth (Meadows et al. 1972). The concern was that scientists unwittingly smuggled contestable value judgments about how the world (supposedly) is, and how it ought to be, in the guise of "objective" and "rational" representations of human biology or the nonhuman world (see, for example, Marshal Sahlins's critique *The Use and Abuse of Biology* [1976]). Given the authority science enjoyed (and still, despite some credibility crises, enjoys), these scholars' interventions were designed to prevent "science imperialism"—that is, a situation where scientific metaphors, findings, and conclusions illicitly structure the framing of societal "problems" by nonscientists and the consequent identification of ethically acceptable "solutions."

Underpinning all this were two convictions. The first was that it is a matter of convention—not necessity—that aspects of the material world come to be labeled "natural." The second was that claims made about "nature" said as much about those doing the representing as about the material world supposedly being made to speak "in its own voice." In the case of science, the suggestion was that it both internalized and presented in its own idioms the cultural norms and desires of the wider society. Its various technical claims about "nature" were thus said to be politics by other means, since all politics involves debatable decisions about the means and ends of any collective endeavor. In this light, the various "natural sciences" were presented as being *social* "all the way down," despite appearances to the contrary.

Four decades ago no one in the social sciences and humanities could have anticipated just how rich and varied subsequent research into nature's "social constitution" would become. Ranging far beyond scientific representations and associated practices, this research has inquired into who speaks for nature, how, and with what effects. Among others, the discourses, images, and artefacts of Safari park owners, genetic counselors, biotechnology companies, nature poets, environmental journalists, and wildlife documentary makers have been subject to critical scrutiny. In the process an extraordinary range of theories and methods has been employed across a myriad of academic disciplines. There have been key debates between interlocutors, especially about whether and how far nature can be said to be a "construction." Even if we should not unreflexively call it "nature" anymore, some critics have worried that the materiality of the physical world has been underplayed in the determination of some analysts to illuminate nature's unnaturalness. As a result, there has been an "ontological turn" of sorts, but one that eschews the resurrection of an old-fashioned "naturalism" or "realism." Certain neologisms have been invented to help us navigate this need to question "nature" epistemologically while respecting the agency or physical intransigence of those phenomena the word names. Technoscience theorist Donna Haraway's "natureculture" is one of several examples, a concept she underpins with arguments and examples that question the idea of *sui generis* "social" and "natural" realms.

Consequently, *Time*'s announcement that "[n]ature is over" no doubt appeared passé to the likes of Haraway and varied fellow travelers such as Bruno Latour, the late Neil Smith, Andrew Ross, or the literary critic Timothy Morton—but for reasons different from those detailed by Bryan Walsh. For these and like-minded critics, what we call "nature" *never existed in the first place.* Instead, what existed was a socially efficacious belief that "natural" things were separable from "social" relations, institutions, and practices. To inquire into what is said and done on the basis of this belief, as well as what is concealed from view and not done, is to shed light on how Western societies are governed—and how they now seek to govern non-Western countries in the name of everything from "planetary management" to brave-new-world promises of healthier, smarter human bodies and minds. Needless to say, not all those aware of these arguments have welcomed them. For instance, in the late 1990s the so-called science wars broke out in the United States between cultural analysts of science and practicing scientists bothered by the suggestion that their representations were anything but metaphorical mirrors of nature's "real" characteristics (see Ashman and Baringer 2001).

Despite their sustained and insightful attempts to "denaturalize nature," the arguments of social scientists and humanities scholars have not registered strongly in the minds of most nonacademic commentators—Walsh being a case in point. There have been momentary exceptions (e.g., the North American debate about "wilderness" inspired by historian Bill Cronon's [1995] essay "The Trouble with Wilderness"); there have been some lasting achievements too (e.g., indirectly fostering activism geared towards advancing the rights of gay, lesbian, and transgender people by rebutting the suggestion that they are somehow "unnatural": see Sandilands, this volume). But, overall, and to offer some examples, the arguments contained in Haraway's *When Species Meet* (2007), Latour's *We Have Never Been Modern* (1993), Smith's *Uneven Development* (1984), Ross's *Chicago Gangster Theory of Life* (1995), or Morton's *Ecology without Nature* (2007) remain alien to most political, commercial, and third-sector actors, let alone ordinary citizens. As a point of contrast, consider the significant impact the early animal rights philosophers had on societal discourse from the early 1970s. For various reasons, an academic like Peter Singer was able to inspire public debates that, in time, had legislative and practical significance for people and nonhumans alike. Meanwhile, those who today emphasize the power of nature to constrain our lives can easily attract wider attention by rehearsing venerable arguments about imminent societal collapse (think of James Lovelock and Jared Diamond).

All this is unfortunate. The capacity to define societal understandings of what "nature" is (and is not), whether nature is over, and what this implies for our sentiments and policies is an extraordinary one. This is the case because nature has long been, and remains, a very special term in the Western lexicon. As Raymond Williams (1976) famously argued in his germinal book *Keywords*, "nature" is an unusually polysemic and polyreferential signifier. Not only does it have several distinct meanings, but these meanings are, by decision and convention, attached to a bewildering array of phenomena (and in ways that are often contradictory). As part of this, Williams emphasized that the meanings of "nature" were often present in discussions where the word itself was not necessarily used. Among nature's "collateral concepts," as we might call them, are "race," "sex," "biodiversity," "genes," "wilderness," "animals," "environment"—among a great many others (see Castree 2013). These concepts are not always synonyms for "nature," but, in certain circumstances, they are vehicles for communicating some of its established meanings. Used together they comprise "formations of meaning" (Williams 1976, 15) that, at any one moment, have a seeming solidity, though they provide some room for semantic maneuver too. This much is obvious from reading several other essays in this book, such as those written by Stacy Alaimo and Vermonja Alston.

What is obvious too is that "nature" and its collateral terms bridge the divide between expert discourse and everyday language. Unlike a neologism such as "posthumanism" (say), they are "ordinary words" used routinely by lay actors as they make sense of themselves and the world. But various professionals (and not just scientists) get to invest them with significance or, as "the Anthropocene" idea suggests, to challenge their "common sense" status. When this happens, ordinary people may well notice. Consequently, to propose that nature is no more is to pull at one hugely significant thread in the tapestry of Western discourse. In effect it is to challenge some or all of the family of collateral terms that Westerners have invented to order both understanding and behavior. There is a crucial difference between this being done in an open and inclusive way and, even if by accident, undemocratically. When Williams famously noted just how complex the semantics of nature have been historically, he was seeking to emphasize the wide range of human projects that had used the term as a means of legitimation and persuasion. It is not the word's complexity *per se* that is interesting, but what it signifies about a society's politics, both "high" and "low." In philosopher Walter Gallie's (1956) terms, if nature and its collateral terms are not, at any given moment in time, "essentially contested concepts," then something is awry. One way of revealing the "unnaturalness" of these keywords is to consider those cultures in which terms like "nature" are not part of the vocabulary.

The writings of all the various authors mentioned earlier are each, in their own way, attempts to preserve, modify, or supersede "nature" (and by implication the other keywords that share some of its principal meanings). These authors aim to derive normative lessons of wide societal application from claims that nature is about to disappear or that it never existed in the first place. Wittingly or not, they are engaged in a discursive contest over the basic terms that ought to guide thinking and action in a world chock-full of pain and possibility, success and failure. But what I am suggesting is that, in the main, social scientists and humanities scholars have been much better at talking to each other than at talking to others (for example, scientists and science popularizers) who are wont to pronounce on nature, "race," biology, evolution, sustainability, and all the rest. If these scholars believed in the importance of their ideas, more might have been done by them to alter the wider discursive climate they have analyzed symptomatically.

This may be about to change. If not, it arguably needs to—and this is where I would like to end this all-too-short essay. Science, in its various forms, is an indispensable component of modern life. Right now, and not for the first time, it is productive of conceptual ferment in the world outside science. But, in shaping societal thinking about the life and death of "nature," it is but one resource among many. In a recent paper about the social uses of the idea of anthropogenic climate change, Marxist geographer Erik Swyngedouw (2010) detects contemporary science imperialism at work in Western environmental policy. He accuses leading politicians of "scientizing politics." For him, they use the scientific "threshold" of two degrees Celsius additional atmospheric warming to forge a wide but thin social consensus based on a rhetoric of urgency ("More nuclear power, more wind power, more 'green investment' *now*!")—yet all the while doing far too little to reform neoliberal capitalism. Many would agree with his analysis. However, as Latour (2004a) has argued, social scientists and humanities scholars need to supplement critique of this sort not only with *invention* (new ideas, new arguments, new evidence) but also with *intervention*.

For forty years, people like Latour (and Swyngedouw) have been highly inventive, but there is another step to take. The question of nature is not simply an issue of pointing to "the evidence" and then deciding what overall balance of conservation, restoration, preservation, and exploitation to aim for. Phrased more explicitly, whether nature is vanishing and what this signifies is not a question that science—or science alone—can tell us much about. This is a thought many nonacademic actors, and not a few scientists, find challenging. People like me arguably need to prove our societal value by discussing nature beyond the comfortable arenas of peer-reviewed journals or university seminar rooms. We may then show people (including future *Time* journalists!) that "nature" is, and should remain, a contested idea because it allows us to disagree about *both* the "questions" and the "remedies" we might otherwise assume to be obvious. Social scientists and humanists are not simply those who speak only to the soft "cultural" questions (like the propriety of gene therapy), with nature an ontological playground where the men and women in white coats discover "hard truths." An excellent example is Andrew Ross, whose book on the unsustainable character of American urban life, *Bird on Fire* (2011), inspired him to engage with people in Phoenix, Arizona—the immediate subject of his monograph.

In this endeavor we (for I am one of them) will not be alone. Many artists, performers, novelists, and filmmakers are today deeply preoccupied with "the question of nature" in ways that are highly inventive. But so too are many scientists: despite my unfortunate tendency to gloss the differences here, there remain significant internal debates within science about the existence, character, and value of "natural kinds." If more of this variety can inform public discussions about the future of nature—"human" and nonhuman—then we stand a chance of avoiding the sort of normative "reading off" that once informed eugenicist policies and other measures that anchor themselves in supposedly "natural" imperatives, deficiencies, limits, or tendencies.

47
Nature Writing
Karla Armbruster

Nature has been "the subject of imaginative literature in every country and in every age" (Finch and Elder 1990, 19), inspiring sustained attention as far and wide as the *Epic of Gilgamesh*, the Hebrew Bible, the *Tao Te Ching*, Aristotle's *Physics*, Virgil's *Georgics*, and the *Bhagavad Gita* (Torrance 1999). However, the term "nature writing" is most often associated with a much more specific genre of nonfiction prose in English that many scholars trace back to British curate Gilbert White's 1789 *Natural History of Selbourne*. White broke away from the conventional practice of natural history, which focused on examining stuffed or dried specimens, by affectionately describing living animals and plants in their natural settings around his country home. Following in White's footsteps by blending lyricism or other literary qualities with scientific facts or observations of nature, nature writing typically also incorporates personal reflection or philosophical interpretation (Finch and Elder 1990; Lyon 1989; Murphy 2000, 2008; Slovic 2003).

Nature writers from White to U.S. marine biologist Rachel Carson (author of *The Sea around Us*, 1951, and *Silent Spring*, 1962) have often achieved wide popularity, but scholarly interest arose only in the late twentieth century with the development of ecocriticism and the concurrent success of U.S. nature writers like Annie Dillard, Barry Lopez, and Terry Tempest Williams. Perhaps because of the convergence of these trends, the canon of nature writing as anthologized and taught has been heavily weighted towards the U.S. tradition: often constructed as beginning with Thoreau's *Walden* and extending through works like Leopold's *Sand County Almanac* and Edward Abbey's *Desert Solitaire*, this tradition reflects the perceived presence, and then gradual loss, of frontier/wilderness territory, with unspoiled nature celebrated as part of the nation's heritage and valued as a place for (often white male) individuals to escape the limitations of society and achieve spiritual transcendence.

For critics, the genre's equation of nature with wild or pastoral landscapes quickly began to seem limited, ignoring the urban and suburban environments where most people live and work. Nature writing has also been critiqued as disproportionately featuring the experiences and epistemologies of the privileged, leaving out the voices of marginalized people who often suffer environmental injustice. Even some writers and editors closely associated with the genre have rejected it as overly reverential, cautious, and increasingly irrelevant to our lives or environmental concerns (Gessner 2004; "From the Editors" 2010).

Running through these critiques is the overarching concern that traditionally defined nature writing reinforces a vision of culture and nature as separate or even opposed, a vision that many environmental thinkers find at the root of environmental crises. While the broad outlines of the genre justify this concern, they also obscure some real diversity. For example, much Native American literature—often claimed as nature writing—communicates a sense of interconnection and interdependence between humans and nonhumans. Without a cultural history of frontier mythology, British nature writers have been far less likely than their U.S. counterparts to fall for the romance of the individual

in the wilderness, instead incorporating "concerns for local human community and character" (Fritzell 1989, 29), and today critics note the emergence of a "new" British nature writing that avoids the trap of nostalgic, escapist fantasies about rural life that once diminished the genre's reputation (Macfarlane 2003; Stenning and Gifford 2013). In Canada, where figures like Margaret Atwood have famously discerned the tendency for nature to be perceived as threatening (Raglon and Scholtmeijer 1998; O'Brien 1998), there is also a rich history of writing representing positive influence between humans and their natural environments, such as the works of western writers Sharon Butala, Sid Marty, and Wallace Stegner (Banting 1998). Undoubtedly, the complex story of nature writing in each cultural tradition remains ongoing and largely untold (e.g., Tredinnick 2003; Banting 2008).

While continuing to pursue and tell these stories remains an important task for ecocritics, more and more have moved "beyond nature writing," exploring how nature is represented and valued in texts ranging from poetry and fiction to digital media (Armbruster and Wallace 2001). Especially promising is work that focuses on cultures outside those that produced the nature writing tradition—cultures that potentially offer not only diverse literary traditions but also different concepts of "nature" and perspectives on human relationships to the environment. For example, Stephanie Posthumus argues that conservation in France emerged from concern about cultural landscapes rather than concern about natural environments (as it did in the United States), yielding a tradition she terms "landscape writing." Studying the "literature of nature" (Murphy 1998) across cultures can also yield a better understanding of how such literature relates to human attitudes and behaviors: a case in point is

Karen Thornber's recent study of East Asian literature, *Ecoambiguity*, in which she dramatically complicates conventional images of Asian ecological harmony, citing literary references pointing to ecological abuse that go back for centuries, as she also teases out the multifaceted, sometimes contradictory relationships between humans and nature that must be understood in order to respond to the daunting environmental challenges that we face today.

48

Pastoral

Sarah Phillips Casteel

Pastoral is often considered to be a dead mode. *The Princeton Encyclopedia of Poetry and Poetics* informs us that Wordsworth's "Michael" (1800) "well marks the end of serious attempts in the genre" ("Pastoral" 1993, 887), while *The Penguin Book of English Pastoral Verse* locates the end of pastoral slightly later, with the early-nineteenth-century "peasant poet" John Clare. Indeed, John Barrell and John Bull, the editors of the Penguin anthology, insist that a contemporary revival of pastoral "is not . . . something that can be looked for with anything other than alarm. For today, more than ever before, the pastoral vision simply will not do" (1975, 432). A survey of recent scholarship, however, suggests that "pastoral"—a term that traditionally designates poetry and drama that offer a nostalgic and idealized portrayal of the life of shepherds and their rural surroundings—is defined not so much by its obsolescence as by its persistence. Such studies track the continuing reinvention of pastoral in a variety of British, American regional, and postcolonial contexts (James and Tew 2009; Rieger 2009; Cella 2010; Barillas 2006; Potts 2011; Huggan and Tiffin 2010). In a second critical trend, other scholars advance "green readings" of canonical examples of the tradition to show, for instance, how Renaissance pastoral aligns with modern ecocriticism (Borlik 2011; Hiltner 2011).

What accounts for these diverging views of the fate of the pastoral mode? In part the divide is one between those who seek to stabilize definitions of "pastoral" and those who emphasize its versatility. Thus Paul Alpers rejects generalized applications of the term to idealized representations of the Golden Age or of nature and rural life, instead restricting pastoral to the "representative anecdote" of "herdsmen and their lives" (1996, 22). Annabel Patterson, by contrast, argues that defining the term is "a cause lost as early as the sixteenth century." In her view, "It is not what pastoral *is* that should matter to us" but rather "how writers, artists, and intellectuals of all persuasions have *used* pastoral for a range of functions and intentions that the *Eclogues* first articulated" (1987, 7).

A second, related debate surrounds the ideological status of pastoral. Implicit in Barrell and Bull's statement that "the pastoral vision simply will not do" is an understanding of pastoral as ideologically tainted. This view was popularized by Raymond Williams's *Country and the City* (1973), which attacked the pastoral for mystifying class relations and masking the exploitation of rural people. Subsequently, many of Williams's readers have found it difficult to think of pastoral except in pejorative terms. Yet what is sometimes forgotten is that Williams's critique had a specific rather than a general target: it was directed primarily against the seventeenth-century country house pastorals of poets such as Ben Jonson and Thomas Carew that naturalized the prevailing social order and excised laborers from the rural landscape. By contrast, in Virgil's *Eclogues* an awareness of the actual conditions of rural life "breaks through into the poetically distant Arcadia" (Williams 1973, 16). In Williams's account, then, pastoral is not a static mode but instead undergoes a series of historical transformations; yet his denunciation of the seventeenth- and eighteenth-century country house pastoral is so persuasive that it is often taken as though it were a pronouncement on pastoral's essential character. Accordingly, one British critic writes that pastoral is "not just a trap to be avoided, but a challenge to be transcended" by contemporary poets, who must break out of the "closed circuit of pastoral and . . . the anti-pastoral" to develop a "post-pastoral" poetry (Gifford 2002, 55, 57).

American critics have tended to be less suspicious of pastoral, favoring a dialectical rather than univocal model that allows for a greater range of both formal and ideological positions. Preeminently, Leo Marx in *The Machine in the Garden* distinguishes between "complex" and "sentimental" pastoral. In Marx's account, while sentimental pastorals promote a "simple, affirmative attitude . . . toward pleasing rural scenery," complex pastorals contain within them a "counterforce" that juxtaposes the ideal with a more realist vision, thereby "embrac[ing] some token of a larger, more complicated order of experience" (1964, 25). As with William Empson's famous definition of "pastoral" as "putting the complex into the simple" (1935 [1974], 22), Marx's account points to a reflective, critical representational mode that nonetheless may be understood as operating within a pastoral framework. Following Marx, Lawrence Buell notes pastoral's ideological suppleness, its capacity to be simultaneously conservative and oppositional. Buell foregrounds pastoral's "multiple frames" and "ideological multivalence" to illustrate the difficulty of reducing pastoral to a single ideological or class position (1995, 36).

The historical rigidities relating to class and land in Britain may help to explain the contrast between British and American responses to pastoral. Notably, American pastoral's "double-edged character" is also shared by other "post-European" pastorals (Buell 1995, 53, 51). Indeed, some of the most striking instances of contemporary pastoral have been produced by (post)colonial writers from the Caribbean, Australia, Ireland, and elsewhere (Casteel 2007; Huggan and Tiffin 2010; Williamson 2012). Some postcolonial critics have echoed

Williams's misgivings about the pastoral, charging that the pastoral sensibility upholds rather than challenges the dominant order (Coetzee 1988, 5–6) and is anathema to a postcolonial poetics (Glissant 1981 [1999], 145). Others, however, have argued that pastoral takes on a critical as well as a conservative function in a postcolonial setting. Rob Nixon, for instance, suggests that V. S. Naipaul's pastoralism in *The Enigma of Arrival* (1987) at once unsettles and extends the tradition, "disturb[ing] a certain notion of Englishness" (1992, 162) while simultaneously enabling Naipaul to ignore the racial tensions that beset England's urban centers. Sidney Burris (1990) maintains that pastoral has from the start lent itself to social commentary and shows how Seamus Heaney's pastoralism works to subvert an idealized vision.

If writers such as Naipaul and Heaney elaborate what Nixon terms a "postcolonial pastoral" (1992, 161), what is the "colonial pastoral" that provokes this alternative vision of landscape? "Colonial pastoral" may refer to the Arcadian or Edenic vision that European explorers imposed on colonial landscapes, as we see in Columbus's diaries and later in works of French romanticism such as Bernardin de Saint-Pierre's *Paul et Virginie* (1788) and Chateaubriand's *Atala* (1801). This tradition of locating Arcadia in the New World is refashioned by Venezuelan writer Andrés Bello, whose *América* (1823) heralds the arrival of an American Virgil at the dawn of Latin American independence in order to articulate a continental American identity. "Colonial pastoral" may also designate the settler colonial practice of pastoralizing the new surroundings in order to domesticate them. To apply a pastoral vocabulary to an unfamiliar landscape, as Oliver Goldsmith does to Nova Scotia in *The Rising Village* (1825), is to imagine that landscape as habitable and to encourage others to do the same. Finally, colonial pastoral may indicate the dissemination in the colonies of images of rural European landscapes.

The profoundly alienating impact of the transmission of such images is dramatized in a scene in Jamaica Kincaid's *Autobiography of My Mother* in which a young Dominican girl smashes a china plate that depicts an English rural vista (1996, 8–9).

In part, then, the prominence of pastoral imagery in postcolonial literature signals an antipastoral critique of an aesthetic tradition that helped to advance the colonial project, assimilating foreign lands to a European gaze and privileging metropolitan over colonial landscapes. Yet in its complex, critical form, pastoral also holds considerable appeal for postcolonial as well as diaspora writers. The larger relevance of the pastoral tradition to postcolonial and diaspora writing is that it speaks to the contemporary preoccupation with displacement. Indeed, if we return to the classical roots of pastoral, we discover that a tension between place and exile has informed the pastoral mode since its inception. Virgil's first Eclogue opens with the theme of exile:

> Tityrus, there you lie in the beech-tree shade,
> Brooding over your music for the Muse,
> While we must leave our native place, our homes,
> The fields we love, and go elsewhere; meanwhile,
> You teach the woods to echo "Amaryllis." (ll. 1–5)

Throughout "Eclogue I," Meliboeus repeatedly attempts to make Tityrus acknowledge his plight, and we become acutely aware of the fragility of the Arcadian world, of the threat of exile and dispossession that lies just outside its borders. This tension between an idealizing and a historicizing vision of nature is also fundamental to postcolonial pastorals such as Derek Walcott's *Tiepolo's Hound* (2000). Such works advance "a postcolonial sense of place" (Handley 2000, 6), exhibiting an understanding of natural and human history as interlocking that challenges the sentimental pastoral reading of colonial nature as outside of history. Diasporic pastorals such as Philip Roth's *American Pastoral* (1997) similarly exploit tensions between the ideal and the real, blending inherited American pastoral myths with Jewish historical memory in a manner that ultimately puts pressure on the myths themselves.

Crucially, while pastoral supports an exploration of displacement in postcolonial and diaspora writing, it does not evacuate the category of place, as many contemporary theories of space are apt to do (Casteel 2007). When employed in its complex rather than sentimental or nostalgic form, the pastoral mode enables contemporary writers to develop alternative spatial imaginaries that retain place as a meaningful category without reinscribing mystifying and exclusionary myths. At the same time, a contemporary idea of pastoral, alert to the relationship between the human and the natural world, offers a framework for responding to our current environmental crisis. Working at the juncture of history and nature, displacement and emplacement, contemporary pastorals testify to the continuing relevance of the pastoral mode. Pastoral, it seems, is far from "dead." Yet, given the power of the issues that it explores, neither can pastoral be confined to purely literary or aesthetic meanings. On the contrary, as writers respond to their own specific histories, pastoral will continue to play its part in shaping contemporary views of land and nature.

49

Place
Wendy Harcourt

Our understanding of the environment is first and foremost informed by our experience of place—the geographic location where we live, work, and interact with nature and people. Our identity, culture, history, and politics are bound up in a sense of place. Though intuitively place may seem inherently conservative, a reading of place as a site of progressive politics allows us to understand more concretely how environment is linked to culture through relations of power, agency, and responsibility to human and nonhuman environments (Harcourt and Escobar 2005).

My reading of place follows Doreen Massey's definition of place as inevitably inflected by the global (Massey 2002, 2004). Understanding place as "a meeting-place" enables us to theorize place as within networks of relations and forms of power that stretch beyond specific places. I explore place with reference to my own engagement in place as a feminist political ecologist (Harcourt 2009), looking at three stories of struggles, networking, and connections around place at the local, national, and global level.

Bolsena, the place from where I write, is a tiny town in north Lazio, Italy, a small Etruscan town on the shores of a volcanic lake founded over two millennia ago. The politics of place is currently being shaped by burgeoning tourism, failing agriculture, and the vested interests of the local church, government, and business. Here environment and culture intersect as Bolsena, only recently made up of poor fishing settlements, a castle, and pilgrims, is now subject to modern life marked by overfishing, tourism, speed boats, and the influx of waste from agricultural chemicals being used in the hills surrounding the lake. The national authorities have declared the water unsafe to drink and citizens now pay for filtered water collected at a central dispensary in the town. Local community groups hold town hall meetings to lobby national or European authorities to provide pump filters to clean the lake, set up eco-islands around the town, promote biological farming, and control the use of speed boats. The emphasis is on care for the peace and beauty of the lake alongside the culture— the Etruscan and Roman histories, pottery, local wine, and food. This understanding of place and the responsibility of nature intertwined with culture is in contrast to the business lobby that wants to promote Bolsena for its boating and fishing, tourism, resurrected medieval religious festivities, and popular musical evenings.

Negotiating around who decides what happens in Bolsena is a continual struggle among the town hall, governmental committees, the church, and progressive environmental groups. It is mostly women who have been leading the progressive environmental groups. They are the local women who are running the book and coffee shops, "eco" wine bars, and "bio" cheese farms and holding workshops on local food and Etruscan cultures. Defending the environmental and cultural integrity of the Lake of Bolsena in order to ensure the sustainability of the environs has become a daily activity. The current austerity measures imposed by the Italian state government since 2012 have heightened the struggle around the town's economic and ecological survival. The local protests are usually ignored by the mayor, who puts economic benefits from tourist fishing competitions, lake-side caravan parks, and hotels ahead of the polluting impacts of such activities on the lake.

The struggles between progressive and business interests in Bolsena show how place is not static but is shaped by the changing experiences of the people living in the place as they connect to other flows of information, economic activities, and social forces (be they tourism, feminism, or religion).

Australia, my birthplace, has in the last years seen a proliferation of progressive narratives around place, environment, and identity. These narratives recognize the violence of colonial history; acknowledge the complexity of forty-five thousand years of indigenous cultures, divergent understandings of land, time, and spirituality; and attempts to deal with the overexploitation of waters, minerals, and soils. Such narratives counter the vision of Australia as a place of great empty spaces and blue skies, a lucky country that offered Europeans wealth and prosperity to be yielded from its mineral wealth and what they considered to be "unpeopled" land.

Since the 1990s, the racial political landscape in Australia has been transformed through various legal and federal decisions related to land rights, equity, and justice: the 1992 Native Titles Act gave back land to traditional owners along with the renunciation of the term "*terra nullius*"; the 1994 national "Going Home Conference" recognized the aboriginal children, "the stolen generation," who had been forcibly removed from their families; and the 1996 apology campaign led to the annual "Sorry Day" (held every May 26 since 1996 in memory of the stolen generation). The most famous symbol of central Australia, Ayers Rock, was renamed "Uluru" and officially returned in 1993. In 2000, indigenous Australian and gold medal Olympic athlete Cathy Freeman, with her vivid smile and arm tattoo saying "cos I'm free," became the heroine of all Australians in the 2000 Olympic Games, celebrating her four-hundred-meter victory by waving an aboriginal land rights flag along with an Australian one (Harcourt 2001, 197–98).

This reclaiming of identities and reconciliation of white and black Australia illustrates how place can be renegotiated. Taking responsibility for creating new understandings of place, is not easy. A sense of belonging in Australia is wrought with an inherent racism of the poverty, ill health, drink, drugs, and violence that indigenous peoples experience. While other Australians can also suffer poverty, the high levels of child and maternal mortality among indigenous Australians in particular belie the First World status of Australia.

It is challenging for Australians to take responsibility and make connections across the racial and cultural differences. Anglo-Celtic Australian Peter Read (2000), in writing about place in Australia, confronts the racism of rural Anglo-Celtic Australians. He speaks of how the contestation between different understandings of place among indigenous and other Australians—one rich with spiritual meaning and a profound sense of collective living together with the land, contrasting with the predominantly capitalist drive to create individual wealth through the ownership and exploitation of land and its resources—needs to be articulated in order for a new sense of Australia to emerge. Such a new understanding would acknowledge and combine different cultural understandings of place, acknowledge the profound power imbalances, and provide a basis for learning to care for the land materially and spiritually.

Places, such as Rio de Janeiro, Brazil, also exist at global levels. In UN and international-development-speak, "Rio" has become a symbolic place for modern "eco-governmentality" or the creation of the environment as a subject of global public discourse around "sustainable" development. The Earth Summit held in 1992 was not only about reframing public policy to be based on stewardship and resource management, but also about creating the first global meeting place of thousands of individuals campaigning on forestry, water, environmental justice, peace, and human rights. The NGO Global Forum in Flamengo Park, Rio, hosted hundreds of tents where civil society passionately engaged with Agenda 21 (the official outcome document). During the Global Forum, ecologists, the peace movement, trade unions, women's and youth organizations, development NGOs, and local community groups intermingled, creating a collective vision for global environmental justice (Dankelman 2012).

This creation of "alternative" places where advocates, NGOs, and movement people could meet alongside the official negotiations has continued through the last twenty years in peoples' struggle to defend the environment and the culture of local places in public global fora. Rio and its prototypes are highly negotiated places that have helped build connections, understanding, and knowledge among communities defending place, moving beyond the local and national space, shaping a global environmental movement. These places of progressive global negotiations (for example, the UN NGOs Forums held in the early 1990s, the protests held in 1999 in Seattle against the World Trade Organization) have emerged as historic moments flowing onto other places. These created places for movement engagement are illustrations of Massey's "meeting-place" forging networks of relations and forms of power that stretch beyond specific places. I would argue that for all the cynicism around engaging in such global shows, these places have been key nodes for networking that strengthens local power in ways that extend and transform place-based politics and connections. These places are holding a sense of global responsibilities for exploitation of environments and culture. In places like Bolsena, the global forums strengthen the ecological organizations, connecting what is happening in the microcosm world of the lake with the larger movement. Utilizing social media, the forum debates create new forms of learning that politicize beyond the mainstream policy hype. These global places have played an important role in our global imaginary of the environment that can be politicized as something to be defended, creating a culture of translocal defense of place in vibrant global networks of connectivity (Escobar 2008).

These three examples illustrate that places, whether experienced at local, national, or global levels, are not natural, nor fixed. Our experience of place flows across spatial scales from the body to the household to the community, national, and global levels. Place extends beyond the physical. People negotiate place as they protect and conserve places, enhance and modify places, create connections with other places at different levels. Our attachments to place are about social, spiritual, and cultural meaning and identity as well as economic need. Political negotiations over the environmental and cultural issues informing place are part of our lived "glocal" realities (Dirlik 1998).

50

Political Ecology

Mario Blaser and Arturo Escobar

Political ecology (PE) deals with the interrelations among nature, culture, and power, and politics, broadly speaking. It emerged as a field of study in the 1970s out of the interweaving of several ecologically oriented frameworks and political economy. By bringing these frameworks together, PE aimed to work through their respective weaknesses, namely, human and cultural ecology's lack of attention to power and political economy's undeveloped conceptualization of nature. Too mired still in structural and dualist ways of thinking, this "first-generation political ecology" (Biersack 2006) gave way over the past two decades to what could be termed a "second-generation" political ecology, variously informed by those theoretical trends marked as "post-" (poststructuralism, post-Marxism, postcolonialism). This second-generation PE has been a vibrant space of inquiry drawing on many disciplines and bodies of theory. What distinguishes it from its predecessor is its engagement with the epistemological debates fostered by constructivism and anti-essentialism. Although very provisionally, given the newness of the trends in question, it could be said that a third-generation political ecology has been in ascension over the past five to ten years. With roots in the second generation, this emerging PE also draws from the most recent academic debates on postrepresentational epistemologies and flat and relational ontologies in geography, anthropology, cultural studies, and science and technology studies (STS). Outside the academy, this tendency is influenced by the increasing visibility acquired by "environmental conflicts" that cannot be easily accommodated in any realist account of struggles over "nature." We believe that this outgrowth of PE2 marks a noticeable shift from political ecology toward what we tentatively call "political ontology." In what follows we provide a (necessarily) partial overview of the trajectory that PE has followed as it grappled with the power-laden relation between culture and the environment; we stress the limits it has encountered and how political ontology may help to push PE's intent further.

First-generation PE grappled with a disjointed relation between culture and the environment in social theory. On the one hand, there was the human and cultural ecology approach that tended to reduce culture to a homeostatic adaptation to the environment (Steward 1955; Rappaport 1968). In other words, culture was an extension of nature insofar as the driving force that shaped the former was the latter. On the other hand, political economy tended to reduce the environment to a background of little import to social dynamics. This was especially the case of dependency and world-system theory, which understood the expansion of capitalism as a process that overruns local specificities (one could say, adaptations to specific environments), ultimately lumping together cultural differences into one single overarching system (Cardoso 1972; Wallerstein 1974; Amin 1976). First-generation PE tackled these problems in two ways: (1) by resituating the question of adaptation into the larger scale in which world-systems and dependency theory operated; and (2) by stressing that, albeit subordinated to the capitalist economy, local adaptations reworked themselves creatively to also meet their own ends (see Wolf 1982). The picture that obtained from these moves was one in which local cultures adapted not only to the "natural environment" but also to the pressures of an encroaching capitalist system. Interestingly, the syncretic forms that emerged from these multiplex adaptations were not necessarily seen as good adaptations; on the contrary, political ecologists tended to highlight how the demands of the capitalist economy forced "maladaptations" that degraded the environment and restricted local control and access to resources (see Nietschmann 1973; Watts 1983; Blaikie 1985; Blaikie and Brookfield 1987; Hecht and Cockburn 1990; Little and Horowitz 1987; O'Connor 1998). A typical example of this would be peasant populations lured into mono-crops by market demands and taxation pressures. In many cases this generates a vicious circle since mono-crops eventually begin to reduce soil productivity, and thus chemical fertilizers are needed in increasing proportion to sustain a large enough yield to cover increasing costs of production. Eventually, the peasant ends up with soils so severely degraded that the equation between input and output becomes unsustainable and the only alternative left is to open new land in the forest. The cycle, once again started, becomes part of a larger process through which the "agricultural frontier" advances over forests and biodiversity.

Although first-generation PE sought to develop a more complex and balanced understanding of the relation between culture and the environment, it retained a strong commitment to ontological dualism and epistemological realism. In other words, the environment remained a domain distinct from culture, and it was assumed that reality was accessible from the standpoint of the sciences (be they hard or soft). Second-generation PE questioned these assumptions. One can distinguish two versions of this challenge, although sometimes they partially overlap. On the one hand, some argue that while nature is a distinct ontological domain, it has become inextricably hybridized with culture and technology and increasingly produced by our knowledge

(Leff 1986, 1993, 2002). On the other hand, there is a stronger epistemological claim that the environment is *always* already cultural since we can only know it through meaning-making practices that are inherently cultural (see Escobar 2010 for a more detailed discussion of PE2). The parting point between these versions tends to be organized around the essentialist/constructivist divide. Essentialism and constructivism are contrasting positions on the relation between knowledge and reality, thought and the real. Succinctly, essentialism is the belief that things possess an unchanging core, independent of context and interaction with other things, which knowledge can progressively know. The world, in other words, is always predetermined from the real. Constructivism, on the contrary, accepts the ineluctable connectedness between subject and object of knowledge and, consequently, the problematic relation between thought and the real. From a certain constructivist perspective (Foucauldian and Deleuzian poststructuralism in particular) language and discourse do not reflect "reality" but constitute it.

Constructivist political ecologists have done a good job in terms of ascertaining the various representations or meanings given to nature by different cultures, the conflicts that these differences beget, and the consequences or impacts that all this has in terms of what is actually done to nature (e.g., Cronon 1995; Hinchcliffe and Woodward 2000; Macnaghten and Urry 1998). This is very important, yet constructivism tends to bypass the question of the ontologically specific character of biophysical reality, implicitly treating material entities as mere "envelopes of meaning" (Pels, Hetherington, and Vandenberghe 2002). In this way, the character of the "real" is left up for grabs, which then contributes to the continued dominance of epistemological realism. In effect, to the constructivist argument that a rock has different meaning for different cultures, realists would

respond that if you throw the rock at the head of people with different cultures, they will all bleed—which in turn shows that there are some core qualities of the rock that are not culturally dependent but constitute an independent reality! Commonsensical observations like this help naturalize the linkages among power, science, production, and technology and make it very hard for radical epistemological constructivism to gain a solid footing in concrete environmental struggles. The shortcomings of this kind of constructivism provide one of the driving impetuses of the emerging political ecology that tackles the question of the "real" head on.

This is not the place to provide a thorough review of the various intellectual sources that nurture the emergent third-generation PE or its various disciplinary expressions (for this see Escobar 2010; Braun 2008). However, we can point out that phenomenology, actor network theory, and Deleuzian philosophy, along with ethnographic research with a host of non-Western groups, are main sources. What gives a family resemblance to the various approaches we label third-generation PE is that rather than seeking to rethink the relation between nature and culture that so concerned previous generations of PE, they challenge the taken-for-granted ontological character of the divide; that is, they challenge the assumption that the divide is universally applicable as if it represented the ultimate reality. This is done in two ways: via conceiving reality as an emergent effect of relations and interdependencies that permanently overflow the boundaries based on modern binaries (nature/culture, subject/object, material/immaterial, and so on); and via a revaluation of non-Western knowledges that emerge from conceiving what exists and make up the world in other ways than in terms of those binaries. Although each "stream" of the challenge has remained somehow privileged within specific disciplines, they strongly resonate in foregrounding "relational ontologies" (see Ingold 2000 [2011]; Castree and Nash 2006; Braun 2008; Whatmore 2002; Descola and Pálsson 1996).

As a category of analysis, the term "relational ontologies" signals various issues. First, it constitutes a way to talk about emergent forms of life and politics that are not based on the nature/culture divide. Second, it is a practice-based concept that calls for ethnographic attention to the distinctions and relations that give shape to the entities populating the worlds of those who do not adhere to modern binaries. Politically, "relational ontologies" points at the fact that while these worlds have been under attack for centuries, their differences persist, not unchanged of course, but as something not reducible to "the same," that is, to a universal modernity. Moreover, by positing, say, the sentience of all beings and mobilizing this notion politically, these worlds unsettle the modern political arrangement by which only science represents nature while other forms of knowledge represent culture (de la Cadena 2010). It is at this point where third-generation political ecology becomes a political ontology, for it is now concerned with the power-laden negotiations involved in bringing into being the entities that make up a particular world or ontology, and delineates a field of study that focuses on these negotiations but also on the conflicts that ensue when different worlds or ontologies strive to sustain their own existence as they interact and mingle with each other (see Blaser 2009, 2010).

51

Pollution
Serenella Iovino

"Garbage hills are alive," Robert Sullivan writes in the travelogue of his explorations along the waste dumps outside Manhattan: "there are billions of microscopic organisms thriving underground in dark, oxygen-free communities" (2006, 96). After metabolizing the trash of New Jersey or New York, these cells will "exhale huge underground plumes of carbon dioxide and of warm moist methane" (2006, 96), soaking through the ground or crawling up into the atmosphere, where they will eventually compost the ozone layer.

Whether organic or not, an alien agency is a constant feature in landscapes of pollution. As in the garbage hills of New Jersey, there is life in Naples's exhausted landfills—life, in the hazardous elements that saturate the pastures in the "Triangle of Death," a former agricultural area comprised by the towns of Acerra, Nola, and Marigliano, now entirely poisoned by dioxin, polychlorinated biphenyls, and other noxious waste (Senior and Mazza 2004). There is life in the asbestos plaques abandoned under the motorway flyovers, and in the barrels of toxic substances spilled overnight in the canals meant to irrigate the lush country at the foothills of Mount Vesuvius. There is life, but not because these things are alive. In fact, only a small portion of the waste discarded in the soil, sea, and watercourses of Campania, Naples's region, is "metabolic." In most cases these are only heaps of inorganic matter, concentrations of nonliving toxicity. Nevertheless, such waste lives in and across the bodies it intersects in its magmatic trajectories of contamination. Epidemiological studies report an alarming increase of toxic-related diseases due to an intermingling of elements such as arsenic in water supplies, lead, cadmium, and mercury in the ground, and dioxin in the residents' blood (De Felip and Di Domenico 2010).

The case of Naples is emblematic of a combination of historical factors that includes decades of misrule, missed social integration, wrong development policies, industrial contamination, poor enforcement of environmental regulation, a chronic need for jobs, and insufficient waste facilities for a growing population. Over the years this dangerous mix has led to a "waste crisis" culminating in 2008 with four hundred thousand tons of uncollected rubbish piled in the streets. Making this picture gloomier is the recent emergence of the "ecomafia." Coined in 1994 by Legambiente, the major Italian environmental NGO, the term "ecomafia" describes a whole series of ecologically lethal criminal activities, organized in networks that extend far beyond Italian borders. Among these crimes, the unlawful disposing of waste stands out as the most profitable and harmful (Pergolizzi 2012). It has been calculated that, from 1993 to 2008, ten million tons of waste have been illegally dumped in the Campania region (Legambiente 2008). The Neapolitan writer Roberto Saviano has pictured this pile of rubbish as "the tallest mountain on earth" (2007, 283), bigger than Mont Blanc and Everest put together, rising 47,900 feet from a base of three hectares. Targeted in detectable, but more often in undetectable ways, the population can hardly avoid the risks of contamination.

As the story shows, there is a form of pollution just as infiltrating and dangerous as the material one. This is the pollution hiding in behaviors and words, in the hazy territory where political discourses and social life intermingle with environmentally lethal activities, some of which are illegal, some others of which are perfectly within the bounds of the law and regulation. To see pollution as a material or "physical" phenomenon is therefore to see it only in part. As Herbert Marcuse said, "Pollution and poisoning are mental as well as physical phenomena, subjective as well as objective phenomena" (2005, 175). Confirming this argument, the Naples case proves how pollution is an interplay of harmful material substances and harmful discourses and practices, whereby inadequate environmental policies, corruption, criminal complicities, and lack of adequate epistemological instruments often cooperate with uncontrolled industrial activities and forms of maldevelopment in ripping the social fabric, creating unequal protection, thwarting citizenship (Iovino 2009), and damaging human and nonhuman life. All these elements merge into the flesh of the body politic, proving how ecological health—namely, the health of biological systems—is strictly connected to political, social, and ethical issues in the forms of a participatory democracy and "cognitive justice" (de Sousa Santos 2007a; Armiero 2008; Armiero and D'Alisa 2011).

One of the major causes of the ecological crisis, pollution, has been, for more than two centuries, the dark outcome of industrial development. While liberating many people from hardships and needs, this development has also created new forms of oppression for human and nonhuman natures. This makes "pollution" a keyword for understanding these oppressions. Considered in its complex framework of material and discursive elements, in fact, pollution has the power to reveal abuses and inequalities. Tracing pollution through the bodies of living organisms and living land as in a litmus test, this keyword signals the stories of political failures, socio-ecological decline, and the discriminatory practices that infiltrate uneven societies.

POLLUTION   SERENELLA IOVINO

52

Queer Ecology

Catriona Sandilands

The term "queer ecology" refers to a loose, interdisciplinary constellation of practices that aim, in different ways, to disrupt prevailing heterosexist discursive and institutional articulations of sexuality and nature, and also to reimagine evolutionary processes, ecological interactions, and environmental politics in light of queer theory. Drawing from traditions as diverse as evolutionary biology, LGBTTIQQ2SA (lesbian, gay, bisexual, transgender, transsexual, intersex, queer, questioning, two-spirited, and allies) movements, and queer geography and history, feminist science studies, ecofeminism, and environmental justice, queer ecology currently highlights the complexity of contemporary biopolitics, draws important connections between the material and cultural dimensions of environmental issues, and insists on an articulatory practice in which sex and nature are understood in light of multiple trajectories of power and matter.

In *The History of Sexuality*, Michel Foucault lays the groundwork for much contemporary queer ecological scholarship in his observation that, beginning in the nineteenth century, modern regimes of biopower came to conceive of sex as a specific object of scientific knowledge, organized through, on the one hand, a "biology of reproduction" that considered human sexual behavior in relation to the physiologies of plant and animal reproduction, and on the other, a "medicine of sex" that conceived of human sexuality in terms of desire and identity (1978, 54). Although Foucault rightly notes the

tenuous early connections between the two discourses, the establishment of sex as a matter of biopolitical *truth* could not help but be connected to ideas of nature, and especially to racialized, sexualized, and other anxieties over hygiene and degeneracy. In this context, the figure of the homosexual came to haunt the margins of emerging discourses in urban development, environmental health, and even wilderness preservation: the effeminate homosexual and the lesbian gender invert were not only seen increasingly as *against nature* but also sometimes considered symptoms of the toxic underside of industrial, urban, and increasingly cosmopolitan modernity.

In this context, a prehistory of queer ecology must include the attempts of such authors as André Gide and Radclyffe Hall to turn these discourses on their heads (famously, of course, Oscar Wilde *embraced* his position "against" nature). Specifically, the placement of sexual desire and orientation in the purview of evolutionary concern gave rise not only to the articulation of homosexuality with degeneracy but also to the possibility that a variety of sexual practices and identities could be understood as "natural" and, therefore, morally neutral; indeed, early sexologist Havelock Ellis wrote, in 1905, that "one might be tempted to expect that homosexual practices would be encouraged whenever it was necessary to keep down the population" (9). Drawing on the rich historical inclusion of same-sex (male) eroticism in pastoral literatures, Gide's *Corydon* (published in four parts from 1911 to 1920) thus pursued the idea that the homosexual activities of boy-shepherds represented a more authentic and innocent sexuality than the heterosexual conventions they needed to learn in order to enter into adult relations of (enforced) heterosexuality (Shuttleton 2000). Similarly, Hall's *Well of Loneliness* (1928) painted a tragic portrait of a lesbian gender invert who was, if anything, by nature nobler and *more* moral than any of the other characters in the book (a portrayal that caused the book to be banned in Britain, see Mortimer-Sandilands 2008b). Although both texts were thoroughly steeped in the class, racialized, and gender politics of the era and are problematic for a host of reasons associated with neopastoral literatures more broadly (*Corydon* is deeply misogynist; *The Well* is profoundly nationalist; both deploy nature as an agent of conservatism), they and others played an important role in setting the conceptual stage for more recent attempts to queer nature and ecology. Indeed, despite their considerable political differences, one can place in productive conversation with Gide and Hall such phenomena as the post-Stonewall emergence of gay and lesbian-feminist anti-urbanisms (Herring 2007; Kleiner 2003; Sandilands 2002) and the more recent articulation of (some) discourses on urban cruising and public sex with organicist renderings of eroticism, freedom, and opposition to heteronormative productions of/restrictions on urban public space (Gandy 2012; Ingram 2010).

From these roots, beginning in the mid-1990s (e.g., *Undercurrents* 1994; Gaard 1997), a distinct body of queer ecological scholarship has emerged that has attempted to develop theoretical and activist connections between sexual and ecological politics, often drawing from ecofeminist and environmental justice perspectives and including concerted attention to the racialized, gendered, colonial, and species politics with which notions of sex and nature are articulated (especially as influenced by the writings of Donna Haraway, e.g., 1991b). As Rachel Stein writes, "by analyzing how discourses of nature have been used to enforce heteronormativity, to police sexuality, and to punish and exclude those persons who have been deemed sexually transgressive, we can begin to understand the deep, underlying commonalities between struggles against sexual oppression and other struggles for environmental justice" (2004, 7). This emphasis on the intersectional politics of sexuality, race, and ecology informs the work of Dayna Scott (2009) and Giovanna Di Chiro (2010), both of whom develop critical responses to the "gender-bending" impacts of endocrine-disrupting hormones in the environment (which particularly affect racialized and indigenous communities) while simultaneously disrupting heterosexist panics about the so-called feminization of organisms and populations of organisms, human and otherwise. Beth Berila (2004), Stacy Alaimo (2010a), and Mel Chen (2012) likewise examine the politics of contamination as toxic affects move in, through, and *as* gendered, sexualized, and racialized bodies: as Chen writes powerfully, toxic substances and discourses organize worlds in which some biopolitical subjects are enabled and their life optimized by medical and environmental interventions at the same time as other subjects are themselves considered toxic, expellable, and expendable. And Eli Clare (1999) treats the intersections of class and disability in his account of (among other things) the varied exclusions of an urban environmental politics and aesthetics that pays scant critical attention to the ways in which people *work* and *move* in natural environments, and also to bodies and landscapes that are abjected in the production and management of the pristine.

Sharing similar roots but on a slightly different trajectory, beginning in the late 1990s (although drawing on a longer line of inquiry, see Terry 2000), research in animal behavior by the likes of Bruce Bagemihl (1999) and Joan Roughgarden (2004) has drawn popular attention to the fact that a large array of animal species include elements of same-sex eroticism in their sexual repertoires; in so doing, this research has attempted to dislodge heterosexual reproduction from its singularly privileged place in evolutionary biology, to connect notions of sexual diversity with biological diversity (for a critique, see McWhorter 2010), and to open the door for a nonteleological reconsideration of animal/human sociality and pleasure (Alaimo 2010b; Johnson 2011). As homonormative popular culture has embraced so-called queer animals in the midst of a larger naturalization of same-sex marriage (especially "gay" male penguins, see Sturgeon 2010), and as images of polymorphously perverse animal and invertebrate sex have come to populate the ideosphere in the service of sexual pluralism (*Green Porno* 2009), several scholars suggest ways in which this queering of the more-than-human could proceed in more critical directions. Elizabeth Wilson (2002) and Myra Hird (2004, 2008) consider the ways in which nonhuman sexual and gender diversity both calls into question human exceptionalism and destabilizes notions of identity, authenticity, and technology on which modern categories of human sexual orientation rest; Karen Barad (2012), deploying a notion of queer performativity to strong posthumanist ends, explores the ways in which the material world can be understood as unfolding according to a process of relational co-constitution in which materialization itself can be understood queerly (see also Morton 2010); and several authors call upon the queer potentialities of Deleuze and Guattari's thought in order to nudge queer and trans theory to think more ecologically, in other words, to consider sexual and gender becomings as complex biological, technological, and political assemblages rather than as either purely discursive or biologically determined processes (Parisi 2008; MacCormack 2009; Chisholm 2010; see also Hayward 2008 and, for a critique, Garrard 2010).

Recent scholarship has drawn from combinations of these theoretical trajectories (and beyond) to generate a proliferating array of queer ecological possibilities, including provocative considerations of cross-species and eco-sexualities as part of an ethico-political opening of love to include more-than-human corporealities (e.g., Kuzniar 2008; Sandilands 2004; Stephens and Sprinkle 2012); of the implications of queer theorist Lee Edelman's critique of reproductive futurism for ecological politics (Giffney 2008; Anderson 2011; Seymour 2013); and of canonical literary texts, reread and restaged in light of queer ecological concerns (Estok 2011; Azzarello 2012). In addition, new forms of specifically ecoqueer *activism*, demonstrating interesting forms of coalitional politics in opposition to homonormative agendas, are also emerging in response to such concerns as violence, space, and food (Hogan 2010; Sbicca 2012). In both theoretical and activist milieux, however, questions remain about the specific meaning of the term "queer," given its historically and politically unstable relations to LGBTTIQQ2SA communities and activisms; the current constellation of queer ecologies—including as it does both more conventional lgbt histories and communities and more recent challenges to these terms and political affinities—may be on the verge of something new.

53

Religion

Mary Evelyn Tucker and John Grim

Religions can be understood in their largest sense as a means whereby humans, recognizing the limitations of phenomenal reality, undertake specific practices to effect self-transformation and community cohesion within cosmological and natural contexts. Religions refer to those cosmological stories, symbol systems, ritual practices, ethical norms, historical processes, and institutional structures that transmit a view of the human as embedded in a world of meaning and responsibility, transformation and celebration. Religions connect humans with a divine presence or numinous force. They bond human communities, and they assist in forging intimate relations with the broader Earth community. In summary, religions link humans to the larger cultural, biological, and material matrices of life.

Most definitions of religion are based on concepts from Western Abrahamic religions of God and salvation. These concepts are presented as universal, but generally do not take into account the varied kinds of religious sensibilities in the world religions, especially from Asian or indigenous traditions. Most Asian religions, for example, do not require belief in a Creator God or in redemption and salvation outside this world. Some, such as Buddhism, are considered to be atheistic, or not focused on God or supernatural beings. Religious sensibilities, thus, are not limited to monotheism or even theism. Nor does religion preclude those inner experiences associated with spirituality. The cultivation of a spiritual path has extensive roots in religious traditions and beyond. Spirituality involves a search for the sacred in which humans experience their authentic being in relation to a larger whole, for example, local bioregions, the Earth, and the cosmos itself.

As the environmental crisis has been well documented, it is increasingly clear that human attitudes and decisions, values and behavior are crucial for the survival and flourishing of ecosystems and life forms on Earth. Religion, ethics, and spirituality are contributing to the shaping of a sustainable future along with the natural and social sciences. In addition, a more comprehensive worldview of the interdependence of life is being articulated along with an ethical responsiveness to the need to care for life for future generations. One telling of this worldview has been formulated as an evolutionary story in *Journey of the Universe*, which is a narrative telling in film and book form by Brian Swimme and Mary Evelyn Tucker. Thomas Berry described this worldview shift as a new cosmological story.

Certain distinctions need to be made between the particularized expressions of religion identified with institutional or denominational forms of religion and those broader worldviews that animate such expressions. By "worldviews" we mean those ways of knowing, embedded in symbols and stories, that find lived expressions, consciously and unconsciously, in the lives of particular cultures. In this sense, worldviews arise from and are formed by human interactions with natural systems or ecologies. Consequently, one of the principal concerns of religions in many communities is to describe the emergence of local geography as a realm of the sacred. This is evident, for example, with mountains such as Wutai in China, rivers such as the Ganges in India, and cities such as Jerusalem, Rome, and Mecca. Worldviews generate rituals and ethics, ways of acting, that guide human behavior in personal, communal, and ecological exchanges. Religions have also helped people to celebrate the gifts of nature, such as air, water, and food, that sustain life.

The exploration of worldviews as they are both constructed and lived by religious communities is the work of religion and ecology scholars who attempt to discover formative and enduring attitudes regarding human-Earth relations. There are many examples of diverse ecological worldviews in religions. Buddhism sees change in nature and the cosmos as a potential source of suffering for the human while Confucianism and Daoism affirm nature's changes as the source of the Dao, and the Western monotheistic traditions of Judaism, Christianity, and Islam view the seasonal cycle of nature as an inspiration for the challenges of human life. Among indigenous traditions, nature is the locus for encountering the power of the sacred, such as *kami* for Shinto practitioners in Japan, *manitou* among Great Lakes Native Americans, and *mana* among Polynesian peoples.

The creative tensions between humans seeking to transcend the anguish of the human condition and at the same time yearning to be embedded in this world are part of the dynamics of world religions. Christianity, for example, holds the promise of salvation in the next life as well as celebration of the incarnation of Christ as a human in the world. Similarly, Hinduism holds up a goal of *moksha*, of liberation from the world of *samsara*, while also highlighting the ideal of devotion to the god Krishna acting in the world. Indigenous traditions also manifest such creative tensions but do not move toward radical transcendence. For example, among the Navajo/Diné of the American Southwest, the Holy People come from beyond the human realm and yet reveal themselves in the inner forms of the natural world.

This realization of creative tensions leads to a more balanced understanding of the possibilities and limitations of religions regarding environmental concerns. Many religions retain orientations toward personal salvation outside this world; at the same time they can foster and have fostered commitments to social justice, peace, and ecological integrity in the world. Historically, religions have contributed to social change in areas such as the abolitionist and civil rights movements. There are new alliances emerging now that are joining social justice with environmental justice.

In alignment with these "eco-justice" concerns, religions are formulating broad environmental ethics that include humans, ecosystems, and other species. Scientists acknowledged this potential of religious communities for encouraging a broadened environmental ethics from their particular traditions. Two key documents were issued in the early 1990s. One is titled "Preserving the Earth: An Appeal for Joint Commitment in Science and Religion." A second is called "World Scientists' Warning to Humanity" and was signed by over two thousand scientists, including more than two hundred Nobel laureates. This document states, "A new ethic is required—a new attitude towards discharging our responsibility for caring for ourselves and for the earth." Moreover, the Earth Charter, inspired by the Earth Summit in 1992, brings ecology, justice, and peace together in a scientific evolutionary framework.

The response to these appeals was slow at first but is rapidly growing. It might be noted that there were some strong voices advocating a religious response to environmental issues over half a century ago. These included Walter Lowdermilk, who in 1940 called for an Eleventh Commandment of land stewardship, and Joseph Sittler, who in 1954 wrote an essay titled "A Theology for the Earth." Likewise, the Islamic scholar Seyyed Hossein Nasr has been calling since the late 1960s for a renewed sense of the sacred in nature that draws on perennial philosophy. Lynn White's essay in 1967 on "The Historical Roots of our Ecologic Crisis" sparked controversy over his assertion that the Judeo-Christian tradition has contributed to the environmental crisis by devaluing nature. In 1972, the theologian John Cobb published a book titled *Is It Too Late?* underscoring the urgency of environmental problems and calling Christians to respond.

Over the last several decades, key movements have taken place among religious communities that have shown growing levels of understanding of the environmental crisis and growing commitment to alleviating it. In particular, the religions have come together in innovative interreligious dialogues, witnessing to their common concern for endangered ecosystems and biodiversity. The Parliament of World Religions held in Chicago in 1993 and in Cape Town, South Africa, in 1999 issued major statements on global ethics that embraced human rights and environmental issues. The Parliament in Melbourne in 2009 had more than twenty sessions on religion and ecology. Indigenous peoples gathered in Bolivia in 2010 to formulate the *Universal Declaration of the Rights of Mother Earth* as an expression of their concern for the degradation of land, religion, and culture. Indigenous peoples continue to report on the evidence of climate change in various regions through the Indigenous Environmental Network (ienearth.org).

Several major international religious leaders have emerged as a result of a renewed appreciation of their tradition's respect for the environment. The Tibetan Buddhist leader, the Dalai Lama, and the Vietnamese Buddhist monk Thich Nat Hahn have spoken on the universal responsibility the human community has toward the environment and toward all sentient species. The Greek Orthodox patriarch Bartholomew has sponsored symposia on water that have brought together scientists, religious leaders, civil servants, and journalists to highlight the problems of marine pollution and fisheries depletion. Many of the world religions have issued statements on the need to care for the Earth and to take responsibility for future generations.

It is within this global context that the field of religion and ecology has emerged within academia over the last two decades, animated by several key questions. Theoretically, how has the interpretation and use of religious texts and traditions contributed to human attitudes regarding the environment? Ethically, how do humans value nature and thus create moral grounds for protecting the Earth for future generations? Historically, how have human relations with nature changed over time, and how has this change been shaped by religions? Culturally, how has nature been perceived and constructed by humans, and conversely, how has the natural world affected the formation of human culture? From a socially engaged perspective, in what ways do the values and practices of a particular religion activate mutually enhancing human-Earth relations? It is at this lively interdisciplinary intersection between theoretical, historical, and cultural research and engaged scholarship that the field of religion and ecology is still emerging.

The values embodied by religious perspectives on the environment provide viable alternative views of nature simply as resources or as services to humans. The shared values of the world religions include *reverence* for the Earth and its profound ecological processes, *respect* for Earth's myriad species and an extension of ethics to include all life forms, *reciprocity* in relation to both humans and nature, *restraint* in the use of natural resources combined with support for effective alternative technologies, a more equitable *redistribution* of economic opportunities, the acknowledgment of human *responsibility* for the continuity of life, and *restoration* of both humans and ecosystems for the flourishing of life. These values are beginning to intersect with other approaches to environmental problems from the perspectives of science, economics, law, and policy.

54

Risk Society

Robert J. Brulle

For most of human experience, the primary societal concern was over scarcity—having adequate food, shelter, and resources to survive. While these concerns still exist for many in the world, a new concern has arisen, that of safety. The rise of technology has created abundance, and thus eliminated the issue of hunger and resource scarcity for many, especially in the developed countries. However, at the same time, these technologies have created anxiety about potential harms. Issues such as radioactivity, genetic engineering, or environmental degradation create apprehension and insecurity in the population. Hence, alongside concern about scarcity, a new concern over risk has emerged. One way to characterize this shift has been to note the emergence of a new form of concern, in which the original concern over scarcity, emblematized by the phrase "I am hungry," has been expanded and partially replaced by the concern "I am afraid." This creates a society in which the politics of eliminating scarcity to provide equity and freedom from want is overlaid by a politics based on minimizing anxiety and eliminating risk by providing safety (Beck 1986 [1992], 49). This shift is characterized as the emergence of a "risk society."

The concept of the "risk society" is primarily based on the work of Ulrich Beck (1986, 1995). For Beck, the continued development of industrial production is based on the dynamic of modernization and industrialization. At the center of the process of modernization is the application of scientific research and knowledge to the expansion of economic growth. The power to define technological development, and thus our future, has become concentrated in the private corporate power that controls and directs much of research and development. This results in a shift in the locus of power from the nation-state to the corporations and their control over the scientific agenda. Driven by the need to maximize profits, these corporations continue to develop new technologies that produce unforeseen risks for the entire society that are unevenly distributed. These economic and technological projects operate in a manner that is "blind and deaf to consequences and dangers" (Beck 1986 [1992], 28).

As a result, industrial risks quantitatively and qualitatively change. They are no longer limited in time or space, aftercare or restoration is not possible, monetary compensation is not possible, and the causality or blame for an accident cannot be defined. Thus accountability for an accident cannot be established, and these industrial projects are not insurable (Beck 1996, 31; 1995, 22–23). This breaks down the ability of society to ensure the safety of its citizens from the production of industrial hazards. Thus the security of citizens against harm from industrial processes is no longer maintained (Beck 1995, 22–23).

In addition, the unquestioned linkage between economic development and progress becomes challenged. This generalization of risks that are not limited in space or time creates a phenomenon labeled by Beck as the "End of the Other." In the course of human history, one group of people inflicted violence on the "other," whether in the form of an enemy, a scapegoat, or a dissident. Today, the harm caused by global environmental problems is inflicted on all persons, regardless of social class or ethnicity (Beck 1995, 27). While the levels of risk exposure vary dramatically, and the powerful and well-off populations are better able to insulate themselves from the immediate effects, there is no population that remains immune from the risks generated by economic and technological development. Property becomes devalued due to ecological destruction (Beck 1995, 60). Ozone depletion creates skin cancers among all classes. Climate change raises sea levels across the globe, flooding both the rich and the poor.

That is not to say that the adverse consequences of environmental risk are evenly distributed or that all peoples have equal ability to insulate themselves from its adverse consequences. In the distribution of environmental risks across society, Beck (1986 [1992], 53) notes that there are winners and losers due to differentials in economic and political power. This results in differential "risk positions" corresponding to variation in the levels of exposure to environmental degradation. The resulting stratification follows the social distribution of power in which "like wealth, risks adhere to the class pattern, only inversely; wealth accumulates at the top, risks at the bottom" (Beck 1986 [1992], 35). Thus although the entire society faces increased risks, some segments confront more intense exposure than others do. Thus Beck (1995, 29) concludes, "What is denied collects itself into geographical areas, into 'loser regions' which have to pay with their economic existence for the damage and its unaccountability." Environmental risk is concentrated in low-income residential neighborhoods near centers of industrial production or in poor underdeveloped countries, where much of the world's toxic waste is dumped. The residents of these communities and nations are disproportionately exposed to industrial pollution and have the least ability to prevent exposure. In addition, the populations in these highly impacted communities have the lowest amount of educational, political, or economic resources to either effectively resist or mitigate their exposures.

This "class pattern" is also complemented by a "race pattern" that influences the unequal distribution of environmental risk. In the United States, the dynamic of racism has created a substantial differentiation in both occupational characteristics and community of residence between white and nonwhite populations. Persistent discrimination in educational opportunities and employment has restricted nonwhites to lower socioeconomic status and thus limits these populations' access to residence in more affluent communities. In addition, people of color—especially African Americans—are restricted in their choice of residence by a series of mechanisms that result in racial segregation. As a result, many people of color are concentrated in highly segregated communities that are significantly more disadvantaged than those of the white population. Racial segregation is a major contributor to the creation and maintenance of environmental inequality because governments and corporations often seek out the path of least resistance when locating polluting facilities in urban and rural settings. Thus polluters can site locally unwanted land uses in such neighborhoods because they are more isolated socially and relatively powerless politically. This distribution of risks is also mirrored in the world system. The less developed and economically poor countries suffer from increased exposure to environmental risks. For example, Indonesia suffers extensive deforestation to supply Japan with its necessary timber supply, and much of the world's toxic waste is dumped in the developing world, revealing a pattern of global environmental inequalities by race, class, and nation.

Given the potentially explosive political threat posed by environmental risks, Beck argues that these threats need to be continually repressed and denied. There is the development of an entire politics that either denies or minimizes the extent and nature of environmental degradation in order to maintain the political and economic power of key economic interests (Beck 1995, 140–42). The politics of a risk society thus has the potential to challenge the fundamental premises on which industrial society is constructed. What is at stake in these conflicts is the question of whether "our concepts of 'progress,' 'prosperity,' 'economic growth,' or 'scientific rationality' are, or ever were correct. In this sense, the conflicts that erupt here take on the character of doctrinal struggles within civilization over the proper road for modernity" (Beck 1986 [1992], 40). First, the impartiality and rationality of science is challenged (Beck 1986 [1992], 29). The supposed ability of science to develop objective knowledge is demystified by the confrontation of the problems that its application has brought about. Science is seen as a social construction that not only solves problems but also creates them. Its legitimacy as an impartial and objective guide to action is delegitimated by the "failure of technoscientific rationality in the face of growing risks and threats from civilization" (Beck 1986 [1992], 59). So the practices and assumptions of the scientific enterprise become the subjects of critique (Beck 1986 [1992], 155–74).

Thus advanced capitalism expands economic production, which also increases technologically based risk. However, concentrating environmental degradation in the neighborhoods of the poor, people of color, the powerless, and the least developed countries minimizes the political challenge to the systematic production of environmental risk. The more economically and politically well-off populations can continue to reap the benefits of increased economic production without having to bear the brunt of its environmental costs. This unequal sharing of environmental risks thus avoids the creation of legitimacy problems related to the production of environmental degradation.

This unequal distribution of environmental risks has given rise to social movements that seek to resist the imposition of environmental risks on disadvantaged communities, and at the global level, on entire nations. Beck sees this movement as part of a process of reflexive modernization. Reflexive modernization is a social process in which the fundamental beliefs and social arrangements are critically examined and reformulated on the basis of these reflections. This brings the unintended consequences of technological development into critical consciousness and allows them to be rationally modified. Reflexive modernization thus involves a self-conscious process where economic development is subjected to a conscious evaluation. A key aspect of this sort of approach is the application of the precautionary principle to economic development. Social movements play a key role in bringing the taken-for-granted assumptions of economic expansion and its risk-distribution consequences into public discussions, and thus allow a full consideration of actions. This enhances the learning capacity of society, and makes its further development subjected to rational public debate.

Overall, the theory of the risk society adds a new perspective to thinking about how industrial risk has changed over the past two hundred years. The scale and impacts of technological disasters have vastly expanded, and our old systems of dealing with accidents such as shipwrecks, factory explosions, or the like is no longer adequate to address our new set of global environmental risks. Beck's conceptual scheme provides social theorists with some new concepts to address these phenomena. As a theoretical scheme, it lacks direct connections to empirical research, and thus suffers from a lack of grounding in the careful empirical work of sociological research on topics of environmental justice or environmental racism. An integration of his theoretical perspective into these literatures would advance both our theoretical and empirical understanding of the way technological risk is developed and distributed in a global society.

55

Scale

Julie Sze

If there is any phrase that represents both environmental problems and the call to action in the last three decades, it's the injunction to "Think Global–Act Local." On the most basic level, this approach connects thinking and acting. For many, this slogan effectively connects individual action with collective impacts or change. Less obvious, but no less important, thinking globally and acting locally also demands that people more fully comprehend the relationship between the local and the global or, in other words, that they consider *scale*.

Scale "maps" onto the fundamental political, environmental, and social problems that have preoccupied environmental activists and scholars over the last three decades. These are primarily as follows: the intersection of the local and the global spheres; intensifying urbanization and pollution around the world, but particularly in the Global South and Asia; and the increasing awareness of environmental pollution on individuals and communities in many locales. Significant environmental problems, from biodiversity conservation to air and water pollution and habitat degradation, require interdisciplinary analysis and methods with which to analyze complex processes at multiple temporal and spatial scales (Sayre 2005, 277).

"Scale" is a—or possibly *the*—key term in geography. Scale is first a *technical* measurement of the ratio between the size of objects in the world and their size on a map (i.e., a cartographic scale of one to one million). Scale functions as a shorthand for an areal unit on a map, or for what geographers call an "analysis scale," as well as a "phenomenon scale," which is the size at which geographic structures exist and over which geographic processes exist in the world (Montello 2001). Physical geographers refer to a "resolution scale" in Geographic Information Systems (GIS) and remote sensing research (Marston 2000). Cultural and political geographers generally focus on scale at the levels of the bodily, urban, regional, national, and global (Herod 2011); while still others focus on local, gender, or household scales (Marston 2000). Political ecologists and radical geographers generally agree that "scale" is not only a technical term to be measured objectively but is also socially constructed (Marston 2000), historically contingent, and politically contested (although scholars disagree about the mechanisms and the role of the state, capital, and labor; see Swyngedouw and Heynen 2003; McCarthy 2005; MacLeavy and Harrison 2010). These debates about scale have political and analytic implications in terms of how best to define or research a problem and, on a more pragmatic level, how to craft policy responses to these problems as a consequence.

Ecologists and geographers use scale in overlapping yet distinctive ways. Ecological scale consists of both temporal and spatial dimensions. Ecologists generally agree that there is no single correct scale for ecological research; that cross-scalar and interdisciplinary analysis is necessary for ecological investigation; that ecological change is historical; and that scale raises metaphysical, epistemological, and ontological issues (Sayre 2005). Ecologists focus on grain and extent, where "grain" refers to the smallest unit of measurement employed to study some phenomenon while "extent" denotes the overall dimensions over which observations are made, including both space (area) and time (duration) in a given study area (Sayre 2010).

The interplay between space and time is also what makes "scale" a key analytic term for humanists and cultural critics. Environmental activists, writers, critics, and artists engage pollution across space and time using imagination and cultural ideas. Shewry writes that "[s]patial arrangements are made in everyday practices among people and other life and are tapped into broader processes such as time, imperialism, global economy, ecology, evolution and *memory*" (LeMenager, Shewry, and Hiltner 2011, 10; italics added). Imagination is a key resource in responding to environmental problems since "literature commonly layers environmental spaces with memories and seek[s]to dream up other worlds" (LeMenager, Shewry, and Hiltner 2011, 10).

Moving from the ways in which particular academic disciplines and fields define and use "scale," the question remains, why has scale become so prominent? Why now? In part, the emphasis on scale is a result of the understanding that environmental problems often "cross" or jump scales, and there is a "spatial mismatch" that can occur in discussions of an environmental problem, as between the scales of environmental pollution and its political regulation. For example, Hilda Kurtz, in discussing a dispute over a toxic facility in Louisiana, argues that environmental injustice is, in part, a result of scalar ambiguity and that scales are also concepts that shape regulation as well as cultural frames of inclusion and exclusion (2003). In this case, a very localized dispute in a predominantly African American neighborhood was connected to the state regulatory agencies. In addition, local activists succeeded in connecting their struggle with a broader social movement for environmental justice within a federal regulatory context. In this same region, other local social-justice-movement actors similarly leveraged cross-scalar analysis (i.e., the Norco community in Louisiana made references to

"Mother Earth," testified at The Hague, and stood in solidarity with the Ogoni in Nigeria, because both lands were polluted by Shell, a multinational oil corporation). Thus, environmental and other social movements (labor) also organize and network across scales, using communication technologies that transcend space and that marshal sophisticated understandings of local, regional, national, and global relationships (Marston 2000; Pellow 2007).

As a result, multiscalar analysis is not just a methodological imperative but also a political one in environmental studies. One cultural scholar argues that "attention to scale leads us to deeply powerful forms of territorial organization that impact the environment in *uneven ways*" (emphases added, LeMenager, Shewry, and Hiltner 2011). A multiscalar lens allows for a deeper political, cultural, and historical analysis across space and time.

Yet another illustration of an environmental problem with scalar dimensions is the elevated pollution in Arctic Native populations as a result of persistent organic pollutants (POPs). POPs are a set of extremely toxic, long-lasting, chlorinated, organic chemicals that can travel long distances from their emissions source (often created through industrial pollution thousands of miles away) and that bioaccumulate in animals and ecosystems. In the 1980s, scientists observed high levels of toxic chemicals (pesticides, insecticides, fungicides, industrial chemicals, and waste combustion) far from their sites of production. A Nunavik midwife in the Arctic collected breast-milk samples as a control from a "clean" environment, but instead, researchers were surprised to find that Arctic indigenous women had POP concentrations in their breast milk five to ten times higher than did women in southern Canada and among the highest ever recorded (health effects include higher rates of infectious diseases and immune dysfunction, as well as negative effects on neurobehavioral development and height). This problem of Arctic Native breast-milk contamination thus has local effects, created in global economic-political and scientific-ecological contexts (Downie and Fenge 2003). A multiscalar analysis is necessary to fully comprehend the concentrated local effects of Arctic Native POP pollution in its historical, spatial, scientific, and political dimensions. Native activists testified at international meetings about their case, to fully illustrate the relationship between the local Arctic realities and global historical and political injustices and the disproportionate impact of this relationship on indigenous bodies.

Scale as an increasingly prominent analytic tool is intimately connected to intensifying conditions of globalization, specifically capitalist economic development and related ideologies of neoliberalism, privatization, and deregulation. Scale and globalization are linked in part because of the way the increasing movement of pollution and peoples and the concomitant weakness of environmental regulation (at multiple scales) are connected. The failures to address climate change are arguably linked to analytic problems related to scale. For example, a significant percentage of industrial pollution in China is a result of production for consumers in the United States, Europe, and Australia, but this pollution crosses the Pacific, landing in the western United States. How should this pollution be measured? In national terms? Global? Pacific Rim? Sayre suggests that the difficulties in confronting global warming are a function of the unique scalar qualities of climate change, including the temporal realm. He writes, "Temporally, the *grain* is likewise infinitesimal: that split second at which a chemical reaction occurs in combustion, photosynthesis, oxidation, decay, etc. But the *extent* is very long: once a molecule of carbon dioxide or nitrous oxide enters the atmosphere, it remains there for more than a century; most other greenhouse gases persist for one-to-several decades" (Sayre 2010; italics added). Nations such as China and India also use temporal dimensions in their political arguments against international treaties to regulate carbon emissions, pointing out that countries in the West have historically contributed greater carbon emissions and have a head start on economic development.

Scalar analysis is spatially and temporally complex, and thus offers useful analytic lenses through which to approach a wide range of environmental topics. In the case of scale, "believing the world to be scaled . . . is likely to shape how we engage with it and so the kind of knowledge about its materiality that we produce" (Herod 2011, 257). As long as complex environmental problems are multiscalar across space and time, scale will remain a key conceptual tool in interdisciplinary environmental studies.

56

Species

Quentin Wheeler

No concept has greater theoretical and practical importance in biology than species. Without reliable knowledge of species, we are incapable of accurately identifying the kinds of organisms we are studying, and without equally reliable scientific names, our efforts to communicate our observations are muted. Show an unfamiliar plant or animal to anyone, adult or child, and her first question is invariably, "What is it?" A crucial part of the answer to that question is a species name, and behind that name are sophisticated scientific hypotheses. In the age of digital databases, scientific names are playing an ever more central role in biodiversity informatics (Patterson et al. 2006).

"Biodiversity" has become a household word and a focus for international treaties and law. We claim to hold the conservation of biodiversity as a high priority, yet we are nearly wholly ignorant of what biodiversity is. We pass laws, like the Endangered Species Act, and pretend or assume that we have put force behind our commitment to conservation, yet without basic data on what species exist, where they live, and what they do, these actions are nearly meaningless. Until we create more or less complete baseline information about species, we are largely powerless to detect or measure the loss of species or the health of ecosystems that are made up of thousands of interacting species (Wheeler et al. 2012).

Are there more species of mosses or of moths? Unless we use the same species definition in botany and entomology, a simple tallying of names may be misleading (Wheeler and Meier 2000). Many other fundamental biological questions are similarly dependent on comparable and accurate treatments of species. Are there more species in a hectare of southern beech forest or in a northern hemisphere maple forest? Why are there only about five thousand species of mammals but more than one million species of insects? What is the average rate of species extinction through geologic time, and what proportion of species were lost in the last mass species extinction event at the Cretaceous-Tertiary boundary? In every case, an even approximately accurate answer depends on good information about species.

We face an uncertain environmental future. Regardless of the extent to which humans are driving climate change, one need only consult ice cores or sediments to realize that the history of Earth is written in climate change. Human civilization happens to have arisen during a postglacial sweet spot, but such relative stability has never been the norm for very long. Over any lengthy span of time, change is the only climatic constant. The important question is not whether there will be climate change. It is, how will we adapt to its impact on our lives? As it turns out, knowledge of species opens a universe of possibilities for promising options. The emerging field of biomimicry studies the evolutionary adaptations of species in search of clues that can guide innovations in design and engineering (Benyus 1997).

Both evolutionary and environmental biology are dependent on knowledge of species to pursue the questions at the heart of their disciplines. The story of the origin and history of life on our planet is written in species and their characters, from genes to anatomy. Were we to send a manned space flight to another planet populated by living organisms, our first priority, by science and common sense, would be to complete an inventory of the kinds of living things in its biosphere. How many species are on Earth? Since 1758 we have described and named nearly two million species, but our best estimate is that ten million additional plants and animals await discovery. And this does not account for the microbial world that is equally or more diverse.

Taxonomists continue to advance our knowledge of species but against incredible odds. Taxonomy enjoyed a central place in biology from the time of Linnaeus through the first half of the twentieth century. The annual rate of species discovery was on a very positive trajectory of increase, in spite of competing formal concepts (Wheeler and Meier 2000), until it peaked just prior to World War II at about twenty thousand species per year. After the understandable hiatus caused by the war, taxonomy immediately regained its annual rate of productivity, but the trend of accelerating species discovery never resumed. In fact, the rate remained more or less flat and is today about eighteen thousand species per year. Given the urgency of the biodiversity crisis (Kolbert 2014) and advances in both theory and technology, it seems inexplicable that we are not making progress more rapidly.

One reason for this self-destructive neglect of taxonomy is the concurrent rise in the 1940s of the field of (nonhuman) population genetics (Wheeler 2008). Population geneticists study species-in-the-making with a focus on populations. Taxonomists only name and classify fully differentiated species. Herein lies the problem. Because population studies are experimental and taxonomy is comparative and historical, the latter is viewed by experimental biologists as stamp collecting rather than science. Any competent philosopher of science instantly sees the folly in this prejudice. Because taxonomists make all-or-nothing claims about the distribution of characters in and among species, its theories are potentially falsified by single observations and are thereby more rigorous than any experiment (Gaffney 1979). Technological advances now offer exciting new ways to revive and accelerate species exploration, particularly cyberinfrastructure and DNA sequencing.

Although *Homo sapiens* is exceptional in its power of reason and sense of ethics, it is nonetheless only one of twelve million species, each unique and exceptional in its own way. We enrich and deepen our understanding of the meaning of our own lives by acknowledging and studying the lives of our relatives, near and distant. All the best evidence suggests that life on Earth is monophyletic, or in other words, all known species seem to share a single distant common ancestor—literally the mother of all species. If humans stand idly by as millions of species go extinct and do not even bother to discover and describe them, we do a great disservice to biodiversity and to humanity. As a more practical matter, we can only recognize, enjoy, and value those species that we have bothered to learn well enough to describe and name. And our best hopes for a bright ecological future demand that we learn the species around us so that we can assess their status and strive to be effective stewards of biodiversity sustainability.

Two great challenges are before us that can fundamentally change the way we think about and value biodiversity. The first is to adopt a unified or general theory of species so that we may ask and answer fundamental questions about biodiversity and its distribution. Such a unified theory is within reach if we focus on the one thing that all species share in common, evolutionary history, and emphasize that single pattern rather than many processes of speciation (Wheeler 2010). The second is rapid completion of an inventory of our planet's species. If we make taxonomy a priority, we could immediately increase the rate of species discovery. A single order of magnitude increase in the rate to two hundred thousand species per year is easily achievable and could result in knowledge of most of the remaining ten million species in just fifty years (Wheeler et al. 2012).

57

Sublime

Patrick D. Murphy

In many areas of common discourse, the word "sublime" has become as vacuous as "awesome." Yet, unlike the latter word, "sublime" carries with it significant meaning not only as an aspect of aesthetics in general but also as a characteristic of engagement with the natural world. It has a long intellectual history as a theoretical construct, and various contemporary environmental literary critics, such as Christopher Hitt and Timothy Morton, have sought to revise it to signify a particular kind of reverence for intense embodied experience. Yet this alleged universal quality of experiencing natural beauty has had a strong masculinist bias. Such a bias works against incorporating the sublime into any ecological ethic that hopes to be egalitarian and ecocentric. That limitation raises questions about its usefulness for a progressive environmental rhetoric.

Its use in Western European aesthetics and literary theory, and more recently in postcolonial analyses (e.g., DeLoughrey and Handley 2011), is based initially on an essay by Longinus translated and popularized in the eighteenth century. He argued that sublime rhetoric is designed to inspire but that it can only do so if based on "the nature of the subject." He also included "terror" as an example of true elevation. In the European tradition, male writers have most strongly associated the sublime with mountain scenery (Nicholson 1997; Koelb 2009). But, mountains as popular symbols for sublime experience come heavily laden with exclusionary significations identified with the three public arenas of religion,

war, and nation building. Gender exclusion intensified when differentiating the sublime from the beautiful became a project of eighteenth-century European aesthetics (Hinnant 2005). John Dennis (1704), Joseph Addison (1712), and Edmund Burke (1757) based their claims for this distinction frequently on male-only experiences linking the sublime with men's recreational utilization and scientific investigation of wild nature. They disregarded men's and women's experiences of nature in more domesticated and work-related settings, reflecting class elitism as well as gender oppression.

Burke made a crucial rhetorical move when he argued for the mutual exclusivity of the sublime and the beautiful in 1757. His argument was widely accepted by Western European and American philosophers, even though women in the eighteenth century were experiencing and portraying a different point of view in letters, stories, and essays. Burke claimed that both the sublime and the beautiful could produce pleasure in the perceiver, but only the sublime could induce the pleasurable feeling of horror. Sublime pleasure, then, comes from the *post facto* relief of having experienced threat or danger, felt fear, awe, and horror, and survived the experience, eliciting a sense of triumph or domination, both of which he mistook for positive attitudes toward the rest of nature. Burke appears to differ from Kant, who follows in his footsteps, by virtue of his emphasizing a realization of physical limitations as part of the sublime, in contradistinction to the Kantian position of the sublime arising from recognition of intellectual transcendence.

One would think that Burke's recognition of physical limitations ought to lead to a sense of humility, a realization of nonandrocentric reality, which could certainly form a type of transcendence, in the sense of crossing over and getting past egotism and illusions of superiority. Such a philosophical recognition of physical limitations could also lead today to an awareness of human interconnectedness with other living entities. In most literary representations of the sublime, however, recognition of limitations and interdependence is not emphasized until authors start to represent an ecological consciousness that shapes interpretations of sensuous and aesthetic experience. Burke's sense of limits seems not to have become nearly as popular as his sexist attitudes, challenged in his lifetime by Mary Wollstonecraft and others, and his promotion of idealist philosophy: "The Burkean sublime, with its emphasis on the psychological effects of terror, proved decisive in shifting the discourse of the sublime away from the study of natural objects and towards the mind of the spectator" (Shaw 2006, 71), in the second half of the eighteenth century just prior to the onset of the Industrial Revolution.

While Kant rejected significant portions of Burke's theories, he embraced the effort to exclude women from experiencing the sublime even as European societies were limiting upper-class women's direct experience of nature (Kofman 1997; Alexander 1999). In comparison to Burke's idea about physical limitations, Kant's idealism on the sublime displays a defensive denial of the material evidence that would support an ecocentric argument that undomesticated nature has dominion over us and not the inverse illusion. Such idealizing today is reflected in the attitude that the rest of nature was made for, and exists for, the benefit and pleasure of human beings. Kant does critique anthropomorphism but only insofar as it leads away from disinterestedness. Kant does not recognize relative positioning within the biotic community and the rest of material existence; instead, he invokes transcendent superiority of abstract reason. In Kantian idealism the fundamental dualisms of nature/culture, man/nature, and mind/matter are also used as part of an androcentrism that subordinates women and nature to an abstract concept of an autonomous male mind. It is noteworthy that two established ecocritical scholars of the Romantics both critique the sublime: Karl Kroeber (1994) and Jonathan Bate (2000). Bate does, however, define a Wordsworthian type of sublime, different from the norm, that may prove compatible with feminist revisions of the sublime. Bate writes of a moment of excess in Wordsworth's poetry that gives rise to an acceptance and embracing of a person's place within worldly vastness rather than an intellectual flight from it, a position also taken by women writers in the nineteenth century (Pipkin 1998).

In the late twentieth century, feminist thinkers have undertaken considerable and highly influential revisions of the sublime in both philosophical and literary circles, including ecofeminism, often through studying women writers of earlier periods. Literary critic Anne K. Mellor concludes that rather than reinscribing the separation of the masculine and the feminine by means of the dichotomy of the beautiful and the sublime, "In this feminine Romantic tradition, the sublime combines with the beautiful to produce . . . an experience of communion between two different people, that very 'sympathy' or *domesticated* sublimity" (1993, 103). Barbara Claire Freeman in her literary criticism sees the stance adopted by female Romantic writers as "taking up a position of respect in response to an incalculable otherness" (1995, 11). Barbara T. Gates in her history of nineteenth- and early twentieth-century women nature writers claims that "the Victorian female sublime emphasized not power *over* nature but the power *of* nature in a given place" (1998, 170). Thus, from the eighteenth through the twentieth century, women were consistently challenging the masculinist sublime with their own versions while male philosophers were codifying the Burkean/Kantian version of the concept for aesthetics and reinforcing philosophical dualism.

Interest in retaining the concept of the sublime raises this question: why should the concept of the sublime as an aesthetic experience of engagement with other aspects of nature be what defines human culture rather than a practical, interrelational, interdependent recognition that can be intuitive, emotional, and rational at different times? Such an effort would require adopting an ideological position different from the confines of the Burke/Kant orientation for the use of the "sublime" as a rhetorical strategy to have any positive function in defining human perceptions of their place in the world, responsibilities toward other entities, and treatment of a particular habitat. Mellor and Gates may actually be describing different types of interpretations of an experience that is internalized and understood from completely different vantage points than that adopted by Burke or Kant.

The sublime may be salvageable as a critical concept if it is defined in terms of a *participatory* or *integrational* sublime. In contrast, a *transcendental* sublime along the lines of Romantic or Kantian idealist philosophy works against ecological values because it places a premium on the human mind separate from the body as a source of immaterial ideas. Rather than an experience of the traditional sublime, many of the writers discussed by Freeman, Mellor, and Gates experience an intellectual interpretation of sensuous engagement that cannot be adequately described by the language available. The *sublime* has a long history of application to engagement with nature, but the interpretation of such experience is fraught with problems and the utility of the term remains in doubt for ecocritical analysis. Other terms might be worth consideration, such as "liminescence" (a sensation of in-betweenness), "transport" (being carried beyond the threshold of ego-identity), or "attendance" (a sense of engagement without a sense of distance) (Murphy 2012). These terms emphasize a heterarchical positioning of the human subject as one among many subjects and entities and reinforce the material basis for emotional response to aesthetically perceived phenomena.

58

Sustainability

Julian Agyeman

The ideas of "sustainability" and "sustainable development" first achieved prominence among academics and international policy makers, together with policy entrepreneurs in NGOs, in the 1980s. They quickly became central concepts in policy, planning, and development discourses, from global to local, especially after the publication in 1987 of the Brundtland Report "Our Common Future," which defined "sustainable development" as "development that meets the needs of the present without compromising the ability of future generations to meet their own needs," and the 1992 Rio Earth Summit, which gave us Agenda 21 and Local Agenda 21—global and local agendas for the twenty-first century. Since then, these terms, and variants such as "sustainable communities," have become pervasive in government at all levels, among business leaders, and in activist and civil society discourses, and there has been a massive increase in published and online material focused on these topics.

The concept of sustainable development put forward by Brundtland, while a contested concept (e.g., Jacobs 1999; Bourke and Meppem 2000; Gunder 2006; Connelly 2007), implies a process in which reasonable material needs are met within ecosystem limits—despite legitimate critiques regarding various and culturally specific definitions of "development" and "needs." Larrain, Leroy, and Nansen (2003) helpfully describe from a Global South perspective the concept of the "dignity line"—a culturally specific minimum level of material consumption needed to allow a life with dignity. The ubiquity of the terms "sustainability" and "sustainable development" has also led to contestations over what is to be sustained, by whom, for whom, and what is the most desirable means of achieving this goal. To some, the discourse surrounding these terms is too all-encompassing to be of any use. To others, the words are usually prefaced by "environmental" and "environmentally," as in "environmental sustainability" or "environmentally sustainable development." The term "ecological sustainability" is also sometimes used to emphasize the interdependence of species within this discourse.

The dominant discourse of sustainable development in Europe is, according to Graham Smith (2003) "ecological modernization," which is

> a discourse of eco-efficiency. Its primary concern is the efficient use of natural resources within a capitalist framework (Hajer 1995, Christoff 1996, Gouldson and Murphy 1997). Criticisms have been leveled at the lack of attention paid to social justice (both within and between nations) and the failure to conceive of nature beyond its value as a resource. (4)

Some see sustainability and sustainable development as trendy, fashionable concepts whose time in the limelight will soon pass. However, it is hard to see this happening when the terms have been around for over thirty years and are still generating a frenzy of interest in academic, activist, and policy and planning circles. To still others, the discourse offers a sense of integrity and holism that is lacking in contemporary, reductionist, silo-based policy making and planning (Davoudi 2001). Indeed, the trend is to talk of sustainable development policy making as "joined up" or "connected" policy making, that is, policy making in specific areas such as housing, economic development, diversity, or environment, with an explicit eye to its intersections, interconnections, and effects on the policy architecture as a whole.

Two major challenges to achieving sustainability, sustainable development, and more sustainable communities are the increasing scientization of sustainability and the need to foreground the issues of equity and social justice. Despite the greater allocation of funding for research on the scientific aspects than on the social scientific aspects, the "science" of sustainability is not our greatest challenge. In almost all domains of sustainability, we know scientifically what we need to do, *and* how to do it. The problem is that all of us, whether in the Global North or in the Global South, are simply not doing it. This is especially so for so-called wicked problems such as climate change: the challenge is not the science but the question of how we shift the paradigm, the political and civic culture, such that the will to act is prized by our politicians—and how we inculcate *public* understanding such that the need for action is both supported and assured.

In part in response to these challenges, a growing number of activists and commentators in both the Global North and the Global South (e.g., Middleton and O'Keefe 2001; Adger 2002; Shiva 2002; McLaren 2003; Buhrs 2004) have commented on issues of equity and social justice, or the "equity deficit" (Agyeman 2005, 44) that still pervades most "green" and "environmental" sustainability theory, rhetoric, *and* practice. Agyeman, Bullard, and Evans (2002) note, "Sustainability cannot be simply a 'green,' or 'environmental' concern, important though 'environmental' aspects of sustainability are. A truly sustainable society is one where wider questions of social needs and welfare, and economic opportunity are integrally related to environmental limits imposed by supporting ecosystems" (78). Integrating social needs and welfare offers us a more "just," rounded,

equity-focused definition of sustainability and sustainable development than Brundtland, while *not* negating the very real environmental threats. A "just" sustainability, which has strong similarities to the concept of environmental justice (see Di Chiro, this volume) is therefore "[t]he need to ensure a better quality of life for all, now and into the future, in a just and equitable manner, whilst living within the limits of supporting ecosystems" (Agyeman, Bullard, and Evans 2003, 5). This definition of "just sustainability" focuses *equally* on four essential conditions for just and sustainable communities of any scale. These conditions are as follows:

1. Improving our quality of life and well-being. Improvement in people's well-being is essential for both justice and sustainability. As is becoming increasingly clear, our current neoliberal model of economic growth cannot be relied upon to deliver this to the majority rather than a minority. Can well-being be delivered without continued economic growth? There is growing interest in the idea that there are emerging economic models, such as *co-production*, that might enable social well-being and flourishing. It refers to involving consumers in the manufacture of the goods and services they consume, thereby blurring the distinction between producer and consumer. One thing is for certain: humanity needs better yardsticks for measuring progress, based on well-being rather than on our current headline indicator—GDP.

2. Meeting the needs of both present and future generations (intra- and intergenerational equity). A key question is, what is the relationship between material consumption and needs, particularly considering the extent to which justice and equity are needs? There is growing evidence (Wilkinson and Pickett 2009) that inequality damages our capability

for flourishing, our ability to meet our needs. Increasingly, human need for social identity is defined by our ability to consume. This is "bling" culture. How do we supplant consumption-based social identity generation within a more just and sustainable framework?

3. Justice and equity in terms of recognition (Schlosberg 1999), process, procedure, and outcome. Justice is not a simple concept. Different ideological foundations can lead to very different conclusions and outcomes: for example, *utilitarian* (justice as the most beneficial outcome for wider society), *egalitarian* (justice as meeting individuals' needs), and *libertarian* (justice as fulfilling merit) perspectives can differ radically. Sen (2009) takes this as reason to argue for a goal of reducing manifest injustice, rather than seeking perfect justice. Both Sen and Nussbaum (2000) suggest a central role of the notion of capabilities for flourishing. Nussbaum's full-capability list includes life; bodily health; bodily integrity; senses, imagination, and thought; emotions; practical reason; affiliation; other species; play; and control over one's environment. Sen, on the other hand, suggests that communities must be involved in listing their own set of capabilities—more because such control over the conditions of life is necessary for justice than because capabilities may be culturally specific, although this latter factor should not be ignored.

4. Living within ecosystem limits (also called one-planet living). Despite several decades of research, the very concept of *environmental limits* remains controversial, especially in the United States. The Club of Rome report (Meadows et al. 1972) and, in the UK, the *Ecologist* magazine's 1972 "Blueprint for Survival" framed debate in terms of "limits to growth" that stimulated very powerful and well-funded counterarguments and rebuttals. By the 1990s, in public and political discourse around the world, the very idea of "limits" had been discredited by the apparent failure of predicted shortages of natural resources to emerge.

However, ecosystem limits are very real (Rockström et al. 2009b). Whether they constitute a fundamental limit to economic growth probably depends more on the nature of the economy than on the economy of nature. What is clear is that as constraints on natural resources have emerged, the capitalist economy has side-stepped them by shifting the crisis around in space, or between environmental domains. For example, in the United States, the approach of Peak Oil has triggered the cry of "Drill Baby Drill" exhorting Americans to exploit oil in yet more remote locations, and to develop unconventional gas and oil sources through "fracking" and tar-sands extraction, both of which involve significantly higher carbon emissions than conventional fossil fuels do. As a result, apparent limits in resource availability have been translated into still greater pressure on the climate system.

The concept of "Greenhouse Development Rights" (Baer et al. 2008) makes an allowance of emissions to meet basic needs, and takes into account the differential capabilities available to reduce emissions (as a function of disposable income) in attempting to determine globally just targets for emissions reduction. Typically, such assessment (e.g., Baer et al. 2008) concludes that rich countries need to make greater reductions in emissions than they currently do. In other words, as well as reducing their own emissions to zero, they need to also take responsibility for financing additional reductions in poor countries, or develop technical means to remove carbon dioxide from the atmosphere, the so-called Negative Emissions Technologies (NETs).

Sustainability, sustainable development, and sustainable communities are contested concepts on many levels. For the past thirty years they have been subject to eager and heated debate. Are they political constructs or rational, technical, or scientifically achievable notions? Are they "destinations," places that we will recognize when we get there, or "journeys" in local civic participation, or both? Should they focus on a "brown" agenda of poverty alleviation, infrastructure development, and public health, as many activists and academics in the Global South have argued, or a "green" agenda of wilderness preservation, greenspace provision, and climate change, as many activists and academics in the Global North would argue? One thing that is becoming increasingly clear is that the potential in these concepts will only be fully realized on two related conditions: first, that progressive activists in different domains should work together to build large-scale, even global "movements" (MoveOn.org [US], 38degrees.org.uk [UK]; Los Indignados [Spain]; MST [Brazil]); and second, that there is a shift from current reformist strategies toward a politics of transformational change. Incremental, linear change based on reform is unlikely to seriously challenge the underlying structures that (re)produce injustice and unsustainability.

59

Translation

Carmen Flys-Junquera and Carmen Valero-Garcés

In academic studies in general, the importance of translation, from one language to another, cannot be overestimated. In environmental studies specifically, what is added or lost when words such as "landscape" or "the environment" are translated into other languages? So, too, in interdisciplinary research, the connotations of a word can shift from one discipline to another and certainly even more so from one language to another. Thus issues of translation necessarily impact environmental studies. As Lynch (2008) remarks, landscapes, in different places in the world, shape and influence the development and use of the words used to describe them. For example, some cultures have developed extensive vocabularies to describe variations of rain, snow, or wind while others lack a precise vocabulary to describe processes that may not occur frequently in another geographic region. This illustrates that translation is often bound up in cultural issues even when addressing terms or processes that may most often be assigned to environmental studies or science. To address these difficulties, the translator must map out cultural differences before attempting translation.

For example, in *Topophilia*, geographer Yi-Fu Tuan has a four-page chapter on "wilderness" (1974) in which the term "wilderness" appears thirty-four times. The Spanish translation of the chapter uses thirteen different words to translate "wilderness," indicating the lack of a clear translation because the concept, with all its denotations and connotations, does not exist in Spanish (Valero-Garcés 2010). Another example can be found in the translation of Antonio Machado's classic literary landmark, *Campos de Castilla*. The title of this book been translated as *Lands of Castile and Other Poems* (2002), *The Landscape of Castile: Poems by Antonio Machado* (2004), and, in a bilingual edition, *Fields of Castile/ Campos de Castilla* (2007), suggesting that the terms "lands," "landscape," and "fields" are not synonymous in all languages.

There are numerous factors that might influence the decisions of a translator: linguistic differences in language systems; use of the language (pragmatics) and rhetorical and stylistic conventions; the translators' background and knowledge; and his/her intentionality or the theoretical framework used. The translator can opt for "a non-fluent or estranging translation style designed to make *visible* the presence of the translator," something that Lawrence Venuti calls "foreignization" (1998, 11). In the more recent translation of *Fields of Castile*, the translator has tried to be literal, forcing the reader of English to perceive the agricultural dimension of the word, rather than the culturally more fluent word "landscape." In the other translations, the translators chose to "domesticate" the translation—in other words, to adapt the original to an idiomatic text, creating the "illusion of transparency" as if no translation process existed (Venuti 1995 [2008], 11). Much depends on the type of landscape or fields the reader imagines. For many non-Spanish readers, "fields" evokes an image of gentle rolling green hills. However, Spanish readers might imagine instead the abrupt, irregular small patches of *pardo*, which is the color of Castile. In Castilian, "*pardo*" designates the color of grayish-brown parched dirt, a color with no adequate translation into other languages (Valero-Garcés 2009, 545–55).

The sciences often rely on Latin or established terminologies to ensure accuracy; in the humanities, where nuances and connotations are inherent in the culture, the challenge is much larger. But even this strategy is not always effective: Latin is not as accessible to modern readers, and translators often opt for the common name, which might actually identify a different species. For example, there are numerous Spanish translations of Henry David Thoreau's classic *Walden* (1854). Thoreau's mention of abundant birds causes problems for the translator and becomes an ornithologist's or birdwatcher's nightmare. The following table illustrates the variety in the different translations of Thoreau's reference to a "veery" (a small North American species) identified by Spanish translator and year of translation:

Walden: Veery
Scientific name: *Catharus fuscescens*
Common name: *Zorzal Dorsirrojo*
Molina y Vedia, 1945: *Tordo canoro*
Gárate, 1949: *Tordo cantor*
Sánchez Rodrigo, 1979: *Pardillo*
Alcoriza y Lastra, 2005: *Tordo*

The reader may assume that the translations correctly identify the species; but in fact, the translations identify different birds: a "*tordo*" is *curaeus curaeus*, a "*pardillo*" is a *cardueles cannabina*, while a "*zorzal*" is *turdus falcklandii*; none of them correspond to the original species that Thoreau identified.

When environmental issues are translated, misinterpretation can also occur. A relatively recent environmental issue in Spain is the repeated flooding of certain areas. In the latter part of the twentieth century, in the midst of the construction boom, many homes were built, legally and illegally, in "*cañadas*." According to

TRANSLATION   CARMEN FLYS-JUNQUERA AND CARMEN VALERO-GARCÉS an English-Spanish dictionary, a *"cañada"* is a "gully, ravine, stream or cattle/sheep track." However, for Spaniards, a *"cañada"* is more like an ancient, shallow, dried riverbed that has been used as a sheep track. Because it is shallow and wide and water no longer runs there, homes have been built there. With climate change and the increasing frequency of intense isolated cold fronts in the Mediterranean in early fall, known as *"gota fria"* ("cold drop"; at times referred to as "cut-off lows" in English), intense rains and floods that occur with minimal warning and stay in one place for several days have washed whole neighborhoods away. For any English reader, building a house in a gully or ravine makes little sense. A "gully" is imagined as too deep a place to sensibly build a house. Thus, the translation must explain not only the word but the environmental history and science of *"cañadas"* in light of climate change, perhaps using metaphors and new images, which, in turn, must also be known to the translator.

Therefore, translation has to be viewed both as a process and as a product. As environmental studies expand internationally, translation will become an increasingly important consideration. Most of the terms explored in *Keywords for Environmental Studies* are quite international by now, but the usage and interpretation of some of the keywords in this volume are necessarily culturally conditioned. How might these keywords change when translated to different languages and cultures? For example, terms such as "nature writing" or "pastoral" have no direct translation to Spanish, and possibly to other languages. In other cases, translation presents different connotations such as "environmentalism" in English and *"ecologismo"* in Spanish. "Translation," then, must be considered a keyword for environmental studies because it raises issues pertinent to both the humanities and the sciences. The purpose behind this essay is to shed light on these factors and point to a complementary project linked to this volume, a website that will present the translation of these keywords as well as abstracts of each essay translated into a number of different languages, signaling what may be lost in translation and what issues emerge with translation of each keyword.

60

Urban Ecology

Nik Heynen

Urban spaces are complex as a result of their socionatural relations and histories, which are shaped by both physical and social processes. The complexity necessitates theoretical frameworks that can demonstrate how urban socionatural outcomes result from the combination of social actions and physical surroundings. Henri Lefebvre suggests,

> Every social space is the outcome of a process with many aspects and many contributing currents, signifying and non-signifying, perceived and directly experienced, practical and theoretical. In short, every social space has a history, one invariably grounded in nature, in natural conditions that are at once primordial and unique in the sense that they are always and everywhere endowed with specific characteristics (site, climate, etc.). (Lefebvre 1991, 110)

Theory about urban ecology can potentially capture this complexity and contingency, but only if its evolution is taken seriously.

Ecology emphasizes site-specific characteristics and interactions between local environs and inhabitants of those environs. Tansley (1935) developed the notion of *ecosystems* and suggested that they were comprised not simply of inherently "natural" dynamics but also of human-made dynamics. He suggests that agriculture is only one example of a "natural entity" that is brought about through human activity and that, therefore, must be analyzed in terms of both its "naturalness" and its "anthropogenic derivates alike" (1935, 304).

However insightful, Tansley's comments do not adequately address the complexity inherent to urban ecology. While urban ecosystems are not necessarily more difficult to understand than others, they are inherently different from others because they are human built, and thus imbued with human politics, culture, and uneven power relations. Hence, although most, if not all, ecological principles apply to urban ecosystems, we must be careful when proceeding to understand the embedded social, political, and economic factors that govern the building, shaping, restructuring, etc., of urban ecosystems.

Notions of urban ecology, distinctly defined, while grounded in other scientific approaches, have come to represent a broad lens through which to investigate urban life. While the study of intrinsically "natural" processes gave way to the study of ecological relations, the inclusion of human activities, especially within urban areas, has contributed to a host of new understandings about human impacts on their local environments. Within social science, and informing cultural studies, the Chicago School's urban ecological model, propounded by scholars such as Robert Park, Ernest Burgess, Roderick McKenzie, and others, used natural ecological processes to help describe and understand social processes. Swapping the socially constructed for the naturally/materially present environment was one of the main contributions of the Chicago School's urban ecological approach (Berry and Kasarda 1977). While many of the systematic organizational components of the basic ecological model, as well as that of human ecology, remained constant within this approach, an individual's relations within, and to, his or her communities became the focal point of investigation.

The field of urban ecology has been critiqued for maintaining too much focus on "ecological" processes that continue to exclude humans. To this end, David Harvey suggests, "It is inconsistent to hold that everything in the world relates to everything else, as ecologists tend to, and then decide that the built environment and the urban structures that go into it are somehow outside of both theoretical and practical consideration. The effect has been to evade integrating understandings of the urbanizing process into environmental-ecological analysis" (1996, 427). However, a study originating in urban ecology necessitates that primacy be given to urban space, form, and process; take for instance the uneven ramifications of urban heat island effects. Alternatively, ecological processes, while implicit in much of urban cultural research, often tend to simply play the role of backdrop for other spatial and social processes.

Along these lines, Mark Gottdiener suggests, "The limitations of contemporary urban ecology are already in evidence. It possesses a biologically reductionist view of human relations which ignores the influence of class, status, and political power" (1985, 40). The recognition that classic urban ecological approaches impede further understanding of urban environmental change necessitates updating and improving the framework to take account of social power relations. While urban ecology provides the necessary context, that is, the urban, it lacks the stringent political-economic foundations necessary for an understanding of those processes that have contributed to the production of cities. Urban political ecology (UPE) has evolved as a response to this lack of inclusion of political-economic factors as contributors to ecological processes of change. The inclusion of political ecology within urban theory provides an appropriate way to discuss the relationship between urban political economy and urban ecology.

The emergence of urban political ecology has deliberately evolved through the theoretical and empirical development of previous urban environmental approaches and is arguably becoming one of the most important areas of urban research. Related to this, Roger Keil (2003) suggests that urban political ecology is an appropriate response to Lefebvre's call for creating an "urban science for an urban world" (2003, 23). Urban political ecology seeks to make up for lack of attention to the way social power relations contributed to uneven development within other urban theoretical/empirical perspectives, by prioritizing the effects that political economy and cultural processes have on urban environments (see Swyngedouw and Heynen 2003). Access to water, food, nonpolluted air, and green space within uneven matrices of class, race, gender, age, and physical ability would all pose the sort of examples urban political ecological investigations would engage.

An increasingly important thread within explicitly Marxist urban political ecology is focusing on the political processes inherent to the "human metabolism with nature" (Hayward 1994, 11). Tim Hayward's translation of Gernot Böhme and Engelbert Schramm suggests that a Marxist approach to these sorts of urban political ecological questions "continually forces reflection back to the material basis, to the concrete interaction between humans and nature . . . embracing not only productive appropriation of nature, but also the consumptive relation of humans to nature; of not just intentional engagement with nature, but also unintended effects" (1994, 8). Notions of metabolism are especially useful for unraveling the uneven socionatural relations inherent to urban ecological change. This is the case because ecologists have tended to address change in too neutral a way, too often. In turn, this often impedes our understanding of the inherent power relations that lead to the transformation of urban ecologies.

Marx modified the idea of metabolism as was first discussed by Justus von Liebig through his insights based on soil chemistry (see Foster 2000). Marx's theoretical incorporation of the idea of metabolization took on an independent meaning, as Erik Swyngedouw (2004) discusses. The initial German word "*Stoffwechsel*" simultaneously means "circulation," "exchange," and "transformation." Marx saw metabolization more historically as the necessary relation through which people interact with nature and themselves as part of nature. To this end, Marx suggested,

> Labour is, first of all, a process between man [*sic*] and nature, a process by which man, through his own actions, mediates, regulates, and controls the metabolism between himself and nature. He confronts the materials of nature as a force of nature. He sets in motion the natural forces which belong to his own body, his arms, legs, head, and hands, in order to appropriate the materials of nature in a form adapted to his own needs. Through this movement he acts upon external nature and changes it, and in this way he simultaneously changes his own nature. (Marx 1976 [1990], 283)

This explanation of metabolism opens up the investigative potential for thinkers working with ideas of urban ecology and/or urban political ecology to approach human/environment interactions and the environmental degradation that tends to always follow such interactions. As the world continues to rapidly urbanize, there will be a pressing need to think not just about urban ecological change in robust ways but also about how social power relations lead to more uneven urban development. The metabolic lens offered through Marxist urban political ecology can be marshaled to consider both who and what suffers and who or what benefits from the interrelated and interdependent processes of urban ecological change marching forward into the twenty-first century.

URBAN ECOLOGY NIK HEYNEN

Bibliography

Abélès, Marc. *The Politics of Survival*. Durham, NC: Duke University Press, 2010.

Aberley, Doug. "Interpreting Bioregionalism: A Story from Many Voices." In *Bioregionalism*, edited by M. V. McGinns, 13–42. New York: Routledge, 1999.

Adams, William Mark. *Green Development: Environment and Sustainability in the South*. New York: Routledge, 2001.

Adamson, Joni. *American Indian Literature, Environmental Justice, and Ecocriticism: The Middle Place*. Tucson: University of Arizona Press, 2001.

———. "Encounter with a Mexican Jaguar: Nature, NAFTA, Militarization, and Ranching in the U.S.-Mexico Borderlands." In *Globalization on the Line: Culture, Capital, and Citizenship at the U.S. Borders*, edited by Claudia Sadowski-Smith, 221–40. New York: Palgrave, 2002.

———. "Indigenous Literatures, Multinaturalism, and *Avatar*: The Emergence of Indigenous Cosmopolitics." *American Literary History (ALH) Special Issue: Sustainability in America* 24, no. 1 (2012a): 143–63.

———. "Medicine Food: Critical Environmental Justice Studies, Native North American Literature, and the Movement for Food Sovereignty." *Environmental Justice* 4, no. 4 (2011): 213–19.

———. "Seeking the Corn Mother: Transnational Indigenous Community Building and Organizing, Food Sovereignty, and Native Literary Studies." In *We the Peoples: Indigenous Rights in the Age of the Declaration*, edited by Elvira Pulitano, 228–49. New York: Cambridge University Press, 2012b.

———. "Source of Life: *Avatar*, *Amazonia*, and an Ecology of Selves." In *Material Ecocriticism*, edited by Serenella Iovino and Serpil Oppermann. Bloomington: University of Indiana Press, 2014.

———. "'¡Todos Somos Indios!' Revolutionary Imagination, Alternative Modernity, and Transnational Organizing in the Work of Silko, Tamez, and Anzaldúa." *Journal of Transnational American Studies* 4, no. 1 (May 2012c): 1–26.

Adamson, Joni, Mei Mei Evans, and Rachel Stein, eds. *The Environmental Justice Reader: Politics, Poetics, and Pedagogy*. Tucson: University of Arizona Press, 2002.

Adamson, Joni, and Kimberly N. Ruffin, eds. *American Studies, Ecocriticism, and Citizenship: Thinking and Acting in the Local and Global Commons*. New York: Routledge, 2013.

Addison, Joseph. "The Spectator, No. 412." In *The Norton Anthology of Theory and Criticism*, 2nd ed., edited by Vincent B. Leitch, et al., 43–45. New York: Norton, 2010.

Adger, Neil W. "Inequality, Environment, and Planning." *Environment and Planning A* 34, no. 10 (2002): 1716–19.

Agamben, Giorgio. *Homo Sacer: Sovereign Power and Bare Life*. Stanford, CA: Stanford University Press, 1998.

Agar, Nicholas. *Humanity's End: Why We Should Reject Radical Enhancement*. Cambridge, MA: MIT Press, 2010.

———. *Liberal Eugenics: In Defence of Human Enhancement*. Malden, MA: Blackwell, 2005.

Agyeman, Julian. *Introducing Just Sustainabilities: Policy, Planning, and Practice*. London: Zed, 2013.

———. *Sustainable Communities and the Challenge of Environmental Justice*. New York: New York University Press, 2005.

Agyeman, Julian, Robert Bullard, and Bob Evans. "Exploring the Nexus: Bringing Together Sustainability, Environmental Justice, and Equity." *Space and Polity* 6, no. 1 (2002): 70–90.

———, eds. *Just Sustainabilities: Development in an Unequal World*. Cambridge, MA: MIT Press, 2003.

Agyeman, Julian, et al., eds. *Speaking for Ourselves: Environmental Justice in Canada*. Vancouver: University of British Columbia Press, 2009.

Akama, John. "The Creation of the Maasai Image and Tourism Development in Kenya." In *Cultural Tourism in Africa: Strat-*

egies for the New Millennium, edited by John S. Akama and Patricia Sterry, 43–54. Arnhem, Netherlands: Association for Tourism and Leisure Education, 2002.

Alaimo, Stacy. *Bodily Natures: Science, Environment, and the Material Self*. Bloomington: Indiana University Press, 2010a.

———. "Discomforting Creatures: Monstrous Natures in Recent Films." In *Beyond Nature Writing: Expanding the Boundaries of Ecocriticism*, edited by Karla Armbruster and Kathleen Wallace, 279–96. Charlottesville: University of Virginia Press, 2001.

———. "Eluding Capture: The Science, Culture, and Pleasure of 'Queer' Animals." In *Queer Ecologies: Sex, Nature, Politics, Desire*, edited by Catriona Mortimer-Sandilands and Bruce Erickson, 51–72. Bloomington: Indiana University Press, 2010b.

Alaimo, Stacy, and Susan Hekman, eds. *Material Feminisms*. Bloomington: Indiana University Press, 2008.

Alcott, Blake. "Jevons' Paradox." *Ecological Economics* 54 (2005): 9–21.

Alexander, Natalie. "Sublime Impersonation: The Rhetoric of Personification in Kant." *Language and Liberation: Feminism, Philosophy, and Language*, edited by Christina Hendricks and Kelly Oliver, 243–69. Albany: State University of New York Press, 1999.

Alexander, Sasha, et al. "Opportunities and Challenges for Ecological Restoration within REDD+." *Restoration Ecology* 19, no. 6 (2011): 683–89.

Alford, John, and John Hibbing. "The New Empirical Biopolitics." *Annual Review of Political Science* 11, no. 1 (2008): 183–203.

Alkon, Alison Hope, and Julian Agyeman. *Cultivating Food Justice: Race, Class, and Sustainability*. Cambridge, MA: MIT Press, 2011.

Allen, Stephen, and Alexandra Xanthaki, eds. *Reflections on the UN Declaration on the Rights of Indigenous Peoples*. Oxford, UK: Hart, 2011.

Alpers, Paul. *What Is Pastoral?* Chicago: Chicago University Press, 1996.

Alston, Vermonja. "Environment." In *Keywords for American Cultural Studies*, edited by Bruce Burgett and Glenn Hendler, 101–3. New York: New York University Press, 2007.

Altman, Rebecca Gasior, et al. "Pollution Comes Home and Pollution Gets Personal: Women's Experience of Household Toxic Exposure." *Journal of Health and Social Behavior* 49 (2008): 417–35.

The American Heritage Dictionary. Boston: Houghton, Mifflin, 2009.

Amin, Samir. *Unequal Development*. New York: Monthly Review Press, 1976.

Anaya, James S. *Indigenous Peoples in International Law*. 2nd ed. New York: Oxford University Press, 2004.

Anderson, E. N., et al., eds. *Ethnobiology*: Hoboken, NJ: Wiley, 2012.

Anderson, Jill. "'Warm Blood and Live Semen and Rich Marrow and Wholesome Flesh!' A Queer Ecological Reading of Christopher Isherwood's *A Single Man*." *Journal of Ecocriticism* 31, no. 1 (2011): 51–66.

Andow, David, et al. "Preserving the Integrity of Manoomin in Minnesota." Wild Rice White Paper (2009). Available at https://www.cfans.umn.edu/sites/cfans.umn.edu/files/WhitePaperFinalVersion2011.pdf.

Annas, George, Lori Andrews, and Rosario M. Isasi. "Protecting the Endangered Human: Toward an International Treaty Prohibiting Cloning and Inheritable Alterations." *American Journal of Law & Medicine* 28, nos. 2–3 (2002): 151–78.

Arendt, Hannah. *The Origins of Totalitarianism*. New York: Harcourt Brace Jovanovich, 1973 (1948).

Armbruster, Karla, and Kathleen R. Wallace, eds. *Beyond Nature Writing: Expanding the Boundaries of Ecocriticism*. Charlottesville: University Press of Virginia, 2001.

Armiero, Marco. *A Rugged Nation: Mountains and the Making of Modern Italy*. Cambridge: White Horse Press, 2011.

———. "Seeing like a Protester: Nature, Power, and Environmental Struggles." *Left History* 13, no. 1 (2008): 59–76.

Armiero, Marco, and Giacomo D'Alisa. "La ciudad de los residuos: Justicia ambiental e incertitumbre en la crisis de los residuos Campania." *Ecología Política* 41 (2011): 97–105.

Armstrong, Rachel. "On Biomimicry as Parametric Snake Oil." Review of *Architecture Follows Nature: Biomimetic Principles for Innovative Design*, by Ilaria Mazzoleni. *Architectural Review* (July 11, 2013). Available at http://www.architectural-review.com/reviews/rachel-armstrong-on-biomimicry-as-parametric-snake-oil/8650000.article.

Arnhart, Larry. *Darwinian Conservatism*. Exeter, UK: Imprint Academic, 2005.

Arnold, Ron. *EcoTerror: The Violent Agenda to Save Nature: The World of the Unabomber.* Bellevue, WA: Free Enterprise Press, 2010 (1997).

———. "Eco-Terrorism: Environmental Extremists Have Declared Guerrilla War on Resource Developers and the Environmental Mainstream Stands by Silently." *Reason* (February 1983): 31–36.

Ashman, Keither, and Phillip Baringer, eds. *After the Science Wars: Science and the Study of Science.* New York: Routledge, 2001.

Athanasiou, Tom. *Divided Planet: The Ecology of Rich and Poor.* Athens: University of Georgia Press, 1998.

Avery, Oswald T., Colin M. MacLeod, and Maclyn McCarty. "Studies on the Chemical Nature of the Substance-Inducing Transformation of Pneumococcal Types: Induction of Transformation by a Desoxyribonucleic Acid Fraction Isolated from Pneumococcus Type III." *Journal of Experimental Mediciine* 79, no. 2 (1944): 137–58.

Azzarello, Robert. *Queer Environmentality: Ecology, Evolution, and Sexuality in American Literature.* Hampshire, UK: Ashgate, 2012.

Baer, Paul, et al. *The Greenhouse Development Rights Framework: The Right to Development in a Climate-Constrained World.* Berlin: Heinrich Böll Foundation, 2008.

Bagemihl, Bruce. *Biological Exuberance: Animal Homosexuality and Natural Diversity.* New York: St. Martin's, 1999.

Bailey, Ronald. *Liberation Biology: The Scientific and Moral Case for the Biotech Revolution.* Amherst, NY: Prometheus, 2005.

Baillie, Jonathan E. M., Craig Hilton-Taylor, and Simon N. Stuart, eds. *2004 IUCN Red List of Threatened Species: A Global Species Assessment.* Gland, Switzerland: IUCN, 2004.

"Bali Principles of Climate Justice." *International Climate Justice Network.* 2002. Available at http://www.indiaresource.org/issues/energycc/2003/baliprinciples.html.

Bang, Megan, et al. "Innovations in Culturally Based Science Education through Partnerships and Community." In *New Science of Learning: Cognition, Computers, and Collaboration in Education,* edited by M. Khine and I. Saleh, 569–92. New York: Springer, 2010.

Banting, Pamela. "Abandoning the Fort: Cultural Difference and Biodiversity in Canadian Literature and Criticism." In *Teaching North American Environmental Literature,* edited by Laird Christensen, Mark C. Long, and Fred

Waage, 112–25. New York: Modern Language Association of America, 2008.

———. "The Land Writes Back: Notes on Four Western Canadian Writers." In *Literature of Nature,* edited by Patrick D. Murphy, 140–46. Chicago: Dearborn, 1998.

Barad, Karen. "Nature's Queer Performativity." In *Kvinder, Køn og forskning / Women, Gender and Research* nos. 1–2 (2012): 25–53.

Barbieri, Marcello. "On the Origin of Language: A Bridge between Biolinguistics and Biosemiotics." *Biosemiotics* 3, no. 2 (2010): 201–23.

Barca, Stefania. *Enclosing Water: Nature and Political Economy in a Mediterranean Valley, 1796–1916.* Cambridge, MA: White Horse Press, 2010.

Barillas, William. *The Midwestern Pastoral: Place and Landscape in Literature of the American Heartland.* Athens: Ohio University Press, 2006.

Barnosky, Anthony D., et al. "Has the Earth's Sixth Mass Extinction Already Arrived?" *Nature* 471 (2011): 51–57.

Barrell, John, and John Bull, eds. *A Book of English Pastoral Verse.* New York: Oxford University Press, 1975.

Barrow, Mark V. *Nature's Ghosts: Confronting Extinction from the Age of Jefferson to the Age of Ecology.* Chicago: University of Chicago Press, 2009.

Barry, Joyce M. *Standing Our Ground: Women, Environmental Justice, and the Fight to End Mountaintop Removal.* Athens: Ohio University Press, 2012.

Bartuska, Tom J. "The Built Environment: Definition and Scope." In *The Built Environment: A Collaborative Inquiry into Design and Planning,* edited by Wendy R. McClure and Tom J. Bartuska, 3–14. Hoboken, NJ: Wiley, 2007.

Bate, Jonathan. *The Song of the Earth.* Cambridge, MA: Harvard University Press, 2000.

Bauman, Whitney A., Richard R. Bohannon II, and Kevin J. O'Brien, eds. *Grounding Religion: A Field Guide to the Study of Religion and Ecology.* New York: Routledge, 2011.

Beck, Ulrich. *Ecological Enlightenment: Essays on the Politics of the Risk Society.* Atlantic Highlands, NJ: Humanities Press, 1995.

———. *Power in the Global Age: A New Political Economy.* Cambridge, MA: Polity Press, 2005.

———. *Risk Society: Towards a New Modernity.* Newbury Park, CA: Sage, 1992 (1986).

———. "World Risk Society as Cosmopolitan Society? Ecological Questions in a Framework of Manufactured Uncertainties." *Theory, Culture, and Society* 13, no. 4 (1996): 1–32.

Bekoff, Marc. *The Emotional Lives of Animals: A Leading Scientist Explores Animal Joy, Sorrow, and Empathy—and Why They Matter*. Novato, CA: New World Library, 2007.

Bekoff, Marc, and Jessica Pierce. *Wild Justice: The Moral Lives of Animals*. Chicago: University of Chicago Press, 2009.

Bello, Andrés. *Selected Writings of Andrés Bello*. Translated by Frances M. Lopez-Morillas. New York: Oxford University Press, 1997.

Benedict, Ruth, and Gene Weltfish. *In Henry's Backyard: The Races of Mankind*. Public Affairs Pamphlet No. 85. New York: Public Affairs Committee, 1943.

Bennett, Jane. *Vibrant Matter: A Political Ecology of Things*. Durham, NC: Duke University Press, 2010.

Benton-Benai, E. *The Mishomis Book: The Voice of the Ojibway*. Hayward, WI: Indian Country Communications, 1988.

Benyus, Janine. *Biomimicry: Innovation Inspired by Nature*. New York: William Morrow, 1997.

Berg, Peter. *Envisioning Sustainability*. San Francisco: Subculture Books, 2009.

Berg, Peter, and Raymond Dasmann. "Reinhabiting California." *Ecologist* 7, no. 10 (1977): 399–401.

Berila, Beth. "Toxic Bodies? ACT UP's Disruption of the Heteronormative Landscape of the Nation." In *New Perspectives on Environmental Justice: Gender, Sexuality, and Activism*, edited by Rachel Stein, 127–36. New Brunswick, NJ: Rutgers University Press, 2004.

Berkes, Fikret. "Traditional Ecological Knowledge in Perspective." In *Traditional Ecological Knowledge: Concepts and Cases*, edited by Julian Inglis, 1–9. Ontario, Canada: International Program on Traditional Ecological Knowledge and International Development Research, 1993.

Berkes, Fikret, Johan Colding, and Carl Folke. "Rediscovery of Traditional Ecological Knowledge as Adaptive Management." *Ecological Applications* 10, no. 5 (2000): 1251–62.

Berry, Brian, and John D. Kasarda. *Contemporary Urban Ecology*. New York: Macmillan, 1977.

Berry, Edward W. "The Term Psychozoic." *Science* 44 (1925): 16.

Berry, Thomas. *The Dream of the Earth*. San Francisco: Sierra Club Books, 1988.

Bérubé, Michael. "The Humanities, Declining? Not According to the Numbers." *Chronicle of Higher Education*. Last modified July 1, 2013. Available at http://chronicle.com/article/The-Humanities-Declining-Not/140093.

Best, Steven. "It's War! The Escalating Battle between Activists and the Corporate-State Complex." In *Terrorists or Freedom Fighters? Reflections on the Liberation of Animals*, edited by Steven Best and Anthony J. Nocella II, 300–339. New York: Lantern Books, 2004.

Biehl, Janet. *Rethinking Eco-Feminist Politics*. Boston: South End Press, 1991.

Biehl, Janet, and Peter Staudenmaier. *Ecofascism Revisited: Lessons from the German Experience*. 2nd ed. Porsgrunn, Norway: New Compass Press, 2011.

Biersack, Aletta. "Introduction." In *Re-imagining Political Ecology*, edited by Aletta Biersack and James Greenberg, 3–40. Durham, NC: Duke University Press, 2006.

Birch, Eugenie Ladner, and Susan M. Wachter, eds. *Growing Greener Cities: Urban Sustainability in the Twenty-First Century*. Philadelphia: University of Pennsylvania Press, 2008.

Blaikie, Piers. *The Political Economy of Soil Erosion in Developing Countries*. London: Longman, 1985.

Blaikie, Piers, and Harold Brookfield. *Land Degradation and Society*. London: Methuen, 1987.

Blaser, Mario. "Political Ontology: Cultural Studies without 'Cultures'?" *Cultural Studies* 23 (2009): 873–96.

———. *Storytelling Globalization from the Chaco and Beyond*. Durham, NC: Duke University Press, 2010.

Boggs, Grace Lee. *The Next American Revolution: Sustainable Activism for the 21st Century*. Berkeley: University of California Press, 2012.

Bogost, Ian. "What Is Object-Oriented Ontology? A Definition for Ordinary Folks." *Ian Bogost Videogame Theory, Criticism, Design*. Last modified December 8, 2009. Available at http://www.bogost.com/blog/what_is_objectoriented_ontolog.shtml.

Bohlen, Jim. *Making Waves: The Origins and Future of Greenpeace*. Montreal: Black Rose Books, 2001.

Böhme, Gernot, and Engelbert Schramm. *Soziale Naturwissenschaft: Wege zu einer Erweiterung der Okologie*. Frankfurt: Fischer, 1984.

Bond, Patrick. *The Politics of Climate Justice: Paralysis Above, Movement Below*. Scottsville, South Africa: University of Kwazulu-Natal Press, 2012.

Bonser, R. H. C., and J. F. V. Vincent. "Technology Trajectories, Innovation, and the Growth of Biomimetics." *Proceedings of the Institution of Mechanical Engineers Part C: Journal of Mechanical Engineering Science* 221 (2007): 1177–80.

Bookchin, Murray. *The Philosophy of Social Ecology: Essays on Dialectical Naturalism.* New York: Black Rose Books, 1990.

———. *Social Ecology and Communalism.* Oakland, CA: AK Press, 2007.

———. "Social versus Deep Ecology: A Challenge for the Ecology Movement." *Green Perspectives: Newsletter of the Green Program Project* 4–5 (Summer 1987): n.p.

Borlik, Todd A. *Ecocriticism and Early Modern English Literature: Green Pastures.* New York: Routledge, 2011.

Bostrom, Nick. "A History of Transhumanist Thought." *Journal of Evolution and Technology* 14, no. 5 (2005). Available at http://jetpress.org/volume14/bostrom.html.

Boswell-Penc, Maia. *Tainted Milk: Breastmilk, Feminisms, and the Politics of Environmental Degradation.* Albany: State University of New York Press, 2006.

Botkin, Daniel. *Discordant Harmonies: A New Ecology for the 21st Century.* New York: Oxford University Press, 1990.

Boulter, Michael Charles. *Extinction: Evolution and the End of Man.* New York: Columbia University Press, 2002.

Bourke, Simon, and Tony Meppem "Privileged Narratives and Fictions of Consent in Environmental Discourse." *Local Environment* 5, no 3. (2000): 299–310.

Bowen, Frances. *After Greenwashing: Symbolic Corporate Environmentalism and Society.* Cambridge: Cambridge University Press, 2014.

Braidotti, Rosi, et al. "The Humanities and Changing Global Environments (Chapter 93)." *Changing Global Environments.* World Social Science Report, ISSC, UNESCO (2013): 506–7.

Braun, Bruce. "Environmental Issues: Inventive Life." *Progress in Human Geography* 32, no. 5 (2008): 667–79.

———. *The Intemperate Rainforest: Nature, Culture, and Power on Canada's West Coast.* Minneapolis: University of Minnesota Press, 2002.

Braun, Bruce, and Noel Castree, eds. *Remaking Reality: Nature at the Millennium.* New York: Routledge, 1998.

Braun, Bruce, and Sarah Whatmore, eds. *Political Matter: Technoscience, Democracy, and Public Life.* Minneapolis: University of Minnesota Press, 2010.

Braungart, Michael, and William McDonough. *Cradle to Cradle: Remaking the Way We Make Things.* New York: North Point Press, 2002.

Breyer, Stephen. *Breaking the Vicious Circle: Toward Effective Risk Regulation.* Cambridge, MA: Harvard University Press, 1993.

Brickman, Ronald, Sheila Jasanoff, and Thomas Ilgen. *Controlling Chemicals: The Politics of Regulation in Europe and the United States.* Ithaca, NY: Cornell University Press, 1985.

Brogan, T. V. F. "Poetics." In *The New Princeton Encyclopedia of Poetry and Poetics,* edited by Alex Preminger, Terry V. F. Brogan, and Frank J. Warnke, 929–38. Princeton, NJ: Princeton University Press, 1993.

Brown, Phil. *Toxic Exposures: Contested Illnesses and the Environmental Health Movement.* New York: Columbia University Press, 2007.

Brown, Phil, and Alissa Cordner. "Lessons Learned from Flame Retardant Use and Regulation Could Enhance Future Control of Potentially Hazardous Chemicals." *Health Affairs* 30, no. 5 (2011): 906–14.

Brown, Phil, and Edwin J. Mikkelsen. *No Safe Place: Toxic Waste, Leukemia, and Community Action.* Berkeley: University of California Press, 1990.

Brown, Phil, Rachel Morello-Frosch, and Steve Zavestoski. *Contested Illnesses: Citizens, Science, and Health Social Movements.* Berkeley: University of California Press, 2011.

Brown, Phil, et al. "The Politics of Asthma Suffering: Environmental Justice and the Social Movement Transformation of Illness Experience." *Social Science and Medicine* 57 (2003): 453–64.

Bryant, Bunyan, and Paul Mohai, eds. *Race and the Incidents of Environmental Hazards: A Time for Discourse.* Boulder, CO: Westview, 1992.

Bryant, Raymond L., and Sinéad Bailey. *Third World Political Ecology.* New York: Routledge, 1997.

Bryson, Scott J. "Introduction." In *Ecopoetry: A Critical Introduction,* edited by Scott J. Bryson. Salt Lake City: University of Utah Press, 2002.

Buckley, Ralf. *Ecotourism: Principles and Practices.* Wallingford, UK: CAB International, 2009.

Buell, Lawrence. *The Environmental Imagination: Thoreau, Nature Writing, and the Formation of American Culture.* Cambridge, MA: Belknap Press of Harvard University Press, 1995.

———. *The Future of Environmental Criticism: Environmental Crisis and Literary Imagination*. Oxford: Blackwell, 2005.

———. "Some Emerging Thoughts on Ecocriticism." *Qui Parle* 19, no. 3 (Spring/Summer 2011): 87–115.

Buhrs, Ton. "Sharing Environmental Space: The Role of Law, Economics, and Politics." *Journal of Environmental Planning and Management* 47, no. 3 (2004): 429–47.

"Bukit Lawang: Natural Disaster or Ecoterrorism?" *Down to Earth*, no. 59. Last modified November 2003. Available at http://www.downtoearth-indonesia.org/story/bukit-lawang-natural-disaster-orecoterrorism.

Bulkeley, Harriet. *Cities and Climate Change*. New York: Routledge, 2013.

Bullard, Robert D. "Confronting Environmental Racism in the Twenty-First Century." In *The Colors of Nature: Culture, Identity, and the Natural World*, edited by Alison H. Deming and Lauret E. Savoy, 90–97. Minneapolis, MN: Milkweed, 2002.

———. *Dumping in Dixie: Race, Class, and Environmental Quality*. 3rd ed. Boulder, CO: Westview, 2000 (1990).

———. "Environmental Justice in the 21st Century." In *Debating the Earth*, edited by John Drysek and David Schlosberg. Oxford: Oxford University Press, 2005.

———. *Growing Smarter: Achieving Livable Communities, Environmental Justice, and Regional Equity*. Cambridge, MA: MIT Press, 2008.

Burawoy, Michael. *Manufacturing Consent: Changes in the Labor Process under Monopoly Capitalism*. Chicago: University of Chicago Press, 1982.

Burgett, Bruce, and Glenn Hendler, eds. *Keywords for American Cultural Studies*. New York: New York University Press, 2007.

Burke, Edmund. *A Philosophical Enquiry into the Origin of Our Ideas of the Sublime and Beautiful*. Edited by Adam Phillips. New York: Oxford University Press, 1990 (1757).

Burnham, Philip. *Indian Country, God's Country: Native Americans and the National Parks*. Washington, DC: Island Press, 2000.

Burris, Sidney. *The Poetry of Resistance: Seamus Heaney and the Pastoral Tradition*. Athens: Ohio University Press, 1990.

Bush, Ray. "Food Riots: Poverty, Power, and Protest." *Journal of Agrarian Change* 10, no. 1 (2010): 119–29.

Byrd, Jodi, and Michael Rothberg. "Between Subalternity and Indigeneity." *Interventions* 13, no. 1 (2011): 1–12.

"C40 Cities." *Copenhagen Climate Communiqué*. Last modified 2009. Available at http://www.c40cities.org/news/news-20091215.jsp.

Cadena, Marisol de la. "Indigenous Cosmopolitics in the Andes: Conceptual Reflections beyond 'Politics.'" *Cultural Anthropology* 25 (2010): 334–70.

Cadena, Marisol de la, and Orin Starn, eds. *Indigenous Experience Today*. Oxford: Berg, 2007.

Caldwell, Lynton. "Biopolitics: Science, Ethics, and Public Policy." *Yale Review* 54, no. 1 (1964): 1–16.

Callicott, J. Baird. "Animal Liberation: A Triangular Affair." *Environmental Ethics* 2 (Winter 1980): 311–38.

———. "Environmental Philosophy IS Environmental Activism: The Most Radical and Effective Kind." In *Environmental Philosophy and Environmental Activism*, edited by Don E. Mariettta Jr. and Lester Embree, 19–35. Lanham, MD: Rowman & Littlefield, 2002 (1995).

Calvino, Italo. *Why Read the Classics?* New York: Pantheon, 1999.

Campbell, Lisa M., Noella J. Gray, and Zoë A. Meletis. "Political Ecology Perspectives on Ecotourism to Parks and Protected Areas." In *Transforming Parks and Protected Areas: Policy and Governance in a Changing World*, edited by Kevin S. Hanna, Douglas A. Clark, and D. Scott Siocombe, 200–221. New York: Routledge, 2008.

Campbell, T. Colin, and Thomas M. Campbell. *The China Study: The Most Comprehensive Study of Nutrition Ever Conducted and the Startling Implications for Diet, Weight Loss, and Long-term Health*. Dallas, TX: BenBella Books, 2006.

Cardoso, Fernando Henrique. "Dependency and Development in Latin America." *New Left Review* 74 (1972): 83–95.

Carmichael, Stokely. "Black Power." In *To Free a Generation: The Dialectics of Liberation*, edited by David Cooper. New York: Collier, 1968.

Carmin, JoAnn, and Julian Agyeman, eds. *Environmental Inequalities beyond Borders: Local Perspectives and Global Injustices*. Cambridge, MA: MIT Press, 2011.

Caro, Tim, et al. "Conservation in the Anthropocene." *Conservation Biology* 26, no. 1 (2011): 185–88.

Carson, Rachel. *Silent Spring*. New York: Houghton Mifflin Harcourt, 1962.

Carter, Christopher. "The Human Genome Is Composed of Viral DNA: Viral Homologues of the Protein Products Cause Alzheimer's Disease and Others via Autoimmune Mechanisms." Available from *Nature Precedings,* http://dx.doi.org/10.1038/npre.2010.4765.1 (2010).

Carter, Majora. "Three Stories of Local Eco-Entrepreneurship." TED Talk. 2010. Available at http://www.ted.com/talks/majora_carter_3_stories_of_local_ecoactivism.html.

Casteel, Sarah Phillips. *Second Arrivals: Landscape and Belonging in Contemporary Writing of the Americas.* Charlottesville: University of Virginia Press, 2007.

Castells, Manuel. *The Information Age: Economy, Society, and Culture.* Vol. 3, *End of Millennium.* Malden, MA: Blackwell, 1997a.

———. *The Information Age: Economy, Society, and Culture.* Vol. 2, *The Power of Identity.* Malden, MA: Blackwell, 1997b.

———. *The Information Age: Economy, Society and Culture.* Vol. 1, *The Rise of the Network Society.* Malden, MA: Blackwell, 1996.

Castree, Noel. *Making Sense of Nature.* Abingdon, UK: Routledge, 2013.

Castree, Noel, and Catherine Nash. "Posthuman Geographies." *Social & Cultural Geography* 7, no. 4 (2006).

Castree, Noel, et al. "Changing the Intellectual Climate." *Nature Climate Change* 4 (September 2014): 763–68.

Ceballos-Lascuráin, Héctor. *Tourism, Ecotourism, and Protected Areas: The State of Nature-based Tourism around the World and Guidelines for Its Development.* Gland, Switzerland: International Union for Conservation of Nature, 1996.

Cederlöf, Gunnel, and K. Sivaramakrishnan. *Ecological Nationalisms: Nature, Livelihoods, and Identities in South Asia.* Seattle: University of Washington Press, 2006.

Cella, Matthew J. C. *Bad Land Pastoralism in Great Plains Fiction.* Iowa City: University of Iowa Press, 2010.

"Center for Responsible Travel." *Handbook 1: A Simple User's Guide to Certification for Sustainable Tourism and Ecotourism.* Center for Ecotourism and Sustainable Development. Accessed May 30, 2013a. Available at http://www.responsibletravel.org/ resources/documents/reports/Ecotourism_Handbook_I.pdf.

"Center for Responsible Travel." *Handbook 3: Practical Steps for Marketing Tourism Certification.* Center for Ecotourism and Sustainable Development. Accessed May 30, 2013b. Available at http://www.responsibletravel.org/resources/documents/reports/Ecotourism_Handbook_III.pdf.

Chakrabarty, Dipesh. "The Climate of History: Four Theses." *Critical Inquiry* 35 no. 2 (Winter 2009): 197–222.

Chalecki, Elizabeth. "A New Vigilance: Identifying and Reducing the Risks of Environmental Terrorism." *Global Environmental Politics* 2, no. 1 (2002): 46–64.

Chateaubriand, François-René, vicomte de. *Atala and René.* Translated by Irving Putter. Berkeley: University of California Press, 1980 (1801).

Chávez, César. "Wrath of Grapes Speech." In *The Words of César Chávez,* edited by Richard J. Jensen and John C. Hammerback. College Station: Texas A&M University Press, 2002.

Chen, Mel. *Animacies: Biopolitics, Racial Mattering, and Queer Affect.* Durham, NC: Duke University Press, 2012.

Chisholm, Diane. "Biophilia, Creative Involution, and the Ecological Future of Queer Desire." In *Queer Ecologies: Sex, Nature, Politics, Desire,* edited by Catriona Mortimer-Sandilands and Bruce Erickson, 359–81. Bloomington: Indiana University Press, 2010.

Chomsky, Noam. "Terror and Just Response." In *Terrorism and International Justice*, edited by James Sterba. New York: Oxford University Press, 2003.

Christ, Carol P. *Rebirth of the Goddess: Finding Meaning in Feminist Spirituality.* Reading MA: Addison-Wesley, 1997.

———. *She Who Changes: Re-Imagining the Divine in the World.* New York: Palgrave Macmillan, 2003.

Christensen, Laird, Mark C. Long, and Fred Waage, eds. *Teaching North American Environmental Literature.* New York: Modern Language Association of America, 2008.

Christoff, Peter. "Ecological Modernization, Ecological Modernities." *Environmental Politics* 5, no. 3 (1996): 476–500.

Clare, Eli. *Exile and Pride: Disability, Queerness, and Liberation.* Cambridge, MA: South End Press, 1999.

Clark, Timothy. *The Cambridge Introduction to Literature and the Environment.* Cambridge: Cambridge University Press, 2011.

Clarke, Jeanne Nienaber, and Daniel McCool. *Staking Out the Terrain: Power Differentials among Natural Resource Management Agencies.* Albany: State University of New York Press, 1985.

Clements, Frederic E. *Plant Succession.* Washington, DC: Carnegie Institute of Washington, 1916.

Clifford, James, and George Marcus. *Writing Culture: The Poetics and Politics of Ethnography*. Berkeley: University of California Press, 1986.

Cobb, John. *Is It Too Late? A Theology of Ecology*. Denton, TX: Environmental Ethics Books, 1972.

Coetzee, J. M. *White Writing: On the Culture of Letters in South Africa*. New Haven, CT: Yale University Press, 1988.

Colchester, Marcus. "Conservation Policy and Indigenous Peoples." *Environmental Science & Policy* 7, no. 3 (2004): 145–53.

Cole, Luke, and Sheila Foster. *From the Ground Up: Environmental Racism and the Rise of the Environmental Justice Movement*. New York: New York University Press, 2001.

Coletta, John W. "Literary Biosemiotics and the Postmodern Ecology of John Clare." *Semiotica* 127, nos. 1–4 (1999): 239–71.

Collectif Argos. *Climate Refugees*. Cambridge, MA: MIT Press, 2010.

Commoner, Barry. *The Closing Circle: Nature, Man, and Technology*. New York: Knopf, 1971.

———. "The Environmental Cost of Economic Growth." In *Population, Resources, and the Environment*, 339–63. Washington, DC: U.S. Government Printing Office, 1972.

Confalonieri, U., et al. "Human Health: Climate Change 2007; Impacts, Adaptation, and Vulnerability." In *Fourth Assessment Report of the Intergovernmental Panel on Climate Change*, edited by M. L. Parry, et al., 391–431. Cambridge: Cambridge University Press, 2007.

Connell, Joseph H. "Citation-Classic: Diversity in Tropical Rain-Forests and Coral Reefs." *Current Contents/Agriculture, Biology & Environmental Sciences* 46 (1987): 16.

———. "Diversity in Tropical Rain Forests and Coral Reefs." *Science* 199, no. 4335 (1978): 1302–10.

Connelly, Steve. "Mapping Sustainable Development as a Contested Concept." *Local Environment* 12, no. 3 (2007): 259–78.

Conzen, Michael P. *The Making of the American Landscape*. New York: Routledge, 1994.

Corbera, Esteve. "Problematizing REDD+ as an Experiment in Payments for Ecosystem Services." *Current Opinion in Environmental Sustainability* 4, no. 6 (2012): 612–19.

Corburn, Jason. *Street Science: Community Knowledge and Environmental Health Justice*. Cambridge, MA: MIT Press, 2005.

Costanza, Robert, ed. *Ecological Economics: The Science and Management of Sustainability*. New York: Columbia University Press, 1991.

Costanza, Robert, et al. "Development: Time to Leave GDP Behind." *Nature* 505, no. 7483 (2014): 283–85.

Costanza, Robert, et al. *An Introduction to Ecological Economics*. Boca Raton, FL: St. Lucie Press, 1997.

Costanza Robert, et al. "Quality of Life: An Approach Integrating Opportunities, Human Needs, and Subjective Well-Being." *Ecological Economics* 61 nos. 2–3 (2007): 267–76.

Crick, Francis, and James Watson. "Molecular Structure of Nucleic Acids." *Nature* 25 (1953): 737.

Croll, Elisabeth, and David Parkin. *Bush Base, Forest Farm: Culture, Environment, and Development*. New York: Routledge, 1992.

Cronon, William. "The Trouble with Wilderness; or, Getting Back to the Wrong Nature." In *Uncommon Ground: Rethinking the Human Place in Nature*, edited by William Cronon. New York: Norton, 1995.

Cruikshank, Julie. *Do Glaciers Listen? Local Knowledge, Colonial Encounters, and Social Imagination*. Seattle: University of Washington Press, 2005.

Crutzen, Paul J. "Geology of Mankind." *Nature* 415 (2002): 23.

Crutzen, Paul J., and Eugene F. Stoermer. The "Anthropocene." *Global Change Newsletter* 41 (2000): 17–18.

Crystal, David. *Language Death*. Cambridge: Cambridge University Press, 2000.

Curtin, Deane. "Toward an Ecological Ethic of Care." *Hypatia* 6, no. 1 (1991): 60–74.

Cutter, Susan L., and William H. Renwick. *Exploitation, Conservation, Preservation: A Geographic Perspective on Natural Resource Use*. 4th ed. Hoboken, NJ: Wiley, 2004.

Dahl, Robert A. *Who Governs? Democracy and Power in an American City*. New Haven, CT: Yale University Press, 1961.

Daily, Gretchen C., et al. "Ecosystem Services: Benefits Supplied to Human Societies by Natural Ecosystems." *Issues in Ecology* 2 (1997): 1–16.

Daily, Gretchen C., et al. "The Value of Nature and the Nature of Value." *Science* 289, no. 5478 (2000): 395–96.

Daly, Herman E., and Joshua C. Farley. *Ecological Economics:*

Principles and Application. Washington, DC: Island Press, 2004.

Dana, Samuel Trask, and Sally K. Fairfax. *Forest and Range Policy: Its Development in the United States.* 2nd ed. New York: McGraw-Hill, 1980.

Dankelman, Irene. "Women Advocating for Sustainable Livelihoods and Gender Equality on the Global Stage." In *Women Reclaiming Sustainable Livelihoods: Spaces Lost, Spaces Gained*, edited by Wendy Harcourt. London: Palgrave, 2012.

Darwin, Charles. *The Descent of Man and Selection in Relation to Sex.* New York: New York University Press, 1989 (1871).

———. *On the Origin of Species by Means of Natural Selection; or, The Preservation of Favoured Races in the Struggle for Life.* London: John Murray, 1859.

———. *On the Origin of Species by Means of Natural Selection and The Descent of Man and Selection in Relation to Sex.* New York: Modern Library, 1936.

David, Peter F. "Wild Rice (Manoomin) Abundance and Harvest in Northern Wisconsin in 2005." In *Great Lakes Indian Fish and Wildlife Commission Administrative Report* 08–22 (2008).

Davion, Victoria. "Ecofeminism." In *A Companion to Environmental Philosophy*, edited by Dale Jamieson, 233–47. Oxford, UK: Blackwell, 2001.

Davis, Ellen F. *Scripture, Culture, and Agriculture: An Agrarian Reading of the Bible.* Cambridge: Cambridge University Press, 2008.

Davoudi, Simin. "Planning and the Twin Discourses of Sustainability." In *Planning for a Sustainable Future*, edited by Antonia Layard, Simin Davoud, and Susan Batty, 81–99. London: Spon, 2001.

Deacon, Terrence W. "Emergence: The Hole at the Wheel's Hub." In *The Re-Emergence of Emergence: The Emergentist Hypothesis from Science to Religion*, edited by Philip Clayton and Paul Davies, 11–50. Oxford, UK: Oxford University Press, 2006.

De Castro, Eduardo Viveiros. "Cosmological Deixis and Amerindian Perpsectivism." *Journal of the Royal Anthropological Institute* 4, no. 3 (1998): 469–88.

De Felip, Elena, and Alessandro Di Domenico. "Studio epidemiologico sullo stato di salute e sui livelli d'accumulo dei contaminanti organici persistenti nel sangue e nel latte materno in gruppi di popolazione a differente rischio d'esposizione nella Regione Campania." Rome: Istituto Superiore di Sanità, 2010. Accessed May 2, 2012. Available at http://speciali.espresso.repubblica.it/pdf/sebiorec2010.pdf.

Deleuze, Giles, and Felix Guattari. *A Thousand Plateaus: Capitalism and Schizophrenia.* Minneapolis: University of Minnesota Press, 1987.

DeLoughrey, Elizabeth M., and George B. Handley, eds. *Postcolonial Ecologies: Literatures of the Environment.* New York: Oxford University Press, 2011.

Dennis, John. *The Grounds of Criticism in Poetry.* Menston, UK: Scholar Press, 1971.

Derrida, Jacques. *The Animal That Therefore I Am*, edited by Marie-Louise Mallet. Translated by David Wills. New York: Fordham University Press, 2008.

Descola, Philippe, and Gisli Pálsson. *Nature and Society: Anthropological Perspectives.* New York: Routledge, 1996.

De-Shalit, Avner. *The Environment: Between Theory and Practice.* Oxford, UK: Oxford University Press, 2000.

Desmarais, Annette. *La Vía Campesina: Globalization and the Power of Peasants.* London: Pluto Press, 2007.

Desmond, Adrian. *The Politics of Evolution: Morphology, Medicine, and Reform in Radical London.* Chicago: University of Chicago Press, 1992.

Devall, Bill, and George Sessions. *Deep Ecology: Living as If Nature Mattered.* Salt Lake City, UT: G.M. Smith, 1985.

De Young, Raymond, and Thomas Princen. *The Localization Reader: Adapting to the Coming Downshift.* Cambridge, MA: MIT Press, 2012.

Dibley, Ben. "'The Shape of Things to Come': Seven Theses on the Anthropocene and Attachment." *Australian Humanities Review* 52 (2012).

Di Chiro, Giovanna. "Beyond Ecoliberal 'Common Futures': Toxic Touring, Environmental Justice, and a Transcommunal Politics of Place." In *Race, Nature, and the Politics of Difference*, edited by Donald Moore, Jake Kosek, and Anand Pandian. Durham, NC: Duke University Press, 2003.

———. "Climate Justice Now! Imagining Grassroots EcoCosmopolitanism." In *American Studies, Ecocriticism, and Citizenship*, edited by Joni Adamson and Kimberly Ruffin. New York: Routledge, 2013.

———. "Living Environmentalisms: Coalition Politics, Social Reproduction, and Environmental Justice." *Environmental Politics* 17, no. 1 (2008): 276–98.

———. "Polluted Politics? Confronting Toxic Discourse, Sex Panic, and Eco-Normativity." In *Queer Ecologies: Sex, Nature, Politics, Desire*, edited by Catriona Mortimer-Sandilands and Bruce Erickson, 199–230. Bloomington: Indiana University Press, 2010.

Dirlik, Arif. "Globalism and the Politics of Place." *Development* 44, no. 2 (1998): 7–13.

Dobson, Andy, et al. "Homage to Linnaeus: How Many Parasites? How Many Hosts?" *PNAS* 105 (2008): 11482–89.

Dobzhansky, T. "A Critique of the Species Concept in Biology." *Philosophy of Science* 2, no. 3 (1935): 344–55.

Donne, John. *Poems of John Donne*, vol. 1, edited by E. K. Chambers. London: Lawrence & Bullen, 1896.

Doremus, Holly D., and Dan A. Tarlock. *Water War in the Klamath Basin: Macho Law, Combat Biology, and Dirty Politics*. Washington, DC: Island Press, 2008.

Douglas, Ian. *Cities: An Environmental History*. London: I.B. Tauris, 2013.

Downey, Liam. "Environmental Racial Inequality in Detroit." *Social Forces* 85 (2006): 771–96.

Downie, David Leonard, and Terry Fenge, eds. *Northern Lights against POPs: Combatting Toxic Threats in the Arctic*. Montreal, Canada: McGill-Queen's University Press, 2003.

Dreiser, Theodore. *The Color of a Great City*. New York: Boni & Liveright, 1923.

Dungy, Camille T. *Black Nature: Four Centuries of African American Nature Poetry*. Athens: University of Georgia Press, 2009.

Dunlap, Riley, and Angela G. Mertig, eds. *American Environmentalism: The U.S. Environmental Movement, 1970–1990*. Philadelphia: Taylor & Francis, 1992.

Eaton, Heather, and Lois Ann Lorentzen, eds. *Ecofeminism and Globalization: Exploring Culture, Context, and Religion*. Lanham, MD: Rowman & Littlefield, 2003.

Edelstein, Michael. *Contaminated Communities: The Social and Psychological Impacts of Residential Toxic Exposure*. Boulder, CO: Westview, 2004.

Ehrlich, Paul. *The Population Bomb*. New York: Ballantine, 1968.

Ehrlich, Paul, and Anne Ehrlich. *Extinction: The Causes and Consequences of the Disappearance of Species*. New York: Ballantine, 1981.

Ehrlich, Paul R., and John P. Holdren. "Impact of Population Growth." *Science* 171 (1971): 1212–17.

Ehrlich, Paul R., and Peter H. Raven. "Butterflies and Plants: A Study in Co-Evolution." *Evolution* 18, no. 4 (1964): 586–608.

Elder, John. *Imagining the Earth: Poetry and the Vision of Nature*. Urbana: University of Illinois Press, 1985.

Eldredge, Niles. "Passionate Lynn Margulis." In *Lynn Margulis: The Life and Legacy of a Scientific Rebel*, edited by Dorion Sagan, 47–49. White River Junction, VT: Chelsea Green Publishing, 2012.

Ellis, Erle C., Erica C. Antill, and Holder Kreft. "All Is Not Loss: Plant Biodiversity in the Anthropocene." *PLoS ONE* 7, no. 1 (2012).

Ellis, Havelock. *Studies in the Psychology of Sex*. Volume 1, Part 4, *Sexual Inversion*. New York: Random House, 1936 (1905).

"Environment." *Oxford English Dictionary*. Accessed Oct. 26, 2012. Available at http://www.oed.com.ezproxy.library.yorku.ca/view/Entry/63089?redirectedFrom=environment#eid.

Empson, William. *Some Versions of Pastoral*. New York: New Directions, 1974 (1935).

Escobar, Arturo. "Beyond the Third World: Imperial Globality, Global Coloniality, and Anti-globalisation Social Movements." *Third World Quarterly* 25, no. 1 (2004): 207–30.

———. "Culture Sits in Places: Reflections on Globalism and Subaltern Strategies of Localization." *Political Geography* 20, no. 2 (2001): 139–74.

———. *Encountering Development: The Making and Unmaking of the Third World*. Princeton, NJ: Princeton University Press, 2011 (1995).

———. "Postconstructivist Political Ecologies." In *International Handbook of Environmental Sociology*, 2nd ed., edited by Michael Redclift and Graham Woodgate, 91–105. Cheltenham, UK: Elgar, 2010.

———. *Territories of Difference: Place, Movements, Life, Redes*. Durham, NC: Duke University Press, 2008.

———. "Whose Knowledge, Whose Nature? Biodiversity, Conservation, and the Political Ecology of Social Movements." Journal of Political Ecology 5, no. 1 (1998): 53–82.

Estok, Simon. *Ecocriticism and Shakespeare: Reading Ecophobia*. Hampshire, UK: Palgrave Macmillan, 2011.

Evanoff, Richard. *Bioregionalism and Global Ethics: A Transactional Approach to Achieving Ecological Sustainability, Social Justice, and Human Well-Being*. New York: Routledge, 2011.

Ezrahi, Yaron. "Science and Utopia in Late-20th-Century Pluralist Democracy." In *Nineteen Eighty-Four: Science between Utopia and Dystopia*, edited by Everett Mendelsohn and Helga Nowotny, 273–90. Dordrecht: Springer Netherlands, 1984.

Faber, Daniel. *The Struggle for Ecological Democracy: Environmental Justice Movements in the United States*. New York: Guilford Press, 1998.

Fabian, Johannes. *Time and the Other: How Anthropology Makes Its Object*. New York: Columbia University Press, 1983.

Fanon, Frantz. *The Wretched of the Earth*. New York: Grove Press, 1963.

Farina, Almo. "The Landscape as a Semiotic Interface between Organisms and Resources." *Biosemiotics* 1, no. 1 (2008): 75–83.

Farley, Joshua, and Robert Costanza. "Envisioning Shared Goals for Humanity: A Detailed Shared Vision of a Sustainable and Desirable USA in 2100." *Ecological Economics* 43 (2002): 245–59.

Favareau, Donald. "Introduction: An Evolutionary History of Biosemiotics." In *Essential Readings in Biosemiotics*, edited by Donald Favareau, 1–77. Berlin: Spinger, 2009.

Federici, Silvia. *Caliban and the Witch: Women, the Body, and Primitive Accumulation*. New York: Autonomedia, 2004.

Felstiner, John. *Can Poetry Save the Earth? A Field Guide to Nature Poems*. New Haven, CT: Yale University Press, 2009.

Ferguson, Kennan. *William James: Politics in the Pluriverse*. Lanham, MD: Rowman & Littlefield, 2007.

Figueroa, Robert Melchior, and Gordon Waitt. "Climb: Restorative Justice, Environmental Heritage, and the Moral Terrains of Uluru-Kata Tjuta National Park." *Environmental Philosophy Special Issue: Ecotourism and Environmental Justice* 7, no. 2 (Fall 2010): 135–64.

Finch, Robert, and John Elder, eds. *The Norton Book of Nature Writing*. New York: Norton, 1990.

Fleming, James Rodger. *Fixing the Sky: The Checkered History of Weather and Climate Control*. New York: Columbia University Press, 2010.

Fletcher, Angus. *A New Theory for American Poetry: Democracy, the Environment, and the Future of Imagination*. Cambridge, MA: Harvard University Press, 2004.

Flys-Junquera, Carmen, Jose Manuel Marrero Henríquez, and Julia Barella Vigal, eds. *Ecocríticas: Literatura y medio ambiente*. Madrid: Iberoamerica, 2010.

Foltz, Bruce V. *Inhabiting the Earth: Heidegger, Environmental Ethics, and the Metaphysics of Nature*. Amherst, NY: Humanity Books, 1995.

Food and Agriculture Organization of the United Nations. "Livestock's Long Shadow: Environmental Issues and Options." Rome: United Nations Publications, 2006.

Foote, Stephanie, and Elizabeth Mazzolini, eds. *Histories of the Dustheap: Waste, Material Cultures, Social Justice*. Cambridge, MA: MIT Press, 2012.

Ford, J. R., et al. "An Assessment of Lithostratigraphy for Anthropogenic Deposits." In *A Stratigraphical Basis for the Anthropocene*, edited by C. N. Waters, et al., 55–89. London: Geological Society of London, Special Publication 395 (2014).

Foster, John Bellamy. *Marx's Ecology: Materialism and Nature*. New York: Monthly Review Press, 2000.

Foucault, Michel. *The Birth of Biopolitics: Lectures at the Collège de France, 1978—1979* (Lectures at the College de France). New York: Picador, 2010.

———. *The Birth of the Clinic: An Archaeology of Medical Perception*. Translated by A. M. Sheridan Smith. New York: Pantheon, 1973.

———. *Discipline and Punish: The Birth of the Prison*. Translated by Alan Sheridan. New York: Pantheon, 1977.

———. *The History of Sexuality*. Vol. 1, *An Introduction*. Translated by Robert Hurley. New York: Pantheon, 1978.

———. *Madness and Civilization: A History of Insanity in the Age of Reason*. Translated by Richard Howard. New York: Pantheon, 1965.

Foushee, Lea, and Renee Gurneau. *Sacred Water: Water for Life*. Lake Elmo, MN: North American Water Office, 2010.

Fox, Steve. *Toxic Work: Women Workers at GTE Lenkurt*. Philadelphia: Temple, 1993.

Fox, Warwick. *Toward a Transpersonal Ecology: Developing New Foundations for Environmentalism*. Boston: Shambalah, 1990.

Freeman, Barbara Claire. *The Feminine Sublime: Gender and Excess in Women's Fiction*. Berkeley: University of California Press, 1995.

Freudenberg, Nicholas, and Carol Steinsapir. "Not in Our Backyards: The Grassroots Environmental Movement." In *American Environmentalism*, edited by Riley E. Dunlap and Angela G. Mertig. Philadelphia: Taylor and Francis, 1992.

Freudenburg, William R., Scott Frickel, and Robert Gramling. "Beyond the Nature/Society Divide: Learning to Think about a Mountain." *Sociological Forum* 10, no. 3 (1995): 361–92.

Fritzell, Peter. *Nature Writing and America: Essays upon a Cultural Type*. Ames: Iowa State University Press, 1989.

"From the Editors." *Orion Magazine*. Last modified April 14, 2010. Available at http://www.orionmagazine.org/index. php /articles/article/5469.

Fukuyama, Francis. *Our Posthuman Future: Consequences of the Biotechnology Revolution*. New York: Farrar, Strauss, Giroux, 2002.

Gaard, Greta. "Global Warming Narratives: A Feminist Ecocritical Perspective." In *The Future of Ecocriticism: New Horizons*, edited by Serpil Oppermann and Ufuk Ozdag. Newcastle upon Tyne, UK: Cambridge Scholars Publishing, 2011.

———. "Milking Mother Nature." *Ecologist* 24, no. 6 (1994): 202–3.

———. "New Directions for Ecofeminism: Toward a More Feminist Ecocriticism." *Interdisciplinary Studies in Literature and Environment* 17, no. 4 (2010a): 643–65.

———. "Reproductive Technology, or Reproductive Justice? An Ecofeminist, Environmental Justice Perspective on the Rhetoric of Choice." *Ethics & the Environment* 15, no. 2 (2010b): 103–29.

———. "Toward a Queer Ecofeminism." *Hypatia* 12, no. 1 (1997): 114–37.

Gaffney, Eugene S. "An Introduction to the Logic of Phylogeny Reconstruction." In *Phylogenetic Analysis and Paleontology*, edited by J. Cracraft and N. Eldredge, 79–111. New York: Columbia University Press, 1979.

Galbraith, Kate. "International Interest Grows in Green-Building Certification." *New York Times*, March 7, 2012. Last visited May 4, 2015. Available at http://www.nytimes. com/2012/03/08/business/global/international-interest-grows-in-green-building-certification.html?_r=0.

Galilei, Galileo. *Dialogue concerning the Two Chief World Systems*. Berkeley: University of California Press, 1953 (1632).

Gallie, Walter. "Essentially Contested Concepts." *Proceedings of the Aristotelian Society* 56, no. 2 (1956): 167–98.

Galtung, Johan. "Violence, Peace, and Peace Research." *Journal of Peace Research* 6, no. 3 (1969): 167–91.

Gandy, Matthew. "Queer Ecology: Nature, Sexuality, and Heterotopic Alliances." *Environment and Planning D: Society and Space* 30, no. 4 (2012): 727–47.

Garrard, Greg. *Ecocriticism*. London: Routledge, 2004.

———. "How Queer Is Green?" *Configurations* 18, nos. 1–2 (2010): 73–96.

Garrett, Paul B. "Dying Young: Pidgins, Creoles, and other Contact Languages as Endangered Languages." In *The Anthropology of Extinction: Essays on Culture and Species Death*, edited by Genese Marie Sodikoff, 143–62. Bloomington: Indiana University Press, 2012.

Gates, Barbara T. *Kindred Nature: Victorian and Edwardian Women Embrace the Living World*. Chicago: University of Chicago Press, 1998.

Geertz, Clifford. *The Interpretation of Cultures*. New York: Basic Books, 1973.

Gessner, David. "Sick of Nature." *Boston Globe*. Last modified April 1, 2004. Available at http://www.boston.com/news/ globe/ ideas/articles/2004/08/01/sick_of_nature.

Gibbs, Jack. "Conceptualization of Terrorism." *American Sociological Review* 54, no. 3 (1989): 329–40.

Gibson-Graham, J. K., Jenny Cameron, and Stephen Healy. *Take Back the Economy: An Ethical Guide for Transforming Our Communities*. Minneapolis: University of Minnesota Press, 2013.

Giddens, Anthony. *Contemporary Critique of Historical Materialism: Power, Property, and the State*. London: Macmillan, 1981.

Giffney, Noreen. "Queer Apocal(o)ptic/ism: The Death Drive and the Human." In *Queering the Non/Human*, edited by Noreen Giffney and Myra Hird, 55–78. Hampshire, UK: Ashgate, 2008.

Gifford, Terry. "Towards a Post-Pastoral View of British Poetry." In *The Environmental Tradition in English Literature,* edited by John Parham, 51–63. Hampshire, UK: Ashgate, 2002.

———. *Understanding Contemporary Nature Poetry*. Manchester, UK: Manchester University Press, 1995.

Gilbert, Scott F, Jan Sapp, and Alfred I. Tauber. "A Symbiotic View of Life: We Have Never Been Individuals." *Quarterly Review of Biology* 87, no. 4 (2012): 325–41.

Gilcrest, David. *Greening the Lyre: Environmental Poetics and Ethics*. Reno: University of Nevada Press, 2002.

Gleason, Henry. "The Individualistic Concept of the Plant Association." *Bulletin of the Torrey Botanical Club* 53, no. 1 (1926): 7–26.

Gleason, William A. *Sites Unseen: Architecture, Race, and American Literature*. New York: New York University Press, 2011.

Glissant, Edouard. *Caribbean Discourse*. Translated by J. Michael Dash. Charlottesville: University Press of Virginia, 1999 (1981).

Glotfelty, Cheryll, and Harold Fromm, eds. *The Ecocriticism Reader: Landmarks in Literary Ecology*. Athens: University of Georgia Press, 1996.

Godfrey, Laurie R., and Emilienne Rasoazanabary. "Demise of the Bet Hedgers: A Case Study of Human Impacts on Past and Present Lemurs of Madagascar." In *The Anthropology of Extinction: Essays on Culture and Species Death*, edited by Genese Marie Sodikoff, 165–99. Bloomington: Indiana University Press, 2012.

Goldman, Michael. *Imperial Nature: The World Bank and Struggles for Social Justice in the Age of Globalization*. New Haven, CT: Yale University Press, 2005.

Goldsmith, Oliver. *The Rising Villlage*, edited by Gerald Lynch. London, Ontario: Canadian Poetry Press, 1989 (1825, 1834).

Goodbody, Axel. "German Ecocriticism: An Overview." In *The Oxford Handbook of Ecocriticism*, edited by Greg Garrard, 547–59. New York: Oxford University Press, 2015.

Goodland, Robert, et al. *Environmentally Sustainable Economic Development: Building on Brundtland*. Paris: UNESCO, 1991.

Gosnell, Hannah, and Erin Clover Kelly. "Peace on the River? Social-Ecological Restoration and Large Dam Removal in the Klamath Basin, USA." *Water Alternatives* 3, no. 2 (2010): 362–83.

Gottdiener, Mark. *The Social Production of Urban Space*. Austin: University of Texas Press, 1985.

Gottlieb, Robert. *Forcing the Spring: The Transformation of the American Environmental Movement*. Washington, DC: Island Press, 1993.

Gottlieb, Robert, and Anupama Joshi. *Food Justice*. Cambridge, MA: MIT Press, 2010.

Gottlieb, Roger S. *A Greener Faith and Our Planet's Future*. New York: Oxford University Press, 2006a.

——, ed. *Oxford Handbook of Religion and Ecology*. New York: Oxford University Press, 2006b.

Gottweis, Herbert. *Governing Molecules: The Discursive Politics of Genetic Engineering in Europe and the United States*. Cambridge, MA: MIT Press, 2000.

Gould, Kenneth A. "Tactical Tourism." *Organization & Environment* 12, no 3. (1999): 245–62.

Gould, Stephen Jay. "Darwinian Fundamentalism." *New York Review of Books* (June 1997): 1–2.

Gouldson, Andrew, and Joseph Murphy. "Ecological Modernization and the Restructuring of Industrial Economies." In *Greening the Millennium: The New Politics of Environment*, edited by Michael Jacobs. Oxford, UK: Blackwell, 1997.

Goulet, Jean-Guy, and Bruce Miller. *Extraordinary Anthropology: Transformations in the Field*. Lincoln: University of Nebraska Press, 2007.

Grant, Gary. *Ecosystem Services Come to Town: Greening Cities by Working with Nature*. Oxford, UK: Wiley, 2012.

Great Ape Project. Accessed April 29, 2012. Available at http://www.greatapeproject.org.

Gregory, Andrew. *Ancient Greek Cosmogony*. London: Duckworth, 2007.

"Green." *Oxford English Dictionary Online*. Oxford, UK: Oxford University Press. Accessed August 20, 2012.

Green Porno. Directed by Isabella Rossellini and Jody Shapiro. Film Series (4 vol.), Sundance Channel, 2009.

Grenoble, Lenore A., and Lindsay J. Whaley, ed. *Endangered Languages: Language Loss and Community Response*. Cambridge: Cambridge University Press, 1998.

Griffiths, Tom. *Seeing "RED"? "Avoided Deforestation" and the Rights of Indigenous Peoples and Local Communities*. Forest Peoples Programme, 2007. Last visited May 10, 2015. Available at http://www.forestpeoples.org/topics/un-redd/publication/2010/seeing-red-avoided-deforestation-and-rights-indigenous-peoples-and-l.

Grossman, Zoltán. "Indigenous Nations' Responses to Climate Change." *American Indian Culture & Research Journal* 32, no. 3 (2008): 5–27.

Gruen, Lori. "Navigating Difference (Again): Animal Ethics and Entangled Empathy." In *Strangers to Nature: Animal Lives and Human Ethics,* edited by Gregory Smulewicz-Zucker, 213–33. Lanham, MD: Lexington Books, 2012.

Guha, Ramachandra. "The Authoritarian Biologist and the

Arrogance of Anti-Humanism: Wildlife Conservation in the Third World." *Ecologist* 27, no. 1 (1997): 14–20.

———. *Environmentalism: A Global History*. New York: Longman, 2000.

———. "Radical American Environmentalism and Wilderness Preservation: A Third World Critique." *Environmental Ethics* 11, no. 1 (1989): 71–83.

———. *The Unquiet Woods: Ecological Change and Peasant Resistance in the Himalaya*. Berkeley: University of California Press, 2000.

Gunder, Michael. "Sustainability: Planning's Saving Grace or Road to Perdition?" *Journal of Planning Education and Research* 26 (2006): 208–21.

Gunn, Eileen. "How America's Leading Science Fiction Authors Are Shaping Your Future." *Smithsonian Magazine* (May 2014). Available at http://www.smithsonianmag.com/arts-culture/how-americas-leading-science-fiction-authors-are-shaping-your-future-180951169/?page=2.

Gupta, Aarti. "Transparency in Global Environmental Governance." *Global Environmental Politics* 10, no. 3 (2010): 1–9.

Hadley, John. "Animal Rights Extremism and the Terrorism Question." *Journal of Social Philosophy* 40, no. 3 (2009): 363–78.

Haeckel, Ernst. *Generelle Morphologie der Organismen: Allgemeine Grundzuege der organischen Formen-wissenschaft, mechanisch begruendet durch die von Charles Darwin reformirte Deszendenz-Theorie*. Berlin: Reimer, 1866.

Hajer, Maarten A. *The Politics of Environmental Discourse: Ecological Modernization and the Policy Process*. Oxford, UK: Oxford University Press, 1995.

Hall, Marcus. *Earth Repair: A Transatlantic History of Environmental Restoration*. Charlottesville: University of Virginia Press, 2005.

Hall, Stuart. "The West and the Rest: Discourse and Power." In *Formations of Modernity*, edited by Bram Gieben and Stuart Hall, 275–332. Cambridge, UK: Polity Press, 1992.

Hancock, G. J., et al. "The Release and Persistence of Radioactive Anthropogenic Nuclides." In *A Stratigraphical Basis for the Anthropocene*, edited by C. N. Waters et al., 265–81. London: Geological Society of London, Special Publication 395 (2014).

Handley, George. "A Postcolonial Sense of Place and the Work of Derek Walcott." *ISLE* 7, no. 2 (Summer 2000): 1–23.

Hanson, David, and Edwin Marty. *Breaking through Concrete: Building an Urban Farm Revival*. Berkeley: University of California Press, 2012.

Haraway, Donna. *The Companion Species Manifesto: Dogs, People, and Significant Otherness*. Chicago: Prickly Paradigm Press, 2003.

———. "A Cyborg Manifesto: Science, Technology, and Socialist-Feminism in the Late Twentieth Century." In *Simians, Cyborgs, and Women: The Reinvention of Nature*, 149–81. New York: Routledge, 1991a.

———. "A Game of Cat's Cradle: Science Studies, Feminist Theory, Cultural Studies." *Configurations* 2, no. 1 (1994): 59–71.

———. *Primate Visions: Gender, Race, and Nature in the World of Modern Science*. New York: Routledge, 1990.

———. *Simians, Cyborgs, and Women: The Reinvention of Nature*. New York: Routledge, 1991b.

———. "Situated Knowledges: The Science Question in Feminism and the Privilege of Partial Perspective." *Feminist Studies* 14, no. 3 (1988): 575–99.

———. *When Species Meet*. Minneapolis: University of Minnesota Press, 2007.

Harcourt, Wendy. *Body Politics in Development: Critical Debates in Gender and Development*. London: Zed, 2009.

———. "Reflections: The Politics of Place and Racism in Australia: A Personal Exploration." *Meridians: Feminism, Race, Transnationalism* 1, no. 2 (2001): 194–207.

Harcourt, Wendy, and Arturo Escobar, eds. *Women and the Politics of Place*. Boulder, CO: Kumarian Press, 2005.

Harcourt, Wendy, and Ingrid Nelson, eds. *Practicing Feminist Political Ecologies: Moving beyond the "Green Economy."* London: Zed, 2015.

Harcourt, Wendy, et al. "A Massey Muse." In *Spatial Politics: Essays for Doreen Massey*, edited by David Featherstone and Joe Painter, 158–77. Malden, MA: Wiley, 2013.

Hardin, Garrett. *Exploring New Ethics for Survival: The Voyage of the Spaceship Beagle*. New York: Viking Press, 1972.

———. "Lifeboat Ethics: The Case against Helping the Poor." *Psychology Today* 8 (1974): 38–43.

———. *The Limits of Altruism: An Ecologist's View of Survival*. Bloomington: Indiana University Press, 1977.

———. "The Tragedy of the Commons." *Science*, New Series (Dec. 13, 1968), 1243–48. Web.

Harding, Sandra. *The Science Question in Feminism*. Ithaca, NY: Cornell University Press, 1986.

Hardt, Michael, and Antonio Negri. *Empire*. Cambridge, MA: Harvard University Press, 2000.

Harris, Dianne. *Little White Houses: How the Postwar Home Constructed Race in America*. Minneapolis: University of Minnesota Press, 2012.

Harris, John. *Enhancing Evolution: The Ethical Case for Making Better People*. Princeton, NJ: Princeton University Press, 2007.

Harrison, David K. *When Languages Die: The Extinction of the World's Languages and the Erosion of Human Knowledge*. New York: Oxford University Press, 2007.

Harrison, Jill. *Pesticide Drift and the Pursuit of Environmental Justice*. Cambridge, MA: MIT Press, 2011.

Hartmann, Betsy. "Conserving Racism: The Greening of Hate at Home and Abroad." *DiffrenTakes* no. 27 (2004). Available at http://popdev.hampshire.edu/projects/dt/27.

Harvey, David. *The Condition of Postmodernity*. Cambridge, UK: Blackwell, 1990.

———. "Contested Cities: Social Process and Spatial Form." In *Transforming Cities: Contested Governance and New Spatial Divisions*, edited by Nick Jewson and Susanne MacGregor, 17–24. London: Routledge, 1997.

———. *Justice, Nature, and the Geography of Differences*. Cambridge, UK: Blackwell, 1996.

———. *The New Imperialism*. New York: Oxford University Press, 2005.

———. *Spaces of Global Capitalism: Towards a Theory of Uneven Geographical Development*. London: Verso, 2006.

Harvey, Graham. *Animism: Respecting the Living World*. New York: Columbia University Press, 2006.

Hatemi, Peter, et al. "Genetic Influences on Political Ideologies: Twin Analyses of 19 Measures of Political Ideologies from Five Democracies and Genome-Wide Findings from Three Populations." *Behavior Genetics* 44, no. 2 (2015): 282–94.

Hawkins, Roberta, and Diana Ojeda. "Gender and Environment: Critical Tradition and New Challenges." *Environment and Planning D: Society and Space* 29, no. 2 (2011): 237–53.

Hay, Peter. *Main Currents in Environmental Thought*. Bloomington: Indiana University Press, 2002.

Hays, Samuel P. *Conservation and the Gospel of Efficiency. The Progressive Conservation Movement, 1890–1920*. Cambridge, MA: Harvard University Press, 1959.

Hayward, Eva. "Lessons from a Starfish." In *Queering the Non/Human*, edited by Noreen Giffney and Myra Hird, 249–63. Hampshire, UK: Ashgate, 2008.

Hayward, Tim. "The Meaning of Political Ecology." *Radical Philosophy* 66 (1994): 11–20.

Head, Dominic. "Beyond 2000: Raymond Williams and the Ecocritic's Task." In *The Environmental Tradition in English Literature*, edited by John Parham. Aldershot, UK: Ashgate, 2002.

Hecht, Susanna, and Alexander Cockburn. *The Fate of the Forest: Developers, Destroyers, and Defenders of the Amazon*. New York: HarperPerennial, 1990.

Heise, Ursula K. "Lost Dogs, Last Birds, and Listed Species: Cultures of Extinction." *Configurations* 18, nos. 1–2 (2011): 39–62.

Held, David, et al. *Global Transformations: Politics, Economics, and Culture*, Stanford, CA: Stanford University Press, 1999.

The Helsinki Group. *Declaration of Rights for Cetaceans*. Accessed April 29, 2012. Available at http://www.cetacean-rights.org.

Herod, Andrew. *Scale*. London: Routledge, 2011.

Herring, Scott. "Out of the Closets, into the Woods: *RFD*, *Country Women*, and the Post-Stonewall Emergence of Queer Anti-Urbanism." *American Quarterly* 59, no. 2 (2007): 341–72.

Hesiod. "Works and Days." Translated by Hugh G. Evelyn-White. *Sacred Texts*, 1914. Available at http://www.sacred-texts.com/cla/hesiod/works.htm.

Hibbard, Michael, et al., eds. *Toward One Oregon: Rural-Urban Interdependence and the Evolution of a State*. Corvallis: Oregon State University Press, 2011.

Hiltner, Ken. *What Else Is Pastoral? Renaissance Literature and the Environment*. Ithaca, NY: Cornell University Press, 2011.

Hilton-Taylor, Craig, et al. "State of the World's Species." In *Wildlife in a Changing World: An Analysis of the 2008 IUCN Red List of Threatened Species*, edited by Jean-Christophe Vié, Craig Hilton-Taylor, and Simon N. Stuart. Gland, Switzerland: IUCN, 2008.

Hinchcliffe, Steve, and Kath Woodward, eds. *The Natural and* the Social: Uncertainty, Risk, Change. London: Routledge, 2000.

Hinnant, Charles. "Shaftesbury, Burke, and Wollstonecraft: Permutations on the Sublime *and* the Beautiful." *Eighteenth Century* 46, no. 1 (2005): 17–35.

Hird, Myra. "Animal Trans." In *Queering the Non/Human*, edited by Noreen Giffney and Myra Hird, 227–47. Hampshire, UK: Ashgate, 2008.

———. "Naturally Queer." *Feminist Theory* 5, no. 1 (2004): 85–89.

Hitt, Christopher. "Toward an Ecological Sublime." *New Literary History* 30 (1999): 603–23.

Hoffmeyer, Jesper, and Claus Emmeche. "Code-Duality and the Semiotics of Nature." In *On Semiotic Modeling*, edited by Anderson Myrdene and Merrell Floyd, 117–66. Berlin: de Gruyter, 1991.

Hogan, Katie. "Undoing Nature: Coalition Building as Queer Environmentalism." In *Queer Ecologies: Sex, Nature, Politics, Desire*, edited by Catriona Mortimer-Sandilands and Bruce Erickson, 231–53. Bloomington: Indiana University Press, 2010.

Hogue, Cynthia. "(Re)Storing Happiness: Toward an Ecopoetic Reading of H.D.'s *The Sword Went Out to Sea (Synthesis of a Dream), by Delia Alton*." *Interdisciplinary Studies in Literature and Environment* 18, no. 4 (2011): 840–60.

Hokowhitu, Brendan. "Indigenous Existentialism and the Body." *Cultural Studies Review* 15, no. 2 (2009): 101–18.

Holtgrieve, Gordon, et al. "A Coherent Signature of Anthropogenic Nitrogen Deposition to Remote Watersheds of the Northern Hemisphere." *Science* 334, no. 6062 (2011): 1545–48.

Honey, Martha. *The ECOCLUB Interview with Martha Honey.* Last modified September 2008a. Available at http://ecoclub.com/news/099/interview.html.

———, ed. *Ecotourism and Certification: Setting Standards in Practice.* Washington, DC: Island Press, 2002.

———. *Ecotourism and Sustainable Development: Who Owns Paradise?* 2nd ed. Washington, DC: Island Press, 2008b.

Hopkinson, Nalo. *Brown Girl in the Ring.* New York: Grand Central Publishing, 1998.

———. *Midnight Robber.* New York: Grand Central Publishing, 2000.

———. *Report from Planet Midnight.* Oakland, CA: PM Press, 2012.

Horkheimer, Max. *The Eclipse of Reason.* New York: Oxford University Press, 1947.

Hornborg, Alf. "Vital Signs: An Ecosemiotic Perspective on the Human Ecology of Amazonia." *Sign Systems Studies* 29, no.1 (2001): 121–52.

Hubbell, Stephen P. *The Unified Neutral Theory of Biodiversity and Biogeography.* Princeton, NJ: Princeton University Press, 2001.

———. "A Unified Theory of Biogeography and Relative Species Abundance and Its Application to Tropical Rain Forests and Coral Reefs." *Coral Reefs* 16, no. 1 Supplement (1997): S9–S21.

Huggan, Graham, and Helen Tiffin. *Postcolonial Ecocriticism: Literature, Animals, Environment.* New York: Routledge, 2010.

Hughes, J. Donald. "Biodiversity in World History." In *The Face of the Earth: Environment and World History*, edited by J. Donald Hughes, 22–46. London: Sharpe, 2000.

———. *An Environmental History of the World: Humankind's Changing Role in the Community of Life.* 2nd ed. New York: Routledge, 2009.

Hughes, James. *Citizen Cyborg: Why Democratic Societies Must Respond to the Redesigned Human of the Future.* Los Angeles: Westview, 2004.

———. "Contradictions from the Enlightenment Roots of Transhumanism." *Journal of Medicine and Philosophy* 35, no. 6 (2010): 622–40.

———. "TechnoProgressive Biopolitics and Human Enhancement." In *Progress in Bioethics*, edited by Jonathan Moreno and Sam Berger, 163–88. Cambridge, MA: MIT Press, 2009.

Hulme, Mike. *Why We Disagree about Climate Change: Understanding Controversy, Inaction, and Opportunity.* Cambridge: Cambridge University Press, 2009.

Humboldt, Alexander von. *Cosmos: A Sketch of the Physical Description of the Universe.* Translated by E. C. Otté. 2 volumes. New York: Harper and Brothers, 1850 (1845–47); facsimile edition, Baltimore, MD: Johns Hopkins University Press, 1997.

Humboldt, Alexander von, and A. Bonpland. *Essai sur la géographie des plantes.* Paris, 1805. Edited and translated as

Essay on the Geography of Plants by S. T. Jackson and S. Romanowski. Chicago: University of Chicago Press, 2009.

Hustak, Carla, and Natasha Myers. "Involutionary Momentum: Affective Ecologies and the Sciences of Plant/Insect Encounters." *Feminist Theory out of Science (Difference: A Journal of Feminist Cultural Studies)* 25, no. 5 (2012): 74–118.

Hutchinson, George Evelyn. *The Ecological Theater and the Evolutionary Play*. New Haven, CT: Yale University Press, 1965.

———. "Homage to Santa Rosalia; or, Why Are There So Many Kinds of Animals?" *American Naturalist* 93, no. 870 (1959): 145–59.

Ibuse, Masuji. *Black Rain*. Translated by John Bester. Tokyo: Kodansha, 1969 (1966).

Imamura, Shōhei. *Black Rain*. New York: Fox Lorber Home Video, 1991 (1989).

Ingold, Tim. *Being Alive: Essays on Movement, Knowledge, and Description*. London: Routledge, 2011.

———. "Culture on the Ground: The World Perceived through the Feet." *Journal of Material Culture* 9, no. 3 (2004): 315–40.

———. *The Perception of the Environment: Essays on Livelihood, Dwelling, and Skill*. London: Routledge, 2011 (2000).

Ingram, Gordon Brent. "Fragments, Edges, and Matrices: Retheorizing the Formation of a So-called Gay Ghetto through Queering Landscape Ecology." In *Queer Ecologies: Sex, Nature, Politics, Desire*, edited by Catriona Mortimer-Sandilands and Bruce Erickson, 254–82. Bloomington: Indiana University Press, 2010.

The International Ecotourism Society. *Global Ecotourism Fact Sheet*. 2006. Accessed May 30, 2013. Available at https://ibgeography-lancaster.wikispaces.com/file/view/TIES+GLOBAL+ECOTOURISM+FACT+SHEET.PDF.

International Federation of Red Cross and Red Crescent Societies. *World Disasters Report, 2001*. Oxford, UK: Oxford University Press, 2001.

Iovino, Serenella. "Naples 2008; or, The Waste Land: Trash, Citizenship, and an Ethic of Narration." *Neohelicon* 36 (2009): 335–46.

Iovino, Serenella, and Serpil Oppermann. "Material Ecocriticism: Materiality, Agency, and Models of Narrativity." *Ecozon@* 3, no. 1 (2012): 75–91.

Isaacs, Harold R. "Color in World Affairs." *Foreign Affairs* 47, no. 2 (1969): 235–50.

Jackson, Cecile. "Women/Nature or Gender/History? A Critique of Ecofeminist Development." *Journal of Peasant Studies* 20, no. 3 (1993): 389–419.

Jacobs, Michael. "Sustainable Development: A Contested Concept." In *Fairness and Futurity. Essays on Environmental Sustainability and Social Justice*, edited by Andrew Dobson. Oxford, UK: Oxford University Press, 1999.

James, David, and Philip Tew, eds. *New Versions of Pastoral: Post-Romantic, Modern, and Contemporary Responses to the Tradition*. Madison, NJ: Farleigh Dickinson University Press, 2009.

Jamieson, Dale, ed. *A Companion to Environmental Philosophy*. Oxford, UK: Blackwell, 2001.

Jantsch, Erich. *The Self-Organizing Universe: Scientific and Human Implications of the Emerging Paradigm of Evolution*. New York: Pergamon, 1980.

Janzen, Daniel H. "Herbivores and the Number of Tree Species in Tropical Forests." *American Naturalist* 104, no. 940 (1970): 501–28.

Jasanoff, Sheila. *Designs on Nature: Science and Democracy in Europe and the United States*. Princeton, NJ: Princeton University Press, 2005.

———. *The Fifth Branch: Science Advisers as Policymakers*. Cambridge, MA: Harvard University Press, 1990.

———. "The Politics of Public Reason." In *The Politics of Knowledge*, edited by Fernando D. Rubio and Patrick Baert, 11–32. Abingdon, NJ: Routledge, 2011.

———. *Science at the Bar: Law, Science, and Technology in America*. Cambridge, MA: Harvard University Press, 1995.

Jasanoff, Sheila, and Sang-Hyun Kim. "Containing the Atom: Sociotechnical Imaginaries and Nuclear Regulation in the U.S. and South Korea." *Minerva* 47, no 2 (2009): 119–46.

Jenkins, Willis. *Ecologies of Grace: Environmental Ethics and Christian Theology*. Oxford, UK: Oxford University Press, 2008.

Joffe, Paul, Jackie Hartley, and Jennifer Preston, eds. *Realizing the UN Declaration on the Rights of Indigenous Peoples: Triumph, Hope, and Action*. Saskatoon, Canada: Purich, 2010.

Johnson, Alex. "How to Queer Ecology: One Goose at a Time." *Orion Magazine* (March/April 2011). Accessed January 20, 2013. Available at http://www.orionmagazine.org/index.php/articles/article/6166.

Johnston, Basil. *Ojibway Tales*. Lincoln: University of Nebraska Press, 1993.

Jones, Patrick. "Permapoesis and Artist as Family." *Philosophy, Activism, Nature Special Issue: Cultures of Sustainability* 7 (2010): 101–3. Available at http://panjournal.heroku.com/issues/35.

Jones, Van. *The Green Collar Economy: How One Solution Can Fix Our Two Biggest Problems*. New York: Harper, 2008.

Kahn, Matthew. *Green Cities: Urban Growth and the Environment*. Washington, DC: Brookings Institution Press, 2006.

Kaiser, Jocelyn. "The Dirt on Ocean Garbage Patches." *Science* 328, no. 5985 (2010): 1506.

Kant, Immanuel. *The Critique of Judgment*. Translated by James Creed Meredith, 1911. Available at http://philosophy.eserver.org/kant/critique-of-judgment.txt.

Karat, Prakash. "Double-Speak" Charge: Maligning The CPI(M)." *People's Democracy* 31, no. 4 (2007).

Kass, Leon, et al. *Beyond Therapy: Biotechnology and the Pursuit of Happiness*. New York: Regan, 2003.

Katz, Cindi. "Whose Nature, Whose Culture? Private Productions of Space and the 'Preservation' of Nature." In *Remaking Reality: Nature at the Millennium*, edited by Bruce Braun and Noel Castree, 45–62. New York: Routledge, 1998.

Kaufman, Les, and Kenneth Mallory, eds. *The Last Extinction*. Cambridge, MA: MIT Press, 1993.

Keil, Roger. "Urban Political Ecology." *Urban Geography* 24 (2003): 723–38.

Keller, David R., ed. *Environmental Ethics: The Big Questions*. Malden, MA: Wiley Blackwell, 2010.

Kelman, Ari. *A River and Its City: The Nature of Landscape in New Orleans*. Berkeley: University of California Press, 2003.

Kerridge, Richard. "Climate Change and Contemporary Modernist Poetry." In *Poetry and Public Language*, edited by Tony Lopez and Anthony Caleshu, 131–48. Exeter, UK: Sherasman Books, 2007.

Kheel, Marti. "From Heroic to Holistic Ethics: The Ecofeminist Challenge." In *Ecofeminism: Women, Animals, Nature*, edited by Greta Gaard, 243–71. Philadelphia: Temple University Press, 1993.

Kimmerer, Robin. "Weaving Traditional Ecological Knowledge into Biological Education: A Call to Action." *BioScience* 52, no. 5 (2002): 432–38.

Kimmerer, Robin, and Frank Kanawha Lake. "The Role of Indigenous Burning in Land Management." *Journal of Forestry* (2001): 36–41.

Kincaid, Jamaica. *The Autobiography of My Mother*. New York: Farrar, Straus, Giroux, 1996.

Kinsella, John. "The School of Environmental Poetics and Creativity." *Angelaki Special Issue: Ecopoetics and Pedagogies* 14, no. 2 (2009): 143–48.

Kintisch, Eli. *Hack the Planet: Science's Best Hope—or Worst Nightmare—for Averting Climate Catastrophe*. Hoboken, NJ: Wiley, 2010.

Kirksey, Eben S., ed. *The Multispecies Salon*. Durham, NC: Duke University Press, 2014.

Kirksey, Eben S., and Stefan Helmreich. "The Emergence of Multispecies Ethnography." *Cultural Anthropology* 25, no. 4 (2010): 545–76.

Klawiter, Maren. *The Biopolitics of Breast Cancer: Changing Cultures of Disease and Activism*. Minneapolis: University of Minnesota Press, 2008.

Klein, Naomi. "Dancing the World into Being: A Conversation with Idle No More's Leanne Simpson." *Yes! Magazine* 65 (2013). Available at http://www.yesmagazine.org/peace-justice/dancing-the-world-into-being-a-conversation-with-idle-no-more-leanne-simpson.

Kleiner, Catherine. "Nature's Lovers: The Erotics of Lesbian Land Communities in Oregon, 1974–1984." In *Seeing Nature through Gender*, edited by Virginia Scharff, 242–62. Lawrence: University Press of Kansas, 2003.

Koelb, Janice Hewlett. "'This Most Beautiful and Adorn'd World': Nicholson's *Mountain Gloom and Mountain Glory* Reconsidered." *ISLE* 16, no. 3 (2009): 443–68.

Kofman, Sarah. "The Economy of Respect: Kant and Respect for Women." Translated by Nicola Fisher. *Feminist Interpretations of Immanuel Kant*, edited by Robin May Schott, 355–71. University Park: Pennsylvania State University Press, 1997.

Kohn, Eduardo. "How Dogs Dream: Amazonian Natures and the Politics of Transspecies Engagement." *American Ethnologist* 34, no. 1 (2007): 3–24.

Kolbert, Elizabeth. *The Sixth Extinction: An Unnatural History*. New York: Holt, 2014.

Krakoff, Sarah. "Radical Adaptation, Justice, and American Indian Nations." *Environmental Justice* 4, no. 4 (2011): 207–12.

Krampen, Martin. "Phytosemiotics." *Semiotica* 36, no. ¾ (1981): 187–209.

Krauss, Celene. "Mothering at the Crossroads: African American Women and the Emergence of the Movement against Environmental Racism." In *Environmental Justice in the New Millennium: Global Perspectives on Race, Ethnicity, and Human Rights*, edited by Filomina Chioma Steady, 65–89. New York: Palgrave MacMillan, 2009.

Kricher, John. *The Balance of Nature: Ecology's Enduring Myth*. Princeton, NJ: Princeton University Press, 2009.

Kroeber, Karl. *Ecological Literary Criticism: Romantic Imagining and the Biology of Mind*. New York: Columbia University Press, 1994.

Kronk, Elizabeth Ann. "Application of Environmental Justice to Climate Change–Related Claims Brought by Native Nations." In *Tribes, Land, and the Environment*, edited by Sarah A. Krakoff and Ezra Rosser, 75–102. Burlington, VT: Ashgate, 2012.

Kuhn, Thomas. *The Structure of Scientific Revolutions*. Chicago: University of Chicago Press, 1962.

Kull, Kalevi. "Biosemiotics in the Twentieth Century: A View from Biology." *Semiotica* 12, no. 1/4 (1999): 385–414.

———. "Vegetative, Animal, and Cultural Semiosis: The Semiotic Threshold Zones." *Cognitive Semiotics* 4 (2009): 8–27.

Kull, Kalevi, Claus Emmeche, and Donald Favareau. "Biosemiotic Questions." *Biosemiotics* 1, no. 1 (2008): 41–55.

Kull, Kalevi, Jesper Hoffmeyer, and Alexei Sharov, eds. *Approaches to Semiosis of Evolution*. Berlin: Springer, forthcoming.

Kull, Kalevi, et al. "Theses on Biosemiotics: Prolegomena to a Theoretical Biology." *Biological Theory* 4, no. 2 (2009): 167–73.

Kurtz, Hilda. "Scales Frames and Counter-Scale Frames: Constructing the Problem of Environmental Injustice." *Political Geography* 22 (2003): 887–916.

Kuzniar, Alice. "'I Married My Dog': On Queer Canine Literature." In *Queering the Non/Human*, edited by Noreen Giffney and Myra Hird, 205–26. Hampshire, UK: Ashgate, 2008.

Kysar, Douglas. *Regulating from Nowhere: Environmental Law and the Search for Objectivity*. New Haven. CT: Yale University Press, 2010.

LaDuke, Winona. *All Our Relations: Native Struggles for Land and Life*. Cambridge, MA: South End Press, 1999.

———. "Manoomin Wild Rice, Biodiversity, and Bio Piracy."

Issue Brief. *Institute for Agriculture and Trade Policy* (April 6, 2003). Available at http://www.iatp.org/documents/issue-brief-manoomin-wild-rice-biodiversity-and-bio-piracy.

———. "Traditional Ecological Knowledge and Environmental Futures." *Colorado Journal of International Environmental Law & Policy* 5, no. 127 (1994).

LaDuke, Winona, and Brian Carlson. *Our Manoomin, Our Life: The Anishinaabeg Struggle to Protect Wild Rice*. Ponsford, MN: White Earth Land Recovery Project, 2003.

Lafferty, Kevin, et al. "Parasites in Food Webs: The Ultimate Missing Links." *Ecology Letters* 11, no. 6 (2008): 533–46.

Lakhtakia, Akhlesh, and Raúl J. Martín-Palma, eds. *Engineered Biomimicry*. Waltham, MA: Elsevier, 2013.

Langston, Nancy. *Toxic Bodies: Hormone Disruptors and the Legacy of DES*. New Haven, CT: Yale University Press, 2010.

Larrain S., J. P. Leroy, and K. Nansen, eds. *Citizen Contribution to the Construction of Sustainable Societies*. 2003. Accessed Jan. 12, 2014. Available at http://www.chilesustentable.net/web/wp-content/plugins/downloads-manager/upload/Comercio_Conosur-ingles.pdf.

Latour, Bruno. "An Attempt at a 'Compositionist Manifesto.'" *New Literary History* 41 (2010): 471–90.

———. *Pandora's Hope: Essays on the Reality of Science Studies*. Cambridge, MA: Harvard University Press, 1999.

———. *Politics of Nature: How to Bring the Sciences into Democracy*. Cambridge, MA: Harvard University Press, 2004a.

———. *Reassembling the Social: An Introduction to Actor-Network Theory*. New York: Oxford University Press, 2005.

———. *We Have Never Been Modern*. Translated by C. Porter. Cambridge, MA: Harvard University Press, 1993.

———. "Why Has Critique Run out of Steam?" *Critical Inquiry* 30, no. 2 (2004b): 225–48.

La Vía Campesina. *La Vía Campesina Open Book: Celebrating 20 Years of Struggle and Hope*. 2013. Available at http://via-campesina.org/downloads/pdf/openbooks/EN-00.pdf.

Law, John. "Notes on the Theory of the Actor-Network: Ordering, Strategy, and Heterogeneity." *Systems Practice* 5, no. 4 (1992): 379–93.

Law, John, and Annemarie Mol, eds. *Complexities: Social Studies of Knowledge Practices*. Durham, NC: Duke University Press, 2002.

Lawton, John H., and Robert M. May, eds. *Extinction Rates*. Oxford: Oxford University Press, 1995.

Layard, Richard. *Happiness: Lessons from a New Science*. New York: Penguin, 2005.

Lazarus, Neil. *The Postcolonial Unconscious*. New York: Cambridge University Press, 2011.

Leakey, Richard, and Roger Lewin. *The Sixth Extinction: Patterns of Life and the Future of Humankind*. New York: Anchor, 1995.

Lee, Charles, ed. *Proceedings: The First National People of Color Environmental Leadership Summit*. New York: United Church of Christ Commission for Racial Justice, 1992.

Lefebvre, Henri. *The Production of Space*. Cambridge, MA: Blackwell, 1991.

———. *The Urban Revolution*. Minneapolis: University of Minnesota Press, 2003.

Leff, Enrique, ed. *Los problemas del conocimiento y la perspectiva ambiental del desarrollo*. México: Siglo XXI, 1986.

———. "Marxism and the Environmental Question." *Capitalism, Nature, Socialism* 4, no. 1 (1993): 44–66.

———. *Saber Ambiental*, Mexico, DF: Siglo XXI, 2002.

Legambiente. *Ecomafia 2008*. Milano: Edizioni Ambiente, 2008.

Leigh, E. G., Jr. "Neutral Theory: A Historical Perspective." *Journal of Evolutionary Biolology* 20, no. 6 (2007): 2075–91.

LeMenager, Stephanie, Teresa Shewry, and Ken Hiltner, eds. "Introduction." In *Environmental Criticism for the Twenty-First Century*. New York: Routledge, 2011.

Lemke, Thomas. *Biopolitics: An Advanced Introduction*. New York: New York University Press, 2011.

Lenin, Vladimir. *Imperialism: The Highest Stage of Capitalism*. New York: Pluto, 1996.

Lerner, Steve. *Diamond: A Struggle for Environmental Justice in Louisiana's Chemical Corridor*. Cambridge, MA: MIT Press, 2005.

———. *Sacrifice Zones: The Front Lines of Toxic Chemical Exposure in the United States*. Cambridge, MA: MIT Press, 2010.

Lestel, Dominique, Florence Brunois, and Florence Gaunet. "Etho-Ethnology and Ethno-Ethology." *Social Science Information* 45, no. 2 (2006): 155–77.

Levine, Adeline G. *Love Canal: Science, Politics, and People*. Lexington, MA: Lexington Books, 1982.

Lewan, Tamar. "As Interest Fades in the Humanities, Colleges Worry." *New York Times*. Last modified October 20, 2013. Available at http://www.nytimes.com/2013/10/31/education/as-interest-fades-in-the-humanities-colleges-worry.html.

Lewontin, Richard, Steven Rose, and Leon Kamin. *Not in Our Genes: Biology, Ideology, and Human Nature*. New York: Pantheon, 1984.

Light, Andrew, and Erick Katz, eds. *Environmental Pragmatism*. New York: Routledge, 1996.

Light, Andrew, and Avner de-Shalit, eds. *Moral and Political Reasoning in Environmental Practice*. Cambridge, MA: MIT Press, 2003.

Light, Andrew, and Jonathan M. Smith, eds. *Philosophies of Place*. Lanham, MD: Rowman & Littlefield, 1998.

Linnaeus, Carl. *Systema Naturae: Sive Regna Tria Naturae Systematice Proposita per Classes, Ordines, Genera, and Species*. Leiden, The Netherlands: Theodorum Haak, 1735.

Little, Peter, and Michael Horowitz, eds. *Lands at Risk in the Third World,* Boulder, CO: Westview, 1987.

Little River Band of Ottawa Indians. *Nmé (Lake Sturgeon) Stewardship Plan for the Big Manistee River and 1836 Reservation*. Natural Resources Department: Special Report 1 Manistee, 2008. Accessed September 8, 2013. Available at http://www.exploretheshores.org/assets/files/78/lrboi_sturgeon.pdf.

Livingston, Morna. *Steps to Water: The Ancient Stepwells of India*. New York: Princeton Architectural Press, 2002.

Locke, John. *Two Treatises of Government*. Edited by Peter Laslett. New York: Cambridge University Press, 1988.

Loh, Penn, et al. "From Asthma to AirBeat: Community-Driven Monitoring of Fine Particles and Black Carbon in Roxbury, Massachusetts." *Environmental Health Perspectives* 110, no. 1 (2002): 297–301.

Longinus. "On the Sublime." *The Critical Tradition: Classic Texts and Contemporary Trends*. 2nd ed., edited by David A. Richter, 81–107. Boston: Bedford/St. Martin's, 1998.

Lotka, Alfred. *Elements of Physical Biology*. Baltimore, MD: Williams & Wilkins, 1925.

Lovitz, Dara. *Muzzling a Movement: The Effects of Anti-Terrorism Law, Money, and Politics on Animal Activism*. Brooklyn, NY: Lantern Books, 2010.

Lowdermilk, Walter. *Lessons from the Old World to the Americas in Land Use*. Annual Report of the Board of Regents of the Smithsonian Institution. Washington, DC: Government Printing Office, 1944.

Lowe, Celia. "Viral Clouds: Becoming H5N1 in Indonesia." *Cultural Anthropology* 25, no. 4 (2010): 625–49.

Lowenthal, David. "Reflections on Humpty-Dumpty Ecology." In *Restoration and History: The Search for a Usable Environmental Past*, edited by Marcus Hall, 13–34. London: Routledge, 2010.

Lynch, Tom. "Literature in the Arid Zone." *The Littoral Zone: Australian Contexts and their Writers*, edited by C. A. Cranston and Robert Zeller, 71–92. Amsterdam: Rodopi, 2007.

———. *Xerophilia: Ecocritical Explorations in Southwestern Literature*. Lubbock: Texas Tech University Press, 2008.

Lynch, Tom, et al., eds. *The Bioregional Imagination: Literature, Ecology, and Place*. Athens: University of Georgia Press, 2012.

Lyon, Thomas J. "A Taxonomy of Nature Writing." *This Incomparable Lande: A Book of American Nature Writing*, edited by Thomas J. Lyon, 3–7. Boston: Houghton Mifflin, 1989.

Lyons, Oren. "Listening to Natural Law." In *Original Instructions: Indigenous Teachings for a Sustainable Future*, edited by Melissa Nelson. Rochester, VT: Bear & Company, 2008.

Maasai Wilderness Conservation Trust. Accessed June 6, 2013. Available at http://www.maasaiwilderness.org/.

Maathai, Wangari. *The Green Belt Movement: Sharing the Approach and the Experience*. Revised edition. New York: Lantern Books, 2004.

———. *Unbowed: A Memoir*. New York: Knopf, 2006.

MacArthur, Robert H. "Population Ecology of Some Warblers of Northeastern Coniferous Forests." *Ecology* 39, no. 4 (1958): 599–619.

MacArthur, Robert H., and Edward O. Wilson. *The Theory of Island Biogeography*. Princeton, NJ: Princeton University Press, 1967.

MacCannell, Dean. *Empty Meeting Grounds: The Tourist Papers*. New York: Routledge, 1992.

MacCormack, Patricia. "Queer Posthumanism: Cyborgs, Animals, Monsters, Perverts." In *The Ashgate Research Companion to Queer Theory*, edited by Noreen Giffney and Michael O'Rourke, 111–26. Hampshire, UK: Ashgate, 2009.

Macer, Darryl. "A Public Ethos of Enhancement across Asia." *American Journal of Bioethics* 14, no. 4 (2014): 45–47.

Macfarlane, Robert. "Call of the Wild." *Guardian*. Last modified December 5, 2003. Available at http://www.theguardian.com/books/2003/dec/06/featuresreviews.guardianreview34.

Machado, Antonio. *Fields of Castile/Campos de Castilla*. Translated by Stanley Appelbaum. New York: Dover, 2007.

———. *Lands of Castile and Other Poems*. Translated by Paul Burns and Salvator Ortiz-Carboneres. New York: Aris & Phillips, 2002.

———. *The Landscape of Castile: Poems by Antonio Machado*. Translated by Mary G. Berg and Dennis Maloney. Buffalo, NY: White Pine Press, 2004.

Maclaurin, James, and Kim Sterelny. *What Is Biodiversity?* Chicago: University of Chicago Press, 2008.

MacLeavy, Julie, and John Harrison. "New State Spatialities: Perspectives on State, Space, and Scalar Geographies." *Antipode* 42, no. 5 (2010): 1037–46.

Macnaghten, Phil, and John Urry. *Contested Natures*. London: Sage, 1998.

Maier, Donald S. *What's So Good about Biodiversity?* Dordrecht, Netherlands: Springer, 2012.

Mandaluyong Declaration. "Mandaluyong Declaration of the Global Conference on Indigenous Women, Climate Change, and REDD Plus." In *Indigenous Women, Climate Change, and Forests*. Baguio City, Philippines: Tebtebba Foundation, 2011.

"Manoomin (Wild Rice)." *Great Lakes Indian Fish and Wildlife Commission*. Accessed Oct. 15, 2013. Available at http://www.glifwc.org/WildRice/wildrice.html.

Maran, Timo. "Are Ecological Codes Archetypal Structures?" In *Semiotics in the Wild: Essays in Honour of Kalevi Kull on the Occasion of his 60th Birthday*, edited by Timo Maran, et al., 147–56. Tartu, Estonia: Tartu University Press, 2012.

———. "Biosemiotic Criticism." In *Oxford Handbook of Ecocriticism*, edited by Greg Garrard, 260–75. Oxford: Oxford University Press, 2014.

———. "An Ecosemiotic Approach to Nature Writing." *PAN: Philosophy Activism Nature* no. 7 (2010): 79–87.

Maran, Timo, Dario Martinelli, and Aleksei Turovski, eds. *Readings in Zoosemiotics (Semiotics, Communication and Cognition 8)*. Berlin: DeGruyter Mouton, 2011.

Marcus, George. "Ethnography in/of the World System: The Emergence of Multi-Sited Ethnography." *Annual Review of Anthropology* 24 (1995): 95–117.

Marcuse, Herbert. "Ecology and Revolution." In *The New Left and the 1960s*, edited by Douglas Kellner, 173–76. London: Routledge, 2005.

————. *One-Dimensional Man: Studies in the Ideology of Advanced Industrial Society.* Boston: Beacon, 1968.

Margulis, Lynn. *Symbiosis in Cell Evolution,* San Francisco: Freeman, 1992.

Margulis, Lynn, and Joel E. Cohen. "Combinatorial Generation of Taxonomic Diversity: Implication of Symbiogenesis for the Proterozoic Fossil Record." In *Early Life on Earth,* 327–33. New York: Columbia University Press, 1994.

Margulis, Lynn, and Dorion Sagan. *Acquiring Genomes: A Theory of the Origins of Species.* New York: Basic Books, 2003.

————. *Microcosmos: Four Billion Years of Microbial Evolution.* Berkeley: University of California Press, 1997.

Markoš, Anton, and Dan Faltýnek. "Language Metaphors of Life." *Biosemiotics* 4, no. 2 (2011): 171–200.

Marsh, George Perkins. *Man and Nature; or, Physical Geography as Modified by Human Action,* edited by David Lowenthal. Cambridge, MA: Belknap Press of Harvard University Press, 1965 (1864).

Marston, Sally. "The Social Construction of Scale." *Progress in Human Geography* 24, no. 2 (2000): 219–42.

Martin, R. M., and D. R. Darr. "Market Responses to the US Timber Demand-Supply Situation of the 1990s: Implications for Sustainable Forest Management." *Forest Products Journal* 47, nos. 11–12 (1997): 27–32.

Martinez-Alier, Joan. "International Biopiracy versus the Value of Local Knowledge." *Capitalism, Nature, Socialism* 11, no. 2 (2000): 59–66.

————. "Scale, Environmental Justice, and Unsustainable Cities." *CNS* 14, no. 4 (2003): 43–63.

Martinez-Alier, Joan, and Klaus Schlupmann, *Ecological Economics: Energy, Environment, and Society.* New York: Basil Blackwell, 1990.

Marx, Karl. *Capital: A Critique of Political Economy.* Vol. 1. London: Penguin, 1990 (1976).

Marx, Leo. *The Machine in the Garden: Technology and the Pastoral Ideal in America.* New York: Oxford University Press, 1964.

Massey, Doreen. "Don't Let's Counterpose Place and Space." *Development* 45, no. 2 (2002): 24–25.

————. "Geographies of Responsibility." *Geografiska Annaler: Series B, Human Geography* 86, no. 1 (2004): 5–18.

Mathews, Freya. "Towards a Deeper Philosophy of Biomimicry." *Organization and Environment* 24, no. 4 (2011): 364–87.

May, Robert M. "How Many Species?" In *The Fragile Environment: The Darwin College Lectures,* edited by Laurie Friday and Ronald Laskey, 61–81. Cambridge, MA: Cambridge University Press, 1989.

————. "How Many Species?" *Philosophical Transactions: Biological Sciences* 330 (1990): 293–304.

————. "How Many Species Are There on Earth?" *Science* 241 (1988): 1441.

————. "Patterns of Species Abundance and Diversity." In *Ecology and Evolution of Communities,* edited by M. L. Cody and J. M. Diamond, 81–120. Cambridge, MA: Harvard University Press, 1975.

————. *Stability and Complexity in Model Ecosystems.* Princeton, NJ: Princeton University Press, 1974.

May, Robert M., et al. "Assessing Extinction Rates." *Extinction Rates,* edited by J. H. Lawton and R. M. May. New York: Oxford University Press, 1995.

Mayer, Brian. "Cross-Movement Coalition Formation: Bridging the Labor-Environment Divide." *Sociological Inquiry* 79, no. 2 (2009): 219–39.

Mayr, Ernst. *Systematics and Origin of Species.* New York: Columbia University Press, 1942.

McCann, Kevin S. *Food Webs (MPB-50).* Princeton, NJ: Princeton University Press, 2011.

McCarthy, James. "Scale, Sovereignty, and Strategy in Environmental Governance." *Antipode* 37, no. 4 (2005): 731–53.

McGill, Brian J., Brian A. Maurer, and Michael D. Weiser. "Empirical Evaluation of Neutral Theory." *Ecology* 87 (2006): 1411–23.

McGinnis, Michael Vincent. *Bioregionalism.* New York: Routledge, 1999.

McGregor, Deborah. "Linking Traditional Ecological Knowledge and Western Science: Aboriginal Perspectives from the 2000 State of the Lakes Ecosystem Conference." *Canadian Journal of Native Studies* 28, no. 1 (2008): 139–58.

McGurty, Eileen. *Transforming Environmentalism: Warren County, PCBs, and the Origins of Environmental Justice.* New Brunswick, NJ: Rutgers University Press, 2009.

McHarg, Ian L. *Design with Nature.* New York: Wiley, 1969.

McKibben, Bill. *Eaarth.* New York: Henry Holt, 2010.

————. *The End of Nature.* New York: Anchor Books, 1989.

————, ed. "Introduction." *American Earth: Environmental Writ-*

ing since Thoreau. New York: Library of America (Book 182), 2009.

McLaren, Duncan. "Environmental Space, Equity, and the Ecological Debt." In *Just Sustainabilities: Development in an Unequal World*, edited by Julian Agyeman, Robert Bullard, and Bob Evans. London: Earthscan, 2003.

McManus, P. "Conservation." In *The Dictionary of Human Geography,* 4th ed., edited by R. J. Johnston, et al., 106–8. Malden, MA: Blackwell, 2000.

McNeill, J. R. *Something New under the Sun. An Environmental History of the Twentieth-Century World*. New York: Norton, 2000.

McWhorter, Ladelle. "Enemy of the Species." In *Queer Ecologies: Sex, Nature, Politics, Desire*, edited by Catriona Mortimer-Sandilands and Bruce Erickson, 73–101. Bloomington: Indiana University Press, 2010.

MEA (Millennium Ecosystem Assessment). *Ecosystems and Human Well-being: Synthesis*. Washington, DC: Island Press, 2005.

Meadows, Donella, et al. *Limits to Growth*. New York: Universe Books, 1972.

Meinig, D. W., ed. *The Interpretation of Ordinary Landscapes: Geographical Essays*. New York: Oxford University Press, 1979.

Mellor, Anne K. *Romanticism and Gender*. New York: Routledge, 1993.

Melosi, Martin V. *Garbage in the Cities: Refuse, Reform, and the Environment*. College Station: Texas A&M University Press, 1981.

Mendel, Gregor. "Experiments on Plant Hybrids" In *The Origin of Genetics: A Mendel Source Book*, edited by C. Stern and E. R. Sherwood, 1–48. San Francisco: Freeman, 1966.

———. *Verhandlungen des naturforschenden Vereines in Brünn*. Abhandlungen 4, 3, (1866).

Merchant, Carolyn. *The Death of Nature: Women, Ecology, and the Scientific Revolution*. San Francisco: Harper & Row, 1980.

———. *Ecological Revolutions: Nature, Gender, and Science in New England*. Chapel Hill: University of North Carolina Press, 2010 (1989).

Michaels, David. *Manufactured Uncertainty: Contested Science and the Protection of the Public's Health and Environment*. New York: Oxford University Press, 2007.

Middleton, Neil, and Phil O'Keefe. *Redefining Sustainable Development*. London: Pluto, 2001.

Mies, Maria, and Vandana Shiva. *Ecofeminism*. London: Zed, 1993.

Mignolo, Walter. "On Pluriversality." Web log comment, Oct. 20, 2013. Available at http://waltermignolo.com/on-pluriversality/.

Mileti, Dennis S. *Disasters by Design: A Reassessment of Natural Hazards in the United States*. Washington, DC: Joseph Henry Press, 1999.

Miller, DeMond Shondell, Jason David Rivera, and Joel C. Yelin. "Civil Liberties: The Line Dividing Environmental Protest and Ecoterrorists." *Journal for the Study of Radicalism* 2, no. 1 (2008): 109–23.

Miller, Paul. *Better Humans? The Politics of Human Enhancement and Life Extension*. London: Demos, 2006.

Miller, Webb, et al. "Sequencing the Nuclear Genome of the Extinct Woolly Mammoth." *Nature* 456 (2008): 387–90.

Millington, Andrew. "Conservation." In *Our Earth's Changing Land: An Encyclopedia of Land-Use and Land-Cover Change*, vol. 1, edited by Helmut Geist, 137–41. Westport, CT: Greenwood, 2005.

Minnesota Department of Natural Resources. "Natural Wild Rice in Minnesota," 1–117. St. Paul: Minnesota Department of Natural Resources, 2008.

Minteer, Ben, ed. *Nature in Common? Environmental Ethics and the Contested Foundations of Environmental Policy*. Philadelphia: Temple University Press, 2009.

———. *Refounding Environmental Ethics: Pragmatism, Principle, and Practice*. Philadelphia: Temple University Press, 2012.

Mira, Alex, et al. "The Bacterial Pan-Genome: A New Paradigm in Microbiology." *International Microbiology* 13, no. 2 (2010): 45–57.

Mitchell, Don. "Historical Materialism and Marxism." In *A Companion to Cultural Geography*, edited by James Duncan, Nuala Johnson, and Richard Schein, 51–65. Malden, MA: Blackwell, 2004.

———. "There's No Such Thing as Culture: Towards a Reconceptualization of the Idea of Culture in Geography." *Transactions of the Institute of British Geographers*, New Series, 20, no. 1. (1995): 102–16.

Mitchell, Thomas W. J. *Landscape and Power*. Chicago: University of Chicago Press, 2002.

———. *The Last Dinosaur: The Life and Times of a Cultural Icon*. Chicago: University of Chicago Press, 1998.

Mohai, Paul, David Pellow, and J. Timmons Roberts. "Environmental Justice." *Annual Review of Environment and Resources* 34, no. 1 (2009): 405–30.

Mol, Arthur P. J. *Environmental Reform in the Information Age: The Contours of Informational Governance.* New York: Cambridge University Press, 2008.

———. *Globalization and Environmental Reform: The Ecological Modernization of the Global Economy.* Cambridge, MA: MIT Press, 2001.

Mol, Arthur P. J., David A. Sonnenfeld, and Gert Spaargaren, eds. *The Ecological Modernization Reader: Environmental Reform in Theory and Practice.* New York: Routledge, 2009.

Mollett, Sharlene. "Esta Listo (Are You Ready)? Gender, Race, and Land Registration in the Río Plátano Biosphere Reserve." *Gender, Place, and Culture* 17, no. 3 (2010): 357–75.

Montello, Daniel. "Scale in Geography." In *International Encyclopedia of the Social and Behavioral Sciences,* edited by N. J. Smelser and P. B. Baltes, 13501–4. Oxford: Pergamon, 2001.

Moore, Donald S., Jake Kosek, and Anand Pandian, eds. *Race, Nature, and the Politics of Difference.* Durham, NC: Duke University Press, 2003.

Moore, Niamh. "Eco/Feminism, Nonviolence, and the Future of Feminism." *International Feminist Journal of Politics* 10, no. 3 (2008): 282–98.

Morello-Frosch, Rachel, and E. Shenassa. "The Environmental 'Riskscape' and Social Inequality: Implications for Explaining Maternal and Child Health Disparities." *Environmental Health Perspectives* 114, no. 8 (2006): 1150–53.

Moreno, Jonathan. *The Body Politic: The Battle over Science in America.* New York: Bellevue Literary Press, 2012.

Morgan, Thomas Hunt. "The Theory of the Gene." *American Naturalist* 51, no. 608 (1917): 513–44.

Morley, David. "Three Poems Plus a Note on New Forms." *Horizon Review* (2008): 1. Accessed August 17, 2012. Available at http://www.saltpublishing.com/horizon/issues/01/text/morley_david.htm.

Mortimer-Sandilands, Catriona. "Eco/Feminism on the Edge." *International Feminist Journal of Politics* 10, no. 3 (2008a): 305–13.

———. "Masculinity, Modernism, and the Ambivalence of Nature: Sexual Inversion as Queer Ecology in *The Well of Loneliness.*" *Left History* 13, no. 1 (2008b): 35–58.

———. "Melancholy Natures, Queer Ecologies." In *Queer Ecologies: Sex, Nature, Politics, Desire,* edited by Catriona Mortimer-Sandilands and Bruce Erickson, 331–58. Bloomington: Indiana University Press, 2010.

Mortimer-Sandilands, Catriona, and Bruce Erickson. *Queer Ecologies: Sex, Nature, Politics, Desire.* Bloomington: Indiana University Press, 2010.

Morton, Timothy. *The Ecological Thought.* Cambridge, MA: Harvard University Press, 2010.

———. *Ecology without Nature: Rethinking Environmental Aesthetics.* Cambridge, MA: Harvard University Press, 2007.

———. "Guest Column: Queer Ecology." *PMLA* 125, no. 2 (2010): 273–82.

Moseley, William G. "Beyond Knee-Jerk Environmental Thinking: Teaching Geographic Perspectives on Conservation, Preservation, and the Hetch Hetchy Valley Controversy." *Journal of Geography in Higher Education* 33, no. 3 (2009): 433–51.

Moyo, Sam, and Paris Yeros, eds. *Reclaiming the Land: The Resurgence of Rural Movements in Africa, Asia, and Latin America.* New York: Zed, 2005.

Muir, John. *John Muir: Nature Writings,* edited by William Cronon. New York: Library of America, 1997.

Mumford, Lewis. *The Culture of Cities.* New York: Harcourt, 1938.

Murphy, Patrick D. "An Ecological Feminist Revisioning of the Masculinist Sublime." *Revista Canaria de Estudios Ingleses* no. 64 (2012): 79–94.

———. *Farther Afield in the Study of Nature-Oriented Literature.* Charlottesville: University of Virginia Press, 2000.

———, ed. *Literature of Nature: An International Sourcebook.* Chicago: Dearborn, 1998.

———. "The Varieties of Environmental Literature in North America." In *Teaching North American Environmental Literature,* edited by Laird Christensen, Mark C. Long, and Fred Waage, 24–36. New York: Modern Language Association of America, 2008.

Myers, Norman. "Environmental Refugees: Our Latest Understanding." *Philosophical Transactions of the Royal Society* 356 (2001).

———. *The Sinking Ark: A New Look at the Problem of Disappearing Species.* Oxford, UK: Pergamon, 1979.

Naess, Arne. *Ecology, Community, and Lifestyle: Outline of an*

Ecosophy. Translated and reviewed by David Rothenberg. Cambridge: Cambridge University Press, 1989.

Naipaul, Vidiadhar Surajprasad. *The Enigma of Arrival*. New York: Knopf, 1987.

Nakashima, Douglas, et al. *Weathering Uncertainty: Traditional Knowledge for Climate Change Assessment and Adaptation*. Paris: UNESCO, United Nations University, 2012.

Nash, June C. *Mayan Visions: The Quest for Autonomy in an Age of Globalization*. New York: Routledge, 2001.

Nasr, Seyyed Hossein. *Man and Nature: The Spiritual Crisis in Modern Man*. Chicago: Distributed by KAZI Publications, 1997.

———. *Religion and the Order of Nature*. New York: Oxford University Press, 1996.

Nederveen Pieterse, J. "Going Global: Futures of Capitalism." *Development and Change* 28, no. 2 (1997): 367–82.

Nee, Sean. "The Neutral Theory of Biodiversity: Do the Numbers Add Up?" *Functional Ecology* 19 (2005): 173–76.

Nelkin, Dorothy, and Michael Pollak. *The Atom Besieged: Extraparliamentary Dissent in France and Germany*. Cambridge, MA: MIT Press, 1981.

Nelson, Melissa K. *Original Instructions: Indigenous Teachings for a Sustainable Future*. Rochester, VT: Bear, 2008.

Netting, Robert McC. "Agrarian Ecology." *Annual Review of Anthropology* 3 (1974): 21–56.

Nettle, Daniel, and Suzanne Romaine. *Vanishing Voices: The Extinction of the World's Languages*. New York: Oxford University Press, 2000.

Neumann, Roderick. *Imposing Wilderness: Struggles over Livelihood and Nature Preservation in Africa*. Berkeley: University of California Press, 1998.

———. "Primitive Ideas: Protected Area Buffer Zones and the Politics of Land in Africa." *Development and Change* 28, no. 3 (1997): 559–82.

Newman, Peter, Timothy Beatley, and Heather Boyer. *Resilient Cities: Responding to Peak Oil and Climate Change*. Washington, DC: Island Press, 2009.

Nicholson, Marjorie Hope. *Mountain Gloom and Mountain Glory: The Development of the Aesthetics of the Infinite*. Seattle: University of Washington Press, 1997.

Nielsen, Soeren Nors. "Towards an Ecosystem Semiotics: Some Basic Aspects for a New Research Programme." *Ecological Complexity* 4, no. 3 (2007): 93–101.

Nietschmann, Bernard. *Between Land and Water*. New York: Seminar Press, 1973.

Niezen, Ronald. *The Origins of Indigenism: Human Rights and the Politics of Identity*. Berkeley: University of California Press, 2003.

———. *The Rediscovered Self: Indigenous Identity and Cultural Justice* (McGill-Queen's Native and Northern Series, vol. 56). Montréal: McGill-Queen's University Press, 2009.

Nightingale, Andrea. "The Nature of Gender: Work, Gender, and Environment." *Environment and Planning D: Society and Space* 24, no. 2 (2006): 165–85.

Nixon, Rob. *London Calling: V. S. Naipaul, Postcolonial Mandarin*. New York: Oxford University Press, 1992.

———. *Slow Violence and the Environmentalism of the Poor*. Cambridge, MA: Harvard University Press, 2011.

Norgaard, Kari Marie. *Living in Denial: Climate Change, Emotions, and Everyday Life*. Cambridge, MA: MIT Press, 2011.

Norgaard, Richard B. "Ecosystem Services: From Eye-Opening Metaphor to Complexity Blinder." *Ecological Economics* 69, no. 6 (2010): 1219–27.

Norton, Bryan G. *Toward a Unity among Environmentalists*. New York: Oxford University Press, 1991.

Nussbaum, Martha C. *Women and Human Development: The Capabilities Approach*. New York: Cambridge University Press, 2000.

Nye, David E., et al. "Background Paper: The Emergence of the Environmental Humanities." *MISTRA: The Swedish Foundation for Strategic Environmental Research*. Stockholm, Sweden, 2013.

O'Brien, Susie. "Nature's Nation, National Natures? Reading Ecocriticism in a Canadian Context." *Canadian Poetry* 42 (1998). Available at http://canadianpoetry.org/volumes/vol42/obrien.htm.

O'Connor, James. *Natural Causes*. New York: Guilford Press, 1998.

Odum, Eugene P., and Howard T. Odum. *Fundamentals of Ecology*. Philadelphia: Saunders, 1953.

Oelschlaeger, Max. *Caring for Creation: An Ecumenical Approach to the Environmental Crisis*. New Haven, CT: Yale University Press, 1994.

———, ed. *Postmodern Environmental Ethics*. Albany: State University of New York Press, 1995.

Oliver-Smith, Anthony. "Theorizing Vulnerability in a Globalized World: A Political Ecological Perspective." In *Mapping Vulnerability: Disasters, Development, and People*, edited by Greg Bankoff, Georg Frerks, and Dorothea Hilhorst, 10–24. London: Earthscan, 2004.

O'Rourke, Dara. "Citizen Consumer." *Boston Review* (2011): 14–19.

Osofsky, Hari M. "The Inuit Petition as a Bridge? Beyond Dialectics of Climate Change and Indigenous Peoples' Rights." *American Indian Law Review* 31, no. 2 (2006): 675–97.

Oxfam International. "Richest 1% Will Own More Than All the Rest by 2016." January 19, 2015. Last accessed February 8, 2015. Available at http://oxf.am/Zi4L.

Pádua, José Augusto. *Um sopro de destruição. Pensamento Político e Crítica Ambiental no Brasil Escravista (1786–1888)*. Rio de Janeiro: Jorge Zahar, 2002.

Parisi, Luciana. "The Nanoengineering of Desire." In *Queering the Non/Human*, edited by Noreen Giffney and Myra Hird, 283–309. Hampshire, UK: Ashgate, 2008.

Park, Geoff. *Ngā Uruora: The Groves of Life; Ecology and History in a New Zealand Landscape*. Wellington: Victoria University Press, 1995.

Park, Lisa Sun-Hee, and David Nabuib Pellow. *The Slums of Aspen: Immigrants vs. the Environment in America's Eden*. New York: New York University Press, 2011.

Parrish, Susan Scott. "Zora Neale Hurston and the Environmental Ethic of Risk." In *American Studies, Ecocriticism, and Ecology: Thinking and Acting in the Local and Global Commons*, edited by Joni Adamson and Kimberly N. Ruffin, 21–36. New York: Routledge, 2013.

Passino, Kevin. *Biomimicry for Optimization, Control, and Automation*. London: Springer-Verlag, 2005.

Pastor, Manuel, et al. "Bridging the Bay: University-Community Collaborations in the San Francisco Bay Area." In *Breakthrough Communities: Sustainability and Justice in the Next American Metropolis*, edited by M. Paloma Pavel. Cambridge, MA: MIT Press, 2009.

"Pastoral." In *The New Princeton Encyclopedia of Poetry and Poetics*, edited by Alex Preminger and T. V. F. Brogan. Princeton, NJ: Princeton University Press, 1993.

Patterson, Annabel. *Pastoral and Ideology: Virgil to Valery*. Berkeley: University of California Press, 1987.

Patterson, David J., et al. "Taxonomic Indexing: Extending the Role of Taxonomy." *Systematic Biology* 55 (2006): 367–73.

Paul, Jay, and Gerald Graff. "Fear of Being Useful." *Inside Higher Education*. Last modified January 5, 2012. Available at http://www.insidehighered.com/views/2012/01/05/essay-new-approach-defend-value-humanities.

Pawlyn, Michael. *Biomimicry in Architecture*. London: Riba, 2011.

Payne, Daniel G. *Voices in the Wilderness: American Nature Writing and Environmental Politics*. Hanover, NH: University Press of New England, 1996.

Peirce, Charles S. *Collected Papers of Charles Sanders Peirce*. Cambridge, MA: Harvard University Press. (Vols. 1–6, Charles Hartshorne and Paul Weiss, eds., 1931–1935; vols. 7–8, A. W. Burks, ed., 1958; in-text references are to CP, followed by volume and paragraph numbers.)

Pellow, David Naguib. *Garbage Wars: The Struggle for Environmental Justice in Chicago*. Cambridge, MA: MIT Press, 2004.

———. *Resisting Global Toxics: Transnational Movements for Environmental Justice*. Cambridge, MA: MIT Press, 2007.

Pellow, David Naguib, and Robert J. Brulle. *Power, Justice, and the Environment: A Critical Appraisal of the Environmental Justice Movement*. Cambridge, MA: MIT Press, 2005.

Pels, Dick, Kevin Hetherington, and Frèdèric Vandenberghe. "The Status of the Object: Performances, Mediations, and Techniques Theory." *Culture and Society* 19, nos. 5–6 (2002): 1–21.

Peña, Devon. *The Terror of the Machine: Technology, Work, Gender, and Ecology on the U.S.-Mexico Border*. Austin: University of Texas Press, 1997.

Pérez-Peña, Richard. "College Classes Use Arts to Brace for Climate Change." *New York Times* (March 31, 2014): A-12. Accessed April 4, 2014.

Pergolizzi, Antonio. *Toxic Italy: Ecomafie e capitalismo. Gli affari sporchi all'ombra del progresso*. Rome: Castelvecchi, 2012.

Peters, Terri. "Nature as Measure: The Biomimicry Guild." *Architectural Design* 81, no. 6 (2011): 44–47.

Pezzoli, Keith, Kerry Williams, and Sean Kriletich. "A Manifesto for Progressive Ruralism in an Urbanizing World." *Progressive Planning* 186 (2011): 16–19.

Pezzullo, Phaedra. *Toxic Tourism: Rhetorics of Pollution, Travel, and Environmental Justice*. Birmingham: University of Alabama Press, 2009.

Phillips, Dana. *The Truth of Ecology: Nature, Culture, and Literature in America*. New York: Oxford University Press, 2003.

Pierotti, Raymond, and Daniel Wildcat. "Traditional Ecological Knowledge: The Third Alternative." *Ecological Applications* 10, no. 5 (2000): 1333–40.

Pipkin, John G. "The Material Sublime of Women Romantic Poets." *SEL: Studies in English Literature, 1500–1900* 38, no. 4 (1998): 597–619.

Plows, Alexandra, Brian Doherty, and Derek Wall. "Covert Repertoires: Ecotage in the U.K." *Social Movement Studies* 3, no. 2 (2004): 199–221.

Plumwood, Val. *Environmental Culture: The Ecological Crisis of Reason*. London: Routledge, 2002.

———. *Feminism and the Mastery of Nature*. New York: Routledge, 1993.

———. "Integrating Ethical Frameworks for Animals, Humans, and Nature: A Critical Feminist Eco-Socialist Analysis." *Ethics & the Environment* 5, no. 2 (2000): 285–322.

Pois, Robert A. *National Socialism and the Religion of Nature*. New York: Palgrave MacMillan, 1986.

Pollan, Michael. *The Botany of Desire: A Plant's-Eye View of the World*. New York: Random House, 2001.

Ponting, Clive. *A Green History of the World*. London: Sinclair-Stevenson, 1992.

Portney, Kent. *Taking Sustainable Cities Seriously: Economic Development, the Environment, and Quality of Life in American Cities*. Cambridge, MA: MIT Press, 2003.

Postel, Sandra. "A River in New Zealand Gets a Legal Voice." *National Geographic Newswatch*, September 4, 2012. Available at http://newswatch.nationalgeographic.com/2012/09/04/a-river-in-new-zealand-gets-a-legal-voice/.

Posthumus, Stephanie. "Re: [asle] Nature Writing outside Anglo-American Tradition?" Association for the Study of Literature and Environment listserv, 2012.

Potts, Donna L. *Contemporary Irish Poetry and the Pastoral Tradition*. Columbia: University of Missouri Press, 2011.

Povinelli, Elizabeth. *Labor's Lot: The Power, History, and Culture of Aboriginal Action*. Chicago: University of Chicago Press, 1993.

Prigogine, Ilya, and Isabelle Stengers. *Order out of Chaos: Man's New Dialogue with Nature*. New York: Bantam, 1984.

"Principles of Working Together." Adopted at the Second People of Color Environmental Leadership Summit. Washington, DC, 2002. Available at http://www.ace-ej.org/ej_resources.

Pulido, Laura. *Environmentalism and Economic Justice: Two Chicano Struggles in the Southwest*. Tucson: University of Arizona Press, 1996.

———. "Rethinking Environmental Racism: White Privilege and Urban Development in Southern California." *Annals of the Association of American Geographers* 90, no. 1 (2000): 12–40.

Quiroz-Martinez, Julie, Diana Pei Wu, and Kristen Zimmerman. *ReGeneration: Young People Shaping Environmental Justice*. Oakland, CA: Movement Strategy Center, 2005.

Radkau, Joakim. *Nature and Power: A Global History of the Environment*. Cambridge: Cambridge University Press, 2008 (2002).

Raglon, Rebecca, and Marion Scholtmeijer. "Canadian Environmental Writing." In *Literature of Nature: An International Sourcebook*, edited by Patrick Murphy, 130–39. Chicago: Dearborn, 1998.

Rappaport, Roy A. *Pigs for the Ancestors*. New Haven, CT: Yale University Press, 1968.

Raup, David M. *Extinction: Bad Genes or Bad Luck?* New York: Norton, 1991.

Read, Peter. *Belonging: Australians, Place, and Aboriginal Ownership*. Cambridge: Cambridge University Press, 2000.

Redclift, Michael. "The Meaning of Sustainable Development." *Geoforum* 23, no. 3 (1992): 395–403.

Regan, Tom. *The Case for Animal Rights*. Berkeley: University of California Press, 1983.

Reich, David, et al. "Genetic History of an Archaic Hominin Group from Denisova Cave in Siberia." *Nature* 468 (2010): 1053–60.

Reidlinger, Dyanna, and Fikret Berkes. "Contributions of Traditional Knowledge to Understanding Climate Change in the Canadian Arctic." *Polar Record* 37, no. 203 (2001): 315–28.

Reill, Peter Hanns. *Vitalizing Nature in the Enlightenment*. Berkeley: University of California Press, 2005.

Reo, Nicholas, and Kyle Powys Whyte. "Hunting and Morality as Elements of Traditional Ecological Knowledge." *Human Ecology* 40, no. 1 (2012): 15–27.

Revelle, Roger. Testimony to US Congress, House 84 H1526–5,

Committee on Appropriations, Hearings on Second Supplemental Appropriation Bill, 1956.

Revkin, Andrew. *Global Warming: Understanding the Forecast.* American Museum of Natural History, Environmental Defense Fund. New York: Abbeville, 1992.

———. "Researchers Propose Earth's 'Anthropocene' Age of Humans Began with Fallout and Plastics." *NYTimes.com* (January 15, 2015). Available at http://dotearth.blogs.nytimes.com/2015/01/15/researchers-propose-earths-anthropocene-age-of-humans-began-with-fallout-and-plastics/.

Ridgwell, Andy, and Daniela N. Schmidt. "Past Constraints on the Vulnerability of Marine Calcifiers to Massive CO2 Release." *Nature Geoscience* 3 (2010): 196–200.

Rieger, Christopher. *Clear-Cutting Eden: Ecology and the Pastoral in Southern Literature.* Tuscaloosa: University of Alabama Press, 2009.

Rifkin, Jeremy. "Fusion Biopolitics." *Nation* (Feb. 18, 2002): 7.

Rigby, Kate. "Earth, World, Text: On the (Im)Possibility of Ecopoiesis." *New Literary History* 35, no. 33 (2004a): 427–42.

———. *Topographies of the Sacred: The Poetics of Place in European Romanticism.* Charlottesville: University of Virginia Press, 2004b.

Ritsma, Nanda, and Stephen Ongaro. "The Commoditisation and Commercialisation of the Maasai Culture: Will Cultural Manyattas Withstand the 21st Century?" In *Cultural Tourism in Africa: Strategies for the New Millennium*, edited by John S. Akama and Patricia Sterry, 127–36. Arnhem, The Netherlands: Association for Tourism and Leisure Education, 2002. Available at http://www.atlas-euro.org/pages/pdf/Cultural%20tourism%20in%20Africa%20Deel%201.pdf.

Roache, Rebecca, and Steve Clarke. "Bioconservatism, Bioliberalism, and the Wisdom of Reflecting on Repugnance." *Monash Bioethics Review* 1, no. 4 (2009): 1–21

Roberts, Dorothy. *Killing the Black Body: Race, Reproduction, and the Meaning of Liberty.* New York: Vintage, 1998.

Roberts, Peter, Joe Ravetz, and Clive George. *Environment and the City.* New York: Routledge, 2009.

Rocheleau, Dianne E. "Political Ecology in the Key of Policy: From Chains of Explanation to Webs of Relation." *Geoforum* 39 (2008): 716–27.

———. "Roots, Rhizomes, Networks, and Territories: Reimagining Pattern and Power in Political Ecologies." In *International Handbook of Political Ecology*, ed. Raymond Bryant, 70–88. London: Edward Elgar, 2015.

Rocheleau, Dianne, and Robin Roth. "Rooted Networks, Relational Webs, and Powers of Connection: Rethinking Human and Political Ecologies." *Geoforum* 38, no. 3 (2007): 433–37.

Rocheleau, Dianne, Barbara Thomas-Slayter, and Esther Wangari, eds. *Feminist Political Ecology: Global Issues and Local Experiences.* London: Routledge, 1996.

Rockström, Johan, et al. "Planetary Boundaries: Exploring the Safe Operating Space for Humanity." *Ecology and Society* 14, no. 2 (2009a): 32.

———. "A Safe Operating Space for Humanity." *Nature* 461, no. 7263 (2009b): 472–75.

Roger, J., ed. *Buffon: Les Époques de la Nature.* Mémoires du Muséum National d'Histoire Naturelle. Nouvelle Série, Série C, Sciences de la Terre, Tome X. Paris: Editions du Muséum, 1962.

Rose, Deborah Bird. *Dingo Makes Us Human: Life and Land in an Australian Aboriginal Culture.* Cambridge: Cambridge University Press, 2009.

Rose, Deborah Bird, and Thom van Dooren, guest eds. "Introduction to 'Unloved Others: Death of the Disregarded in the Time of Extinctions.'" *Australian Humanities Review* 50 (2011): 1–4.

Rose, Deborah Bird, et al. "Thinking through the Environment, Unsettling the Humanities." *Environmental Humanities* 1 (2012): 1–5.

Rose, Nikolas. *The Politics of Life Itself: Biomedicine, Power, and Subjectivity in the Twenty-First Century.* Princeton, NJ: Princeton University Press, 2007.

Ross, Andrew. *Bird on Fire: Lessons from the World's Least Sustainable City.* New York: Oxford University Press, 2011.

———. *The Chicago Gangster Theory of Life: Nature's Debt to Society.* London: Verso, 1995.

Rosset, Peter, Raj Patel, and Michael Courville, eds. *Promised Land: Competing Visions of Agrarian Reform.* San Francisco: Food First, 2006.

Rossini, Manuela, ed. *Energy Connections: Living Forces in Inter/Intra-Action.* Open Humanities Press, 2012. Last modified October 14, 2012. Available at http://www.livingbooksaboutlife.org/books/Energy_Connections.

Roth, Philip. *American Pastoral*. New York: Vintage, 1997.

Rothenberg, David. *Is It Too Painful to Think? Conversations with Arné Naess*. Minneapolis: University of Minnesota Press, 1993.

Roughgarden, Joan. *Evolution's Rainbow: Diversity, Gender, and Sexuality in Nature and People*. Berkeley: University of California Press, 2004.

Roy, Arundhati. *The Greater Common Good*. Bombay: India Book Distributor, 1999.

Rozzi, Ricardo, et al. "Galapagos and Cape Horn: Ecotourism or Greenwashing in Two Iconic Latin American Archipelagoes?" *Environmental Philosophy Special Issue: Ecotourism and Environmental Justice*, Robert Melchior Figueroa, ed., 7, no. 2 (Fall 2010): 1–32.

Runte, Alfred. *National Parks: The American Experience*. Lincoln: University of Nebraska Press, 1987.

Saelens, Brian E., and Susan L. Handy. "Built Environment Correlates of Walking: A Review." *Medicine and Science in Sports and Exercise* 40, no. 7 (2008): Supplement 1: S550–S566.

Saez, Emmanuel. "Striking It Richer: The Evolution of Top Incomes in the United States (Updated with 2013 Preliminary Estimates)." January 25, 2015. Last accessed February 8, 2015.Available at eml.berkeley.edu/~saez-UStopincomes-2013.pdf.

Sagan, Carl. *Cosmos*. New York: Random House, 1980.

Sagan, Dorion. "The Human Is More Than Human: Interspecies Communities and the New 'Facts of Life.'" In *Cosmic Apprentice: Dispatches from the Edges of Science*, 17–32. Minneapolis: University of Minnesota Press, 2013.

Sagan, Dorion, and Lynn Margulis. "'Wind at Life's Back': Toward a Naturalistic, Whiteheadian Teleology; Symbiogenesis and the Second Law." In *Beyond Mechanism: Putting Life Back into Biology*, edited by Brian Henning and Adam Scarfe. Lanham, MD: Lexington Books/Rowman & Littlefield, 2013.

Sahlins, Marshal. *The Use and Abuse of Biology*. Ann Arbor: University of Michigan Press, 1976.

Saint-Pierre, Jacques-Henri-Bernardin de. *Paul et Virginie*. Paris: Garnier-Flammarion, 1992 (1788).

Sale, Kirkpatrick. *Dwellers in the Land: The Bioregional Vision*. Athens: University of Georgia Press, 2000.

Salleh, Ariel. *Ecofeminism as Politics: Nature, Marx, and the Postmodern*. New York: Zed, 1997.

Samways, Michael. "Translocating Fauna to Foreign Lands: Here Comes the Homogenocene." *Journal of Insect Conservation* 3, no. 2 (1999): 65–66.

Sanchez-Azofeifa, G. A., et al. "Dynamics of Tropical Deforestation around National Parks: Remote Sensing of Forest Change on the Osa Peninsula of Costa Rica." *Mountain Research and Development* 22, no. 4 (2002): 352–58.

Sanders, Douglas E. *The Formation of the World Council of Indigenous Peoples*. Copenhagen, Denmark: International Secretariat of International Work Group for Indigenous Affairs, 1977.

Sandilands, Catriona. "Eco Homo: Queering the Ecological Body Politic." *Social Philosophy Today* 19 (2004): 17–39.

———. "Lesbian Separatists and Environmental Experience: Notes Toward a Queer Ecology." Organization and Environment 15.2 (2002): 131–163.

Sandler, Ronald, and Philip Cafaro, eds. *Environmental Virtue Ethics*. Lanham, MD: Rowman & Littlefield, 2005.

Sapp, Jan. *Evolution by Association: A History of Symbiosis*. New York: Oxford University Press, 1994.

———. "Too Fantastic for Polite Society: A Brief History of Symbiosis Theory." In *Lynn Margulis: The Life and Legacy of a Scientific Rebel*, edited by Fotion Sagan, 66–67. White River Junction, VT: Chelsea Green, 2012.

Sassen, Saskia. *Territory, Authority, Rights: From Medieval to Global Assemblages*. Princeton, NJ: Princeton University Press, 2006.

Saviano, Roberto. *Gomorrah: A Personal Journey into the Violent International Empire of Naples' Organized Crime System*. Translated by V. Jewiss. New York: Farrar, Straus, Giroux, 2007.

Sayre, Nathan. "Climate Change, Scale, and Devaluation: The Challenge of Our Built Environment." *Washington and Lee Journal of Climate, Energy, and the Environment* 1, no. 1 (2010): 93–105.

———. "Ecological and Geographical Scale: Parallels and Potential for Integration." *Progress in Human Geography* 29, no. 3 (2005): 276–90.

Sbicca, Joshua. "Eco-queer Movement(s): Challenging Heteronormative Space through (Re)imagining Nature and Food." *European Journal of Ecopsychology* 3 (2012): 33–52

Schabas, Margaret. *The Natural Origins of Economics*. Chicago: University of Chicago Press, 2005.

Schilhab, Theresa, Frederik Stjernfelt, and Terrence Deacon, eds. *The Symbolic Species Evolved*. Berlin: Springer, 2012.

Schlosberg, David. *Environmental Justice and the New Pluralism: The Challenge of Difference for Environmentalism*. New York: New York University Press, 1999.

———. "Theorising Environmental Justice: The Expanding Sphere of a Discourse." *Environmental Politics* 22, no. 1 (2013): 37–55.

Schnaiberg, Allan. *The Environment: From Surplus to Scarcity*. New York: Oxford University Press, 1980.

Schneider, Eric D., and Dorion Sagan. *Into the Cool: Energy Flow, Thermodynamics, and Life*. Chicago: University of Chicago Press, 2006.

Schrag, Daniel. "Foreword." In *Ecology and the Environment: Perspectives from the Humanities*, edited by Donald K. Swearer. Cambridge, MA: Harvard University Press, 2009.

Scott, Dayna, "'Gender-Benders': Sex and Law in the Constitution of Polluted Bodies." *Feminist Legal Studies* 17 (2009): 241–65.

Sebeok, Thomas A. *Essays in Zoosemiotics* (Monograph Series of the TSC 5). Toronto: Toronto Semiotic Circle, 1990.

———. "In What Sense Is Language a 'Primary Modeling System'?" In *A Sign Is Just a Sign*, edited by Thomas A. Sebeok, 49–58. Bloomington: Indiana University Press, 1991.

———. *Perspectives in Zoosemiotics*. The Hague: Mouton, 1972.

Segerstråle, Ullica. *Defenders of the Truth: The Sociobiology Debate*. Oxford, UK: Oxford University Press, 2001.

Seligman, Martin. *Flourish: A Visionary New Understanding of Happiness and Well-Being*. New York: Free Press, 2012.

Sen, Amartya. *The Idea of Justice*. London: Allen Lane, 2009.

Senior, Kathryn, and Alfredo Mazza. "Italian 'Triangle of Death' Linked to Waste Crisis." *Lancet Oncology* 5, no. 9 (2004): 525–27.

Sepkoski, David. *Rereading the Fossil Record: The Growth of Paleobiology as an Evolutionary Discipline*. Chicago: University of Chicago Press, 2012.

Seppelt, Ralf, et al. "A Quantitative Review of Ecosystem Service Studies: Approaches, Shortcomings, and the Road Ahead." *Journal of Applied Ecology* 48, no. 3 (2011): 630–36.

Sessions, George. *Deep Ecology for the Twenty-First Century*. Boston: Shambalah, 1995.

Seymour, Nicole. *Strange Natures: Futurity, Empathy, and the Queer Ecological Imagination*. Champaign: University of Illinois Press, 2013.

Shah, Anup. "Poverty Facts and Stats." *Global Issues* (Jan. 7, 2013). Last accessed Feb. 8, 2015. Available at http://www.globalissues.org/article/26/poverty-facts-and-stats.

Shapiro, James Alan. *Evolution: A View from the 21st Century*. Upper Saddle River, NJ: FT Press Science, 2011.

"A Shared Fate." *Blue Voice.org*. Last modified June 23, 2008. Available at http://www.bluevoice.org/webfilms_shared-fate.php.

Shaw, Philip. *The Sublime*. New York: Routledge, 2006.

Shearer, Christine. *Kivalina: A Climate Change Story*. Chicago: Haymarket Books, 2011.

Shelley, Mary. *Frankenstein; or, The Modern Prometheus*. Oxford, UK: Oxford University Press, 1998 (1818).

Shepard, Peggy M., and Cecil Corbin-Mark. "Climate Justice." *Environmental Justice* 2, no. 4 (2009): 163–66.

Shepard, Peggy M., et al. "Preface: Advancing Environmental Justice through Community-Based Participatory Research." *Environmental Health Perspectives* 110, Supplement 2 (2002): 139–40.

Sherlock, Robert. *Man as a Geological Agent: An Account of his Action on Inanimate Nature*. London: H. F. & G. Witherby, 1922.

Shiva, Vandana. *Biopiracy: The Plunder of Nature and Knowledge*. Boston: South End Press, 1997.

———. "The Real Reasons for Hunger." *Observer Newspaper*. Last modified June 22, 2002. Available at http://www.guardian.co.uk/world/2002/jun/23/1.

———. *Soil Not Oil: Environmental Justice in an Age of Climate Crisis*. Cambridge, MA: South End Press, 2008.

———. *Staying Alive: Women, Ecology, and Development*. London: Zed, 1988 (1989).

———. *The Violence of the Green Revolution: Third World Agriculture, Ecology, and Politics*. New York: Zed, 1992.

Shiva, Vandana, and Ingunn Moser, eds. *Biopolitics: A Feminist and Ecological Reader on Biotechnology*. London: Zed, 1995.

Shonkoff, Seth, et al. "The Climate Gap: Environmental Health and Equity Implications of Climate Change and Mitigation Policies in California: A Review of the Literature." *Climatic Change* 109, no. 1 (2012): 485–503.

Shukin, Nicole. *Animal Capital: Rendering Life in Biopolitical Times*. Minneapolis: University of Minnesota Press, 2009.

Shuttleton, David. "The Queer Politics of Gay Pastoral." In *De-Centring Sexualities: Politics and Representation beyond the Metropolis*, edited by Richard Phillips, Diane West, and David Shuttleton, 125–46. London: Routledge, 2000.

Siewers, Alfred K. "Pre-Modern Ecosemiotics: The Green World as Literary Ecology." In *The Space of Culture: The Place of Nature in Estonia and Beyond*, edited by Tiina Peil, 39–68. Tartu, Estonia: Tartu University Press, 2011.

Simpson, Leanne. *Dancing on Our Turtle's Back: Stories of Nishnaabeg Re-Creation, Resurgence, and a New Emergence*. Winnipeg, Canada: Arbeiter Ring Publishing, 2011.

Singer, Peter. *Animal Liberation*. 2nd ed. New York: Random House, 1990.

———. *A Darwinian Left: Politics, Evolution, and Cooperation*. New Haven, CT: Yale University Press, 2000.

Sittler, Joseph. "A Theology for the Earth." *Christian Scholar* 37 (1954): 367–74.

Skigaj, Leonard M. *Sustainable Poetry: Four American Ecopoets*. Lexington: University Press of Kentucky, 1999.

Skinner, Jonathan. "Why Ecopoetics?" *Ecopoetics* 1 (2001): 105–6.

Slovic, Scott. "Nature Writing." In *Encyclopedia of World Environmental History*, edited by Shepard Krech, J. R. McNeill, and Carolyn Merchant, 886–91. Great Barrington, MA: Berkshire, 2003.

Smith, Graham. *Deliberative Democracy and the Environment*. London: Routledge, 2003.

Smith, Linda Tuhiwai. *Decolonizing Methodologies: Research and Indigenous Peoples*. London: Zed, 2012.

Smith, Mick. *An Ethics of Place: Radical Ecology, Postmodernism, and Social Theory*. Albany: State University of New York Press, 2001.

Smith, Neil. *Uneven Development*. Oxford, UK: Blackwell, 1984.

Smith, Ted, David Sonnenfeld, and David Pellow. *Challenging the Chip: Labor Rights and Environmental Justice in the Global Electronics Industry*. Philadelphia: Temple University Press, 2006.

Snow, Charles Percy. *The Two Cultures*. London: Cambridge University Press, 2001 (1959).

Snyder, Gary. *A Place in Space: Ethics, Aesthetics, and Watersheds; New and Selected Prose*. Washington, DC: Counterpoint, 1995.

Somit, Alfred, and Steven A. Peterson. "Biopolitics after Three Decades: A Balance Sheet." *British Journal of Political Science* 28 (1998): 559–71.

Sonea, Sorin, and Maurice Panisset. *The New Bacteriology*. Boston: Jones and Bartlett, 1983.

Soper, Kate. *What Is Nature?* Oxford, UK: Blackwell, 1995.

de Sousa Santos, Boaventura. "Beyond Abyssal Thinking: From Global Lines to Ecologies of Knowledges." *Review* 30, no. 1 (2007a): 45–89.

———, ed. *Cognitive Justice in a Global World: Prudent Knowledges for a Decent Life*. Lanham, MD: Lexington Books, 2007b.

———. *Descolonizar el saber, reinventar el poder*. Montevideo: Ediciones Trilce, 2010.

———. "A Discourse on the Sciences." In *Cognitive Justice in a Global World: Prudent Knowledges for a Decent Life*, edited by Boaventura de Sousa Santos, 13–45. Lanham, MD: Lexington Books, 2007c.

Spaargaren, Gert, Arthur P. J. Mol, and Frederick H. Buttel, eds. *Governing Environmental Flows: Global Challenges for Social Theory*. Cambridge, MA: MIT Press, 2006.

Spretnak, Charlene. *The Resurgence of the Real: Body, Nature, and Place in a Hypermodern World*. Reading, MA: Addison-Wesley, 1997.

———. *The Spiritual Dimension of Green Politics*. Santa Fe, NM: Bear & Co., 1986.

Starhawk. *The Spiral Dance: A Rebirth of the Ancient Religion of the Great Goddess*. San Francisco: Harper & Row, 1979.

Starn, Orin. "Writing Culture at 25: Special Editor's Introduction." *Cultural Anthropology* 27, no. 3 (2012): 411–16.

Steady, Philomena Chioma. *Environmental Justice in the New Millennium: Global Perspectives on Race, Ethnicity, and Human Rights*. New York: Palgrave MacMillan, 2009.

Stein, Rachel. "Introduction." In *New Perspectives on Environmental Justice: Gender, Sexuality, and Activism*, edited by Rachel Stein et al., 1–17. New Brunswick, NJ: Rutgers University Press, 2004.

———, ed. *New Perspectives on Environmental Justice: Gender, Sexuality, and Activism*. New Brunswick, NJ: Rutgers University Press, 2004.

Steingraber, Sandra. *Having Faith: An Ecologist's Journey to Motherhood*. New York: Berkeley Books, 2003.

Stengers, Isabelle. "The Cosmopolitical Proposal." In *Making Things Public: Atmospheres of Democracy*, edited by Bruno Latour and Peter Weibel, 994–1003. Cambridge, MA: MIT Press, 2005.

———. *Cosmopolitics*. 2 vols. Translated by Robert Bononno. Minneapolis: University of Minnesota Press, 2010–11.

———. *The Invention of Modern Science*. Minneapolis: University of Minnesota Press, 2000.

Stenning, Anna, and Terry Gifford. "Twentieth-Century Nature Writing in Britain and Ireland." *Green Letters: Studies in Ecocriticism* 17, no. 1 (2013): 1–4.

Stephens, Elizabeth, and Annie Sprinkle. "On Becoming Appalachian Moonshine." *Performance Research* 17, no. 4 (2012): 61–66.

Stern, Nicholas. *Stern Review on the Economics of Climate Change*. New York: Cambridge University Press, 2006.

Steward, Julian. *The Concept and Method of Cultural Ecology: Theory of Culture Change*. Urbana: University of Illinois Press, 1955.

Stone, Christopher D. *Should Trees Have Standing?* Los Altos, CA: William Kaufmann, 1974.

Stoppani, Antonio, *Corso di Geologia*. Milan: Bernardoni and Brigola, 1873.

Strathern, Marilyn. "Out of Context: The Persuasive Fictions of Anthropology." *Current Anthropology* 28, no. 3 (1987): 251–81.

Sturgeon, Noël. "Penguin Family Values: The Nature of Planetary Environmental Reproductive Justice." In *Queer Ecologies: Sex, Nature, Politics*, edited by Catriona Mortimer-Sandilands and Bruce Erickson, 102–33. Bloomington: Indiana University Press, 2010.

Sullivan, Robert. *The Meadowlands: Wilderness Adventures on the Edge of New York City*. London, UK: Granta Books, 2006.

Sundberg, Juanita. "Identities in the Making: Conservation, Gender, and Race in the Maya Biosphere Reserve, Guatemala." *Gender, Place & Culture* 11, no. 1 (2004): 43–66. Available at http://dx.doi.org/10.1080/0966369042000188549.

Sunstein, Cass. *Laws of Fear*. New York: Cambridge University Press, 2005.

Sutti, Stefano. *Biopolitica: il nuovo paradigma*. Colchester, UK: Barbarossa Books, 2005.

Suzuki, David. "Imagining a Sustainable Future: Foresight over Hindsight." Keynote address, University of South Wales, Sydney, Australia. September 22, 2013. Available at http://www.youtube.com/watch?v=UC2XeS1aTYk.

Swimme, Brian, and Mary Evelyn Tucker. *Journey of the Universe*. New Haven, CT: Yale University Press, 2011.

Swyngedouw, Erik. "Apocalypse Forever: Post-Political Populism and the Spectre of Climate Change." *Theory, Culture & Society* 27, nos. 2–3 (2010): 213–32.

———. *Social Power and the Urbanization of Water: Flows of Power*. Oxford, UK: Oxford University Press, 2004.

Swyngedouw, Eric, and Nikolas Heynen. "Urban Political Ecology, Justice, and the Politics of Scale." *Antipode*. 35, no. 5 (2003): 899–918.

Sylvan, Richard, and David Bennett. *The Greening of Ethics*. Tucson: University of Arizona Press, 1994.

Szasz, Andrew. "Is Green Consumption Part of the Solution?" In *The Oxford Handbook of Climate Change and Society*, edited by John S. Dryzek, Richard B. Norgaard, and David Schlosberg, 594–608. New York: Oxford University Press, 2011.

———. *Shopping Our Way to Safety: How We Changed from Protecting the Environment to Protecting Ourselves*. Minneapolis: University of Minnesota Press, 2007.

Sze, Julie. *Noxious New York: The Racial Politics of Urban Health and Environmental Justice*. Cambridge, MA: MIT Press, 2007.

Sze, Julie, and Jonathan London. "Environmental Justice at the Crossroads." *Sociology Compass* 2, no. 4 (2008): 1331–54.

Takacs, David. *The Idea of Biodiversity: Philosophies of Paradise*. Baltimore, MD: Johns Hopkins University Press, 1996.

TallBear, Kim. "Why Interspecies Thinking Needs Indigenous Standpoints." 2012. Last modified April 24, 2011. Available at http://www.culanth.org/?q=node/510; vimeo on Cultural Anthropology site: http://www.culanth.org/?q=node/509. (Link also includes Myra Hird and Augustin Fuentes responses to Essay 1.)

Tansley, A. "The Use and Abuse of Vegetational Concepts and Terms." *Ecology* 16, no. 3 (1935): 284–307.

Tarlo, Harriet. "Recycles: the Eco-Ethical Poetics of Found Text in Contemporary Poetry." *Journal of Ecocriticism Special Issue: Poetic Ecologies* 1, no. 2 (2009): 114–30.

Taylor, Bron, ed. *Encyclopedia of Religion and Nature*. London: Continuum, 2005.

Taylor, Charles. *A Secular Age*. Cambridge, MA: Harvard University Press, 2007.

Taylor, Dorceta E. *The Environment and the People in American Cities, 1600s–1900s: Disorder, Inequality, and Social Change*. Durham, NC: Duke University Press, 2009.

Te Aho, Linda. "Indigenous Challenges to Enhance Freshwater Governance and Management in Aotearoa New Zealand–The Waikato River Settlement." *Water Law* 20, no. 5 (2010): 285–92.

Tebtebba. *Indigenous Women, Climate Change, and Forests*. Baguio City, Philippines: Tebtebba Foundation, 2011.

Tedlock, Dennis, trans. *Popol Vuh: The Definitive Edition of the Mayan Book of the Dawn of Life and the Glories of Gods and Kings*. New York: Touchstone, 1985.

———. *The Spoken Word and the Work of Interpretation*. Philadelphia: University of Pennsylvania Press, 1983.

Terry, Jennifer. "'Unnatural Acts' in Nature: The Scientific Fascination with Queer Animals." *GLQ: A Journal of Lesbian and Gay Studies* 6, no. 2 (2000): 151–93.

Thayer, Jr., Robert L. *LifePlace: Bioregional Thought and Practice*. Berkeley: University of California Press, 2003.

Thomashow, Mitchell. *Bringing the Biosphere Home: Learning to Perceive Global Environmental Change*. Cambridge, MA: MIT Press, 2002.

Thoreau, Henry David. "Walking." In *Excursions*, edited by Joseph J. Moldenhauer, 185–222. Princeton, NJ: Princeton University Press, 2007 (1862).

Thornber, Karen Laura. *Ecoambiguity: Environmental Crises and East Asian Literatures*. Ann Arbor: University of Michigan Press, 2012.

Tokar, Brian. *Toward Climate Justice: Perspectives on the Climate Crisis and Social Change*. Porsgrun, Norway: Communalism Press, 2010.

Torrance, Robert. *Encompassing Nature: Nature and Culture from Ancient Times to the Modern World*. Berkeley, CA: Counterpoint, 1999.

Transition United States. "The Transition Town Movement." 2013. Last accessed February 8, 2015. Available at http://transitionus.org/transition-town-movement.

Tredinnick, Mark. "Belonging to Here: An Introduction." *A Place on Earth: An Anthology of Nature Writing from Australia and North America*, edited by Mark Tredinnick, 25–47. Lincoln: University of Nebraska Press, 2003.

———. *The Land's Wild Music*. San Antonio, TX: Trinity University Press, 2005.

Tsing, Anna. "Arts of Inclusion; or, How to Love a Mushroom." *Australian Humanities Review* 50 (2011): 5–21.

———. *Friction: An Ethnology of Global Connection*. Princeton, NJ: Princeton University Press, 2005.

———. "Unruly Edges: Mushrooms as Companion Species." *Environmental Humanities* 1, no. 1 (2012): 141–54.

Tsosie, Rebecca. "Climate Change, Sustainability, and Globalization: Charting the Future of Indigenous Environmental Self-Determination." *Environmental & Energy Law & Policy Journal* 2 (2009): 188–255.

Tuan, Yi-Fu. *Topofilia*. Madrid: Melusina, 2007.

———. *Topophilia*. New York: Columbia University Press, 1974.

Tucker, Richard. "Environmentally Damaging Consumption: The Impact of American Markets on Tropical Ecosystems in the Twentieth Century." In *Confronting Consumption*, edited by Thomas Princen, Michael Maniates, and Ken Conca, 177–95. Cambridge, MA: MIT Press, 2002.

Tully, John. *The Devil's Milk: A Social History of Rubber*. New York: Monthly Review Press, 2011.

Turner, Dale. *This Is Not a Peace Pipe: Towards a Critical Indigenous Philosophy*. Toronto: University of Toronto Press, 2006.

Turner, Nancy, et al. "Edible and Tended Wild Plants, Traditional Ecological Knowledge, and Agroecology." *Critical Reviews in Plant Sciences* 30, no. 1 (2011): 198–225.

Tüür, Kadri. "Bird Sounds in Nature Writing: Human Perspective on Animal Communication." *Sign Systems Studies* 37, nos. 3–4 (2009): 580–613.

Uekoetter, Frank. *The Green and the Brown: A History of Conservation in Nazi Germany*. Cambridge: Cambridge University Press, 2006.

Uexküll, Jakob von. *A Foray into the Worlds of Animals and Humans*. Translated by Joseph D. O'Neil. Minneapolis: University of Minnesota Press, 2010 (1934).

———. "The Theory of Meaning." *Semiotica* 4, no. 1 (1982): 25–82.

Undercurrents: Journal of Critical Environmental Studies Special Issue: Queer/Nature. Toronto: Faculty of Environmental Studies, York University, 1994.

United Nations. *United Nations Declaration on the Rights of Indigenous Peoples*. Last modified September 13, 2007. Available at http://www.un.org/esa/socdev/unpfii/documents/ DRIPS_en.pdf.

United Nations Department of Economic and Social Affairs. Economic and Social Council Resolution 1998/40: 46th Plenary Meeting (July 30)—Declaring the Year 2002 as the International Year of Ecotourism. Last modified July 30, 1998. Available at http://www.un.org/documents/ecosoc/ res/1998/eres1998-40.htm.

United Nations Division for Sustainable Development. *Agenda 21*, United Nations Conference on Environment & Development, Rio de Janeiro, Brazil (June 3-14, 1992). Available at .http://sustainabledevelopment.un.org/content/documents/Agenda21.pdf.

United Nations Economic and Social Council. "Crisis of Human Environment." *Report of the Secretary General on Problems of the Human Environment.* 47th Session, Agenda Item 10 (1969): 4-6.

United Nations Environment Programme. *Sustainable, Resource-Efficient Cities: Making It Happen!* New York: UNEP, 2012.

United Nations Population Fund. *State of World Population 2007: Unleashing the Potential of Urban Growth.* New York: UNFPA, 2007.

United Nations World Tourism Organization. *Tourism Highlights 2013 Edition.* Accessed July 22, 2013. Available at http://dtxtq4w60xqpw.cloudfront.net/sites/all/files/pdf/ unwto_highlights13_en_hr_0.pdf.

Universal Declaration on the Rights of Mother Earth and Climate Change. "Preamble" 2010. Available at http://pwccc.wordpress.com/programa/.

University of Wisconsin Extension. "Wild Rice: Ecology, Harvest, Management." Madison: University of Wisconsin Extension, 2007.

Urry, John. *Sociology beyond Society.* London: Routledge, 2000.

U.S. National Research Council (NRC). *Risk Assessment in the Federal Government: Managing the Process.* Washington, DC: National Academies Press, 1983.

Valero-Garcés, Carmen. "Reflexiones en torno a la Ecocrítica, traducción y terminología." In *Ecocríticas: Literatura y medio ambiente,* edited by Carmen Flys-Junquera et al., 121-34. Madrid: Iberoamerica/Vervuert, 2010.

———. "Walden & Campos de Castilla. Translated. Landscapes and 'eco-translation'?" In *Proceedings of the 3rd EASCLE Conference "Cultural Landscapes: Heritage and Conservation,"* edited by Carmen Flys-Junquera et al., 545-55. Alcalá de Henares: Servicio de Publicaciones de la Universidad, 2009.

Van Dam, C. "Indigenous Territories and REDD in Latin America: Opportunity or Threat?" *Forests* 2, no. 1 (2011): 394-414.

Vanderheiden, Steve. "Eco-Terrorism or Justified Resistance? Radical Environmentalism and the 'War on Terror.'" *Politics & Society* 33 (2005): 425-47.

Veblen, Thorstein. *The Theory of the Leisure Class.* New York: Macmillan, 1899.

Vennum, Thomas. *Wild Rice and the Ojibway People.* St. Paul: Minnesota Historical Society Press, 1988.

Venter, Craig. *Life at the Speed of Light.* New York: Viking, 2013.

Venuti, Lawrence. *The Scandals of Translation.* New York: Routledge, 1998.

———. *The Translator's Invisibility: A History of Translation.* New York: Routledge, 2008 (1995).

Vernadsky, Vladimir. "The Biosphere and the Noosphere." *American Scientist* 33 (1945): 1-12.

Vidas, Davor. "The Anthropocene and the International Law of the Sea. *Philosophical Transactions of the Royal Society-A* 369, no. 1938 (2011): 909-25.

Virgil. *The Eclogues of Virgil.* Translated by David Ferry. New York: Farrar, Straus, Giroux, 1999.

Volk, Tyler. *CO2 Rising: The World's Greatest Environmental Challenge.* Cambridge, MA: MIT Press, 2008.

———. *Gaia's Body: Toward a Physiology of the Earth.* Cambridge, MA: MIT Press, 2003.

Volterra, Vito. "Fluctuations in the Abundance of a Species Considered Mathematically." *Nature* 118 (1926): 558-60.

Walcott, Derek. *Tiepolo's Hound.* New York: Farrar, Straus, Giroux, 2000.

Wallace, Alfred Russel. "On the Zoological Geography of the Malay Archipelago." *Journal of the Proceedings of the Linnean Society of London: Zoology* 4, no. 16 (1860): 172-84.

Wallerstein, Immanuel. *The Modern World System: Capitalist Agriculture and the Origins of the European World Economy in the Sixteenth Century.* New York: Academic Press, 1974.

———. "The Structures of Knowledge; or, How Many Ways May We Know?" In *Cognitive Justice in a Global World: Prudent Knowledges for a Decent Life,* edited by Boaventura de Sousa Santos, 129-34. Lanham, MD: Lexington Books, 2007.

Walls, Laura Dassow. *Passage to Cosmos: Alexander von Humboldt and the Shaping of America.* Chicago: University of Chicago Press, 2009.

Walsh, Bryan. "Nature is Over." *Time Magazine.* Last modified March 12, 2012. Available at http://content.time.com/time/magazine/article/0,9171,2108014,00.html.

Walsh, Catherine. "Afro and Indigenous Life: Visions in/and Politics: (De)Colonial Perspectives in Bolivia and Ecuador." *Bolivian Studies Journal* 18 (2011): 47–67.

Waring, Marilyn. *If Women Counted: A New Feminist Economics.* San Francisco: HarperCollins, 1988.

Warming, Eugenius. *Plantesamfund: Grundtræk af den økologiske Plantegeografi.* Copenhagenb: P.G. Philipsens Forlag, 1895. Edited and translated as *Oecology of Plants: An Introduction to the Study of Plants* by P. Groom and I. B. Balfour. Oxford: Clarendon Press, 1909.

Warren, Karen, ed. *Ecological Feminism.* London: Routldege, 1994.

Washington, Haydn, and John Cook. *Climate Change Denial: Heads in the Sand.* London: Earthscan, 2011.

Waskow, Arthur. "What Is Eco-Kosher?" In *This Sacred Earth: Religion, Nature, Environment,* edited by Roger S. Gottlieb, 297–300. New York: Routledge, 1996.

Waters, Colin, et al. "Can Nuclear Weapons Fallout Mark the Beginning of the Anthropocene Epoch?" *Bulletin of the Atomic Scientists* 71, no. 3 (2015): 46–57.

Watts, Michael. *Silent Violence: Food, Famine, and Peasantry in Northern Nigeria.* Berkeley: University of California Press, 1983.

Weaver, Jace. "Indigeneity and Indigenousness." In *A Companion to Postcolonial Studies,* edited by H. Schwarz and S. Ray, 221–35. New York: Wiley-Blackwell, 2000.

Westra, Laura, and Peter Wenz, eds. *The Faces of Environmental Racism: The Global Equity Issues.* Lanham, MD: Rowman & Littlefield, 1995.

Whatmore, Sarah. "Humanism's Excess: Some Thoughts on the 'Post-human/ist' Agenda." *Environment and Planning A* 36, no. 8 (2004): 1360–63.

———. *Hybrid Geographies: Natures, Cultures, Spaces,* London: Routledge, 2002.

———. "Materialist Returns: Practising Cultural Geography in and for a More-Than-Human World." *Cultural Geographies* 13 (2006): 600–609.

Wheeler, Quentin D. "Introductory." In *The New Taxonomy,* edited by Q. D. Wheeler, pp. 1–17. London: Taylor and Francis, 2008.

———. "What Can We Learn from 20th-Century Concepts of Species? Lessons for a Unified Theory of Species." In *For a Philosophy of Biology,* edited by Ilse Jahn and Andreas Wessel, 43–60. Munich: Kleine Verlag, 2010.

Wheeler, Quentin D., and Rudolf Meier, eds. *Species Concepts and Phylogenetic Theory: A Debate.* New York: Columbia University Press, 2000.

Wheeler, Quentin D., et al. "Mapping the Biosphere: Exploring Species to Understand the Origin, Organization, and Sustainability of Biodiversity." *Systematics and Biodiversity* 10 (2012): 1–20.

Wheeler, Wendy. "Postscript on Biosemiotics: Reading beyond Words—and Ecocriticism." *New Formations* 64, no. 1 (2008): 137–54.

Whitehead, Alfred North. *Concept of Nature.* Cambridge: Cambridge University Press, 1920.

Whyte, Kyle. P. "An Ethics of Recognition for Environmental Tourism Practices." *Environmental Philosophy Special Issue: Ecotourism and Environmental Justice,* edited by Robert Melchior Figueroa, 7, no. 2 (2010): 75–92.

———. "Justice Forward: Tribes, Climate Adaptation, and Responsibility." *Climatic Change* 120, no. 3 (2013): 117–30.

Wilkinson Richard, and Kate Pickett. *The Spirit Level: Why Equality Is Better for Everyone.* London: Allen Lane, 2009.

Williams, Joy. *Ill Nature.* New York: Vintage, 2002.

Williams, Mark, et al. "The Anthropocene Biosphere." *Anthropocene Review* (2015) DOI: 10.1177/2053019615591020.

Williams, Mark, et al., eds. "The Anthropocene: A New Epoch of Geological Time?" *Philosophical Transactions of the Royal Society A* 369, no. 1938 (2011): 833–1112.

Williams, Raymond. *The Country and the City.* New York: Oxford University Press, 1973.

———. "Culture." In *Keywords: A Vocabulary of Culture and Society,* 87–93. New York: Oxford University Press, 1983 (1976).

———. "Ideas of Nature." In *Problems in Materialism and Culture.* London: Verso, 1980.

———. *Keywords: A Vocabulary of Culture and Society.* London: Fontana, 1976.

Williams, Terry, and Preston Hardison. "Culture, Law, Risk, and Governance: Contexts of Traditional Knowledge in

Climate Change Adaptation." *Climatic Change* (2013): 1–14.

Williamson, Karina. "From Arcadia to Bunyah: Mutation and Diversity in the Pastoral Mode." In *A Companion to Poetic Genre,* edited by Erik Martiny, 568–83. Hoboken, NJ: Wiley-Blackwell, 2012.

Willis, Margaret M., and Juliet B. Schor. "Does Changing a Light Bulb Lead to Changing the World? Political Action and the Conscious Consumer." *The Annals of the American Academy of Political and Social Science* 644, no. 1 (Nov. 2012): 160–90.

Wilson, Edward O., ed. *Biodiversity.* Washington, DC: National Academy Press, 1988.

———. *The Future of Life.* New York: Vintage, 2002.

———. *The Social Conquest of Earth.* New York: Liveright, 2012.

———. *Sociobiology: The New Synthesis.* Cambridge, MA: Harvard University Press, 1975.

Wilson, Edward O., and Frances M. Peter, eds. *Biodiversity: Papers from the 1st National Forum on Biodiversity, September 1986.* Washington, DC: National Academy Press, 1988.

Wilson, Elizabeth. "Biologically Inspired Feminism: Response to Helen Keane and Marsha Rosengarten, 'On the Biology of Sexed Subjects.'" *Australian Feminist Studies* 17, no. 39 (2002): 283–85.

Winkler, Hans. *Verbreitung und Ursache der Parthenogenesis im Pflanzen- und Tierreiche.* Jena: Fischer, 1920.

Winner, Langdon. *The Whale and the Reactor.* Chicago: University of Chicago Press, 1986.

Winright, Tobias, ed. *Green Discipleship: Catholic Theological Ethics and the Environment.* Winona, MN: Anselm Academic, 2011.

Wirzba, Norman, ed. *The Art of the Commonplace: The Agrarian Essays of Wendell Berry.* San Francisco: Continuum, 2002.

Wolf, Eric. *Europe and the People without History.* Berkeley: University of California Press, 1982.

Wolfe, Cary. "Before the Law: Animals in a Biopolitical Context." *Law, Culture, and the Humanities* 6, no. 1 (2010): 8–23.

Wood, Gillen D'Arcy. "The Volcano That Changed the Course of History." *Slate.* Last modified April 9, 2014. Available online.

Worster, Donald. *Naure's Economy: A History of Ecological Ideas.* Cambridge: Cambridge University Press, 1985.

Wright, Jan, and Valerie Harwood, eds. *Biopolitics and the "Obesity Epidemic": Governing Bodies.* New York: Routledge, 2009.

Wylie, John W. *Landscape.* New York: Routledge, 2007.

York, Richard F., Eugene A. Rosa, and Thomas Dietz. "STIRPAT, IPAT, and ImPACT: Analytic Tools for UnPacking the Driving Forces of Environmental Impacts." *Ecological Economics* 46, no. 3 (2003): 351–65.

Yudice, George. "Culture." In *Keywords for American Cultural Studies,* edited by Bruce Burgett and Glenn Hendler, 71–75. New York: New York University Press, 2007.

Zalasiewicz, Jan, Colin N. Waters, and Mark Williams. "Human Bioturbation, and the Subterranean Landscape of the Anthropocene." *Anthropocene* 6 (2014): 3–9.

Zalasiewicz, Jan, et al. "The New World of the Anthropocene." *Environmental Science and Technology* 44 (2010): 2228–31.

Zalasiewicz, Jan, et al. "The Technofossil Records of Humans." *Anthropocene Review* 1, no. 1 (2014): 34–43.

Zalasiewicz, Jan, et al. "When Did the Anthropocene Begin? A Mid-Twentieth Century Boundary Level Is Stratigraphically Optimal." *Quaternary International* (2015), doi.org/10.1016/j.quaint.2014.11.0450.

Zavestoski, Stephen. "The Struggle for Justice in Bhopal: A New/Old Breed of Transnational Social Movement." *Global Social Policy* 9, no. 3 (2009): 383–407.

Zimmerer, Karl. "The Reworking of Conservation Geographies: Nonequilibrium Landscapes and Nature–Society Hybrids." *Annals of the Association of American Geographers* 90, no. 2 (2000): 356–69.

Zimmerman, Michael E. "The Threat of Ecofascism." *Social Theory and Practice* 21 (Summer 1995): 207–38.

———. "Toward a Heideggerian Ethos for Radical Environmentalism." *Environmental Ethics* 5 no. 2 (1983): 99–131.

Zimmerman, et al., eds. *Environmental Philosophy: From Animal Rights to Radical Ecology.* Englewood Cliff, NJ: Prentice Hall, 1993.

About the Contributors

Joni Adamson is Professor of Environmental Humanities in the Department of English and Senior Sustainability Scholar in the Julie Ann Wrigley Global Institute for Sustainability at Arizona State University. She served as President of the Association for the Study of Literature and the Environment (ASLE) in 2012 and co-leads Humanities for the Environment, an Andrew W. Mellon Foundation seed-funded networking project (hfe-observatories.org). She is the author of *American Indian Literature, Environmental Justice, and Ecocriticism* (2001) and peer-reviewed articles and chapters on environmental justice, food justice, global indigenous studies, and cosmopolitics. She is coeditor of five collections, including *Ecocriticism and Indigenous Studies: Conversations from Earth to Cosmos* (forthcoming), *American Studies, Ecocriticism, and Citizenship* (2013), and *The Environmental Justice Reader* (2002).

Julian Agyeman is a Professor in Urban and Environmental Policy and Planning at Tufts University. His research interests are in the complex and embedded relations between humans and the environment and the effects of this on policy and planning processes and outcomes in relation to notions of justice and equity. His books include *Just Sustainabilities: Development in an Unequal World*, *Sustainable Communities and the Challenge of Environmental Justice*, and *Cultivating Food Justice: Race, Class, and Sustainability*.

Stacy Alaimo is Professor of English and Distinguished Teaching Professor, and Director of the Environmental and Sustainability Studies program at the University of Texas at Arlington. Her publications include *Undomesticated Ground: Recasting Nature as Feminist Space* (2000); *Bodily Natures: Science, Environment, and the Material Self* (2010), which won the Association for the Study of Literature and the Environment (ASLE) book award for ecocriticism; and *Material Feminisms* (2008), which she coedited. She has served on the MLA Division of Literature and Science and the inaugural committee of the new MLA forum for Ecocriticism and the Environmental Humanities. She is currently editing the Matter for the Gender series of Macmillan Interdisciplinary Handbooks, and is writing two books: "Protest and Pleasure: New Materialism, Environmental Activism" and "Feminist Exposure and Blue Ecologies: Science, Aesthetics, and the Creatures of the Abyss."

Vermonja R. Alston is Associate Professor in the Departments of English and Equity Studies at York University in Toronto. She has published articles on environmental justice, poetry and poetics, and cosmopolitanism in journals and edited volumes. Alston has completed a manuscript on twentieth-century African American and Caribbean cosmopolitanism. She teaches postcolonial literary studies, ecocriticism, indigenous literature of the Americas, and Caribbean poetry and poetics.

Karla Armbruster is a Professor in the English Department at Webster University in St. Louis, Missouri. A past president of the Association for the Study of Literature and the Environment, she is coeditor of *Beyond Nature Writing: Expanding the Boundaries of Ecocriticism* and *The Bioregional Imagination: New Perspectives on Literature, Ecology, and Place*. Her current project is a book of narrative criticism about the wildness of dogs in literature, popular culture, and everyday life.

Stefania Barca is Senior Research Associate at the Center for Social Studies of the University of Coimbra, Portugal. She is the author of *Enclosing Water: Nature and Political Economy in a Mediterranean Valley, 1796–1916*, winner of the Turku book prize for 2011 (jointly awarded by the Rachel Carson Center for Environment and Society [LMU Munich] and the European Society for Environmental History).

Kamaljit S. Bawa is a Distinguished Professor of Biology at the University of Massachusetts–Boston, and Founder-President of the Bangalore-based Ashoka Trust for Research in Ecology and the Environment (ATREE). He has published more than 200 scientific papers and ten authored or edited books and monographs. He is a recipient of the Gunnerus Prize in Sustainability Science, the MIDORI international Prize in Biodiversity, and a Fellow of the Royal Society. His latest book, *Himalaya: Mountains of Life*, a sequel to *Sahyadris: India's Western Ghats*, was published in 2013.

Mario Blaser is the Canada Research Chair in Aboriginal Studies at Memorial University of Newfoundland, the author of *Storytelling Globalization from the Chaco and Beyond* (2010), and coeditor of *Indigenous Peoples and Autonomy: Insights for the Global Age* (2010) and *In the Way of Development: Indigenous Peoples, Life Projects, and the Environment* (2004).

Phil Brown is University Distinguished Professor of Sociology and Health Sciences and Director of the Social Science Environmental Health Research Institute at Northeastern University. He is the author of *No Safe Place: Toxic Waste, Leukemia, and Community Action* and *Toxic Exposures: Contested Illnesses and the Environmental Health Movement*, and coeditor of *Social Movements in Health* and *Contested Illnesses: Citizens, Science, and Health Social Movements*. He studies biomonitoring and household exposure, social policy concerning flame retardants, reporting back data to participants, and health social movements.

Robert J. Brulle is a Professor of Sociology and Environmental Science in the Department of Sociology at Drexel University in Philadelphia, Pennsylvania. His research focuses on the U.S. environmental movement, critical theory, and public participation in environmental policy making. He is the author of over seventy articles in these areas and of *Agency, Democracy, and the Environment: The U.S. Environmental Movement from the Perspective of Critical Theory*, as well as coeditor, with David Pellow, of *Power, Justice, and the Environment* and coeditor, with Riley Dunlap, of *Climate Change and Society*. He was a fellow at the Center for Advanced Study in the Behavioral Sciences in 2012.

Lawrence Buell is Powell M. Cabot Research Professor of American Literature at Harvard. His books include *The Environmental Imagination* (1995), *Writing for an Endangered World* (2001), and *The Future of Environmental Criticism* (2005). He has held fellowships from the Mellon and Guggenheim foundations and the National

Endowment for the Humanities. In 2007 he received the Modern Language Association's Jay Hubbell Award for lifetime contributions to American literature scholarship.

Sarah Phillips Casteel is an Associate Professor of English at Carleton University. She is the author of *Second Arrivals: Landscape and Belonging in Contemporary Writing of the Americas* (2007) and the coeditor with Winfried Siemerling of *Canada and Its Americas: Transnational Navigations* (2010).

Noel Castree is Professor of Geography at the University of Wollongong, Australia, and the University of Manchester, England. Among other things, he is interested in the construction, circulation, and reception of ideas about "nature." He is author, most recently, of *Making Sense of Nature: Representation, Politics, and Democracy* (2014).

Alissa Cordner is Assistant Professor of Sociology at Whitman College. Her research focuses on environmental sociology, risks and disasters, environmental health and justice, and public engagement in science and policy making. Her book *Toxic Safety: Flame Retardants, Chemical Controversies, and Environmental Health* will be published in 2016.

Robert Costanza is Professor and Chair in Public Policy at the Crawford School of Public Policy, Australian National University. His transdisciplinary research integrates the study of humans with the study of the rest of nature to address research, policy, and management issues at multiple time and space scales, from small watersheds to the global system. He is cofounder of the International Society for Ecological

Economics and founding editor-in-chief of *Solutions* (www.thesolutionsjournal.org).

Ashley Dawson is Professor of English at the CUNY Graduate Center. He is the author of the *Routledge Concise History of Twentieth-Century British Literature* (forthcoming) and *Mongrel Nation: Diasporic Culture and the Making of Postcolonial Britain* (2007), and coeditor of three essay collections: *Democracy, the State, and the Struggle for Global Justice* (2009); *Dangerous Professors: Academic Freedom and the National Security Campus* (2009); and *Exceptional State: Contemporary U.S. Culture and the New Imperialism* (2007).

Giovanna Di Chiro is the Lang Professor for Issues of Social Change at Swarthmore College, and Policy Advisor for Environmental Justice at Nuestras Raíces, Inc. She has published widely on the intersections of environmental science and policy, with a focus on social and economic disparities and human rights. She is coeditor of the volume *Appropriating Technology: Vernacular Science and Social Power* and is completing a book titled "Embodied Ecologies: Science, Politics, and Environmental Justice." Di Chiro's research, teaching, and activism focus on community-based approaches to sustainability and the intersections of social justice and sustainability.

Andy Dobson is an ecologist whose interests are focused on the role that parasites and infectious diseases play in natural ecosystems. His work uses a mixture of mathematical models, long-term field work, and collaborations with parasitologists and wildlife veterinarians; the principle research sites are Serengeti National Park in Tanzania, Yellowstone National Park in Wyoming, the coastal salt marshes of California, and the high Canadian Arctic.

Arturo Escobar is Professor of Anthropology at the University of North Carolina–Chapel Hill. His main interests are political ecology, design, and the anthropology of development, social movements, and science and technology. Over the past twenty years, he has worked closely with several Afro-Colombian social movements in the Colombian Pacific. His main books are *Encountering Development: The Making and Unmaking of the Third World* (2nd ed. 2011) and *Territories of Difference: Place, Movements, Life, Redes* (2008).

Robert Melchior Figueroa is Associate Professor in the School of History, Philosophy, and Religion at Oregon State University and Director of the Environmental Justice Project for the Center for Environmental Philosophy. He is coeditor, with Sandra Harding, of *Science and Other Cultures: Issues in Philosophies of Science and Technology* (2003) and editor of "Ecotourism and Environmental Justice," a special issue of *Environmental Philosophy* (2010). He collaborates on the Uluru Project (Australia) and on the Mesa Verde Project (United States), both on tourism and cultural-political reconciliation through environmental heritage.

Carmen Flys-Junquera is an Associate Professor of American Literature at the University of Alcalá, Madrid, Spain. She founded and coordinates Grupo de Investigación en Ecocrítica (GIECO) and was President of the European Association for the Study of Literature, Culture, and Environment from 2010 to 2012. She is the General Editor of *Ecozon@: European Journal of Literature, Culture, and Environment*. Her research is focused on ecocriticism, ecofeminism, environmental justice, and sense of place, mostly in American contemporary ethnic literature as well as contemporary Spanish literature.

Stephanie Foote is Professor of English and Gender and Women's Studies at the University of Illinois at Urbana-Champaign. In addition to numerous essays, she is the author of *Regional Fictions: Culture and Identity in Nineteenth-Century American Literature* (2001) and *The Parvenu's Plot* (2014); with Elizabeth Mazzolini, the editor of *Histories of the Dustheap: Waste, Material Cultures, Social Justice* (2012); and, with Stephanie LeMenager, the editor of *Resilience: A Journal of the Environmental Humanities*.

Greta Gaard is Professor of English and Coordinator of the Sustainability Program at the University of Wisconsin–River Falls. Her work emerges from the intersections of feminism, environmental justice, queer studies, and critical animal studies, exploring a wide range of issues, including interspecies justice, material perspectives on fireworks and space exploration, postcolonial ecofeminism, and the eco-politics of climate change. She is author or editor of five books and over fifty refereed articles, and her most recent volume is *International Perspectives in Feminist Ecocriticism* (2013), coedited with Simon Estok and Serpil Oppermann. She is currently at work on a manuscript titled "Critical Ecofeminism," and her creative nonfiction eco-memoir, *The Nature of Home* (2007), is being translated into Chinese and Portuguese.

Greg Garrard is Sustainability Professor at the University of British Columbia. A founding member and former Chair of the Association for the Study of Literature and the Environment (UK and Ireland), he is the author of *Ecocriticism* (2004, 2011 2nd ed.) and numerous essays on eco-pedagogy, animal studies, and environmental criticism. He has recently edited *The Oxford Handbook of Ecocriticism* (2014) and become coeditor of *Green Letters: Studies in Ecocriticism*.

William A. Gleason is Professor and Chair of English at Princeton University, where he is also affiliated with the Program in American Studies, the Center for African American Studies, the Program in Urban Studies, and the Princeton Environmental Institute. He is the author of *The Leisure Ethic: Work and Play in American Literature, 1840–1940* (1999) and *Sites Unseen: Architecture, Race, and American Literature* (2011), a runner-up for the 2012 John Hope Franklin Publication Prize in American Studies.

John Grim is currently a Senior Lecturer and Research Scholar at Yale University and Environmental Ethicist-in-Residence at Yale's Center for Bioethics. With Mary Evelyn Tucker, he codirects Yale's Forum on Religion and Ecology, a project arising from a series of conferences held from 1996 to 1998 at Harvard's Center for the Study of World Religions. Grim is the author of *The Shaman* (1983) and an edited volume, *Indigenous Traditions and Ecology* (2001). With Mary Evelyn Tucker, he has coedited *Religion and Ecology: Can the Climate Change?* (2001) and a volume of Thomas Berry's essays, *The Christian Future and the Fate of Earth* (2009).

Wendy Harcourt is Associate Professor in Critical Development and Feminist Studies at the International Institute of Social Studies of the Erasmus University Rotterdam. She joined ISS/EUR in 2011, after twenty years at the Society for International Development in Rome, as Editor of Development and Director of Programmes. Her most recent edited collections are *Practicing Feminist Political Ecology: Beyond the Green Economy* (2015) and *OUP Handbook on Transnational Feminist Movements* (2015). Her monograph *Body Politics in Development: Critical Debates in Gender and Development* (2009) received the 2010 FWSA Book Prize.

Ursula K. Heise is the Marcia Howard Professor of Environmental Humanities in the Department of English and at the Institute of the Environment and Sustainability at UCLA. She is a 2011 Guggenheim Fellow and served as President of the Association for the Study of Literature and the Environment the same year. Her books include *Chronoschisms: Time, Narrative, and Postmodernism* (1997), *Sense of Place and Sense of Planet: The Environmental Imagination of the Global* (2008), and *Nach der Natur: Das Artensterben und die moderne Kultur* (After Nature: Species Extinction and Modern Culture, 2010). She is editor of the book series *Literatures, Cultures, and the Environment* with Palgrave-Macmillan and coeditor of the series *Literature and Contemporary Thought* with Routledge. Her book *Imagining Extinction: The Cultural Meanings of Endangered Species* will appear in 2016.

Nik Heynen is a Professor in the Department of Geography at the University of Georgia. His research interests include urban political ecology and social movement theory with specific interests in environmental and food politics. His main research foci relate to the analysis of how social power relations, including class, race, and gender, are inscribed in the transformation of nature/space, and how in turn these processes contribute to interrelated connections among nature, space, and vulnerable populations.

James J. Hughes is a bioethicist and sociologist at Trinity College in Hartford, Connecticut, where he teaches health policy and serves as Director of Institutional Research and Planning. He is also the Executive Director of the Institute for Ethics and Emerging Technologies. Dr. Hughes is author of *Citizen Cyborg: Why Democratic Societies Must Respond to the Redesigned Human of the Future* and is working on a second book tentatively titled "Cyborg Buddha."

Serenella Iovino is Professor of Ethics at the University of Turin, Research Fellow of the Alexander-von-Humboldt Foundation, and past President of the European Association for the Study of Literature, Culture, and Environment (EASLCE). She has written extensively on ecocriticism, environmental philosophy, and German philosophical literature of the Age of Goethe. You can find more information about her work at http://unito.academia.edu/serenellaiovino.

Basia Irland, author, poet, sculptor, installation artist, and activist, creates international water projects featured in her book *Water Library*. She works with scholars from diverse disciplines restoring riparian zones; filming and producing water documentaries; connecting communities along the lengths of rivers; building rainwater harvesting systems; and creating global waterborne disease projects. She is Professor Emerita in the Department of Art and Art History at the University of New Mexico, where she established the Arts and Ecology Program.

Sheila Jasanoff is Pforzheimer Professor of Science and Technology Studies at Harvard University's John F. Kennedy School of Government. Her research centers on the role of science and technology in democratic governance, with particular focus on the use of science in legal and political decision making. Her books include *The Fifth Branch: Science Advisers as Policymakers*, *Science at the Bar: Law, Science, and Technology in America*, and *Designs on Nature: Science and Democracy in Europe and the U.S.*

Stephanie LeMenager is Barbara and Carlisle Moore Distinguished Professor in English and American Literature and Professor of Environmental Studies at the University of Oregon. Her latest book, *Living Oil: Petroleum Culture in the American Century*, was published by Oxford University Press. She is coeditor of *Resilience: A Journal of the Environmental Humanities*.

Timo Maran is a Senior Research Fellow in the Department of Semiotics at the University of Tartu, Estonia. His publications include *Mimikri semiootika* (Semiotics of Mimicry, 2008), *Readings in Zoosemiotics* (coedited with D. Martinelli and A. Turovski, 2011), and *Semiotics in the Wild: Essays in Honour of Kalevi Kull on the Occasion of His 60th Birthday* (coedited with K. Lindström, R. Magnus, and M. Toennessen 2012).

Joan Martinez-Alier is Emeritus Professor at ICTA, Universitat Autonoma de Barcelona, and at FLACSO, Ecuador; author of *Ecological Economics: Energy, Environment, and Society* (1987) and *The Environmentalism of the Poor: A Study of Ecological Conflicts and Valuation* (2002); coeditor of *Ecological Economics from the Ground Up* (2012); past President of the International Society for Ecological Economics; and Director of the EJOLT project (Environmental Justice Organizations, Liabilities, and Trade), 2011 to 2015.

Arthur P. J. Mol is Rector Magnificus and Vice-Chairman of the Board of Wageningen University and Research, the Netherlands; Professor of Environmental Policy at Wageningen University; joint editor of the journal *Environmental Politics*; and editor of the book series *New Horizons in Environmental Politics*. He has published extensively on environmental social theory, environmental politics and policy, globalization, the information age, and China's struggles to cope with environmental challenges.

Rachel Morello-Frosch is Professor in the Department of Environmental Science, Policy, and Management and the School of Public Health at the University of California–Berkeley. Her scientific work examines the combined, synergistic effects of social and environmental factors in environmental health disparities. She also studies the ways in which health social movements (re)shape scientific thinking about environmental health issues. She is coauthor of the book *Contested Illnesses: Citizens, Science, and Health Social Movements*.

William G. Moseley is a Professor of Geography at Macalester College in Saint Paul, Minnesota. His books include four editions of *Taking Sides: Clashing Views on African Issues* (2004, 2006, 2008, 2011), *Hanging by a Thread: Cotton, Globalization, and Poverty in Africa* (2008), *The Introductory Reader in Human Geography: Contemporary Debates and Classic Writings* (2007), and *African Environment and Development: Rhetoric, Programs, Realities* (2004).

Patrick D. Murphy is Professor and Chair of the Department of English at the University of Central Florida. He has authored *Ecocritical Explorations in Literary and Cultural Studies* (2009), *Farther Afield in the Study of Nature-Oriented Literature* (2000), *A Place for Wayfaring: The Poetry and Prose of Gary Snyder* (2000), and *Literature, Nature, and Other: Ecofeminist Critiques* (1995). He teaches critical theory, modern and contemporary American literature, comparative literature, ecocriticism, and ecofeminism.

Gary Paul Nabhan is the Kellogg Endowed Chair in Sustainable Food Systems at the University of Arizona Southwest Center. A conservation biologist, ethnobotanist, and agroecologist, he worked at the first

Earth Day headquarters in 1970. He has served on the boards of the Society for Conservation Biology, the U.S. National Park System, Wild Farm Alliance, and Seed Savers Exchange. Cofounder of Native Seeds/SEARCH and a MacArthur Fellow, he was a pioneer in the food relocalization movement and the global initiative to save heirloom seeds. He farms in Patagonia, Arizona.

Padini Nirmal is a doctoral student at Clark University. Her doctoral research focuses on the dispossession of indigenous peoples by the development-capitalism-modernity complex and the resistance movements that emerge at its juncture in Kerala, India. Broadly, her research interests lie within political ecology, feminism, and critical development studies.

David N. Pellow is Dehlsen Chair of Environmental Studies and Director of the Global Environmental Justice Project at the University of California–Santa Barbara. His teaching and research focus on environmental and ecological justice in the United States and globally. He has served on the Boards of Directors for the Center for Urban Transformation, Greenpeace USA, and International Rivers.

Keith Pezzoli is Director of the Urban Studies and Planning Program, and a Professor of Teaching in the Communications Department at the University of California–San Diego. He teaches courses on Community-Based Action Research, food justice, environmental movements, and globalization. Pezzoli's research and publications examine science and technology, and human-nature relations in the development of cities and regions, including a book, *Human Settlements and Planning for Ecological Sustainability: The Case of Mexico City* (2000).

Kate Rigby is Professor of English and Comparative Literature at Monash University. Among her publications in the area of literature and environment are *Topographies of the Sacred: The Poetics of Place in European Romanticism* (2004) and *Ecocritical Theory: New European Approaches* (2011). Rigby is a Fellow of the Australian Academy of the Humanities and was the founding President of the Association for the Study of Literature, Environment, and Culture (Australia–New Zealand).

Dianne Rocheleau is a Professor of Geography at Clark University. She is a feminist political ecologist who has worked on emergent ecologies including humans and other beings, and their artifacts, technologies, and territories. She has studied with, for, and about social movements and rural people's ecologies of resistance in farmlands, forests, and regional agroforests in the Dominican Republic, Kenya, Mexico, and the United States. She has coauthored and coedited four books: *Feminist Political Ecology* (1996); *Gender, Environment, and Development in Kenya* (1995); *Power, Process, and Participation: Tools for Change* (1995); and *Agroforestry in Dryland Africa* (1988). She is also coeditor with Arturo Escobar of the Duke University Press series New Ecologies for the 21st Century.

Deborah Bird Rose is Professor of Social Inclusion at Macquarie University and a Visiting Professorial Fellow at the University of New South Wales (Sydney). Her research focuses on how we humans include and exclude other members of the family of life on Earth in this era of extinctions, and her most recent book is *Wild Dog Dreaming: Love and Extinction* (2011).

Andrew Ross is a Professor of Social and Cultural Analysis at New York University. A contributor to the *Nation, Village Voice, New York Times*, and *Artforum*, he is the author of many books, including *Bird on Fire: Lessons from the World's Least Sustainable City, Nice Work If You Can Get It: Life and Labor in Precarious Times, Fast Boat to China: Lessons from Shanghai, No-Collar: The Humane Workplace and Its Hidden Costs*, and *The Celebration Chronicles: Life, Liberty, and the Pursuit of Property Value in Disney's New Town*. His most recent book is *Creditocracy and the Case for Debt Refusal*.

Dorion Sagan's interests include philosophy, science, and literature. He is sole author or coauthor of twenty-nine books translated into thirteen languages. His most recent works include *Lynn Margulis: The Life and Legacy of a Scientific Rebel, Cosmic Apprentice: Dispatches from the Edges of Science, Biospheres: Metamorphosis on Planet Earth*, and *The Sciences of Avatar*.

David E. Salt is interested in understanding the molecular mechanisms that control the way plants acquire the mineral nutrients they require from the soil, along with the evolutionary forces that shape these mechanisms. Professor Salt has held faculty positions in the United States at Rutgers University, Northern Arizona University, and Purdue University, and is currently a Sixth Century Chair at the University of Aberdeen in the United Kingdom. He has published over 110 peer-reviewed papers, which have over seven thousand citations. These include papers published in such journals as *Nature, Proceedings of the National Academy of Sciences, Plant Cell,* and *PLoS Genetics*.

Catriona (Cate) Sandilands is Professor in the Faculty of Environmental Studies at York University, where she teaches and writes at the intersections of environmental literatures and histories, social and political theory, and feminist and queer studies. She is the author of over sixty chapters and articles, and recently the coeditor of *Queer Ecologies: Sex, Nature, Politics, Desire* (2010) and *Green Words, Green Worlds: Environmental Literatures and Politics* (forthcoming).

Bryony Schwan is the cofounder of the Biomimicry 3.8 Institute and served as the founding Executive Director for eight years. Prior to that, Schwan worked for eleven years as the Executive Director and then as the National Campaigns Director for Women's Voices for the Earth (WVE), a nonprofit environmental justice organization that she founded in 1995. She is an affiliate faculty member at the University of Montana, where she teaches in the Environmental Studies program.

Reinmar Seidler is Research Assistant Professor in Environmental Biology at the University of Massachusetts–Boston, where he teaches evolutionary biology, conservation biology, and sustainability science. He has published widely on aspects of environmental management, with a particular focus on South Asia. Most recently, he edited and coauthored P. S. Ashton's *On the Forests of Tropical Asia,* a comprehensive study of the ecology, biogeography, evolutionary history, and human history of Asian tropical forests (2015). With K. S. Bawa, he is currently preparing a volume of essays on the past, present, and future of climate change in the Himalayas.

Teresa Shewry is Associate Professor of English at the University of California–Santa Barbara. She is the author of *Hope at Sea: Possible Ecologies in Oceanic Literature* (2015) and is coeditor of *Environmental Criticism for the Twenty-First Century* (2011).

Andrew Szasz has written books and articles on the toxics movement, green consuming, environmental

regulation, and environmental justice. Szasz is currently Professor and Chair of the Department of Environmental Studies at the University of California–Santa Cruz. He teaches courses on Environmental Justice, Sociology of Climate Change, and Sociological Theory.

Julie Sze is a Professor at the University of California–Davis and the founding director of the Environmental Justice Project for the John Muir Institute for the Environment. Sze's research investigates environmental justice and inequality; culture and environment; race, gender, and power; and urban/community health and activism. Sze's book, *Noxious New York: The Racial Politics of Urban Health and Environmental Justice,* won the 2008 John Hope Franklin Prize. She has authored and coauthored thirty peer-reviewed articles and book chapters and is the author of *Fantasy Islands: Chinese Dreams and Ecological Fears in an Age of Climate Crisis* (2015).

Dorceta E. Taylor is James E. Crowfoot Collegiate Professor at the University of Michigan's School of Natural Resources and Environment, where she is the Coordinator of the Environmental Justice Field of Studies. She also holds a joint appointment with the Program in the Environment. She is a past Chair of the American Sociological Association's Environment and Technology Section. Taylor received doctorates in Sociology and Forestry & Environmental Studies from Yale University in 1991. She is the author of *The Environment and the People in American Cities, 1600s–1900s: Disorder, Inequality, and Social Change* (2009) and *Toxic Communities: Environmental Racism, Industrial Pollution, and Residential Mobility* (2014).

Hava Tirosh-Samuelson is Irving and Miriam Lowe Professor of Modern Judaism and Director of the Center for Jewish Studies at Arizona State University in Tempe, Arizona. She is the author of *Happiness in Premodern Judaism: Knowledge, Virtue, and Well-Being* (2003) and the editor of *Judaism and Ecology: Created World and Revealed Word* (2002), *The Legacy of Hans Jonas: Judaism and the Phenomenon of Life* (2008), and *Building Perfect Humans? Refocusing the Debate on Transhumanism* (2012).

Mitchell Thomashow is Director of the Second Nature Presidential Fellows Program, designed to assist the executive leadership of colleges and universities in promoting a comprehensive sustainability agenda on their campuses. Previously (2006–2011), he was the President of Unity College in Maine. He is the author of *Ecological Identity: Becoming a Reflective Environmentalist* (1995) and *Bringing the Biosphere Home* (2001). *The Nine Elements of a Sustainable Campus* (2014) provides a framework for advancing sustainable living and teaching in a variety of campus environments.

Teresa A. Toulouse is Professor of English at the University of Colorado–Boulder. She is the author of *The Art of Prophesying: New England Sermons and the Shaping of Belief*; *The Captive's Position: Female Captivity, Male Identity, and Royal Authority in Colonial New England*; and the coeditor, with Andrew Delbanco, of volume 2 of *The Complete Sermons of Ralph Waldo Emerson*. As Director of American Studies at Tulane University, she taught and continues to teach courses on American literature and the environment.

Mary Evelyn Tucker is currently a Senior Lecturer and Research Scholar at Yale University and an Environmental Ethicist-in-Residence at Yale's Center for Bioethics. With John Grim, she codirects Yale's Forum on Religion and Ecology, a project arising from a series of conferences held from 1996 to 1998 at Harvard's Center for the Study of World Religions. Tucker is author and coeditor of many books, including *Confucianism and Ecology* (1998), *Confucian Spirituality* (2003), and *The Philosophy of Qi* (2007). With Brian Swimme, she created a multimedia project titled *Journey of the Universe* (2011). With John Grim, she has coedited *Religion and Ecology: Can the Climate Change?* (2001) and a volume of Thomas Berry's essays, *The Christian Future and the Fate of Earth* (2009).

Carmen Valero-Garcés is a Full Professor of Translation and Interpreting at the University of Alcalá, Madrid, Spain, and member of Grupo de Investigación en Ecocrítica (GIECO). She studies the relationships among ecocriticism, environmental studies, and translation. She is the guest editor of a special issue of *Ecozon@: European Journal of Literature, Culture, and Environment* 5, no. 1 (2014), titled "Translating the Environmental Humanities."

Tyler Volk is Professor of Biology and Environmental Studies at New York University. For more than twenty years, his research has focused on the global carbon cycle, the dynamics of the biosphere, and systems at all scales. Volk's books include *CO2 Rising: The World's Greatest Environmental Challenge*; *Gaia's Body: Toward a Physiology of Earth*; and *Metapatterns across Space, Time, and Mind*.

Priscilla Wald is Professor of English and Chair of the Program in Women's Studies at Duke University. She is the author of *Constituting Americans: Cultural Anxiety and Narrative Form* (1995) and *Contagious: Cultures, Carriers, and the Outbreak Narrative* (2008). She is currently at work on a book entitled "Human Being after Genocide," which explores the convergence of science and politics in the production of new creation stories about humanity following the Second World War.

Laura Dassow Walls is the William P. and Hazel B. White Professor of English at the University of Notre Dame and the author of *The Passage to Cosmos: Alexander von Humboldt and the Shaping of America* (2009), *Emerson's Life in Science* (2003), and *Seeing New Worlds: Henry David Thoreau and Nineteenth-Century Natural Science* (1995), as well as numerous essays, and the coeditor of several volumes, including *The Oxford Handbook of Transcendentalism*.

Colin N. Waters is a geologist with twenty-seven years' experience working at the British Geological Survey, with particular interest in Carboniferous and Anthropocene stratigraphy. He is Secretary of the Geological Society Stratigraphy Commission and Anthropocene Working Group. Relevant publications include Zalasiewicz et al., "Are We Living in the Anthropocene?" *GSA Today* 18, no. 2 (2008): 4–8; and Zalasiewicz et al., "Stratigraphy of the Anthropocene," *Philosophical Transactions of the Royal Society A* 369 (2011): 1036–55. He is Senior Editor of "A Stratigraphical Basis for the Anthropocene," Geological Society, London, Special Publications 395 (2014).

Quentin Wheeler is the fourth President of ESF, the College of Environmental Science and Forestry, in Syracuse, New York, and is Founding Director of the International Institute for Species Exploration. He was previously Professor of taxonomy at Cornell University, Director of the division of environmental biology at the National Science Foundation, Keeper and Head of Entomology in the Natural History Museum, London, and Virginia M. Ullman Professor of Natural History and the Environment, Vice-President, and Dean of the

College of Liberal Arts and Sciences at Arizona State University. He writes a weekly column on new species for London's *Observer* newspaper. His most recent book is *What on Earth?* (2013).

Kyle Powys Whyte is an Assistant Professor of Philosophy at Michigan State University. He is an enrolled member of the Citizen Potawatomi Nation, a federally recognized tribe in Oklahoma. Kyle's most recent research addresses moral and political issues concerning climate-change impacts on indigenous peoples.

Mark Williams is a geologist with a particular interest in the reconstruction of ancient climate. One area of his research has focused on the climate of the Pliocene world, some three million years ago, when CO_2 levels were similar to their levels in the present. Williams is a former geologist with the British Geological Survey and British Antarctic Survey, and his geological expertise has taken him from the Cambrian to the Anthropocene, and from the tropics to the Antarctic. He teaches palaeoclimates and micropalaeontology at the University of Leicester in the United Kingdom.

Jan Zalasiewicz is a field geologist, stratigrapher, and palaeontologist formerly with the British Geological Survey and now at the University of Leicester, United Kingdom. His research covers geological processes and environmental change from the Precambrian to the present day, with particular interests in the early Palaeozoic and the late Cenozoic, and in present-day and future geological change. He currently chairs the Anthropocene Working Group of the International Commission on Stratigraphy. He has written the books *The Earth after Us* (2008), *The Planet in a Pebble* (2010), and, with Mark Williams, *The Goldilocks Planet* (2012) and Ocean Worlds (2014).

Michael E. Zimmerman is Professor of Philosophy at the University of Colorado at Boulder. He is the author of *Eclipse of the Self: The Development of Heidegger's Concept of Authenticity*; *Heidegger's Confrontation with Modernity*; *Contesting Earth's Future*; *Integral Ecology* (coauthored with Sean Esbjorn-Hargens); and more than one hundred articles and chapters.

Michael Ziser is the author of *Environmental Practice and Early American Literature* (2013) and Associate Professor of English at the University of California–Davis, where he codirects the Environments and Societies Research Initiative.